A-LEVEL BIOLOGY

COURSE COMPANION

A. G. Toole BSc, MIBiol

Vice Principal,
Pendleton Sixth Form College, Salford
Examiner for the London Board

S. M. Toole BSc

Formerly Head of Biology, North London Collegiate School
Open University Course Tutor
Assistant Chief Examiner for the London Board

Charles Letts & Co Ltd
London, Edinburgh & New York

First published 1982
by Charles Letts & Co Ltd
Diary House, Borough Road, London SE1 1DW

Revised 1984, 1986, 1988
Reprinted 1985

Illustrations: Tek-Art

British Library Cataloguing in Publication Data
Toole, A. G.
  A-level biology : course companion—
  4th ed.—(Letts study aids).
  1. Biology—Examinations, questions, etc.
  I. Title—II. Toole, S. M.
  574′.076   QH316

  ISBN 0 85097 818 1

Printed and bound in Great Britain by
Charles Letts (Scotland) Ltd

Preface to the Fourth Edition

This book is designed to be used throughout the A-level course and not as a last-minute revision aid. It should be used in conjunction with lesson or lecture notes and text books. It is expected that the most useful aspect of the book will be the detailed question analysis provided for each topic. Actual examination questions are used wherever possible and the student is shown what the question means, what the examiner is seeking, the key parts of the question, what knowledge should be used to answer it and how this should be organized into an effective answer plan. Although the book is primarily designed as a student aid it is envisaged that teachers will find it useful; particularly those who have very small A-level groups and who may have no experience of A-level examining techniques.

It is inevitable that with modern technical and scientific advances the nature of a subject such as biology is constantly changing. New discoveries expand our knowledge of some topics and cause traditional ideas associated with others to be questioned. In turn examination syllabuses and questions are revised to incorporate these changes. This fourth edition has been written in order to keep pace with these advances. The analysis of examination syllabuses has been updated to include all changes, and most questions prior to 1980 have been replaced by more recent examples. The suggested further reading sections have been revised to include the latest editions and new references. Small alterations have been made to all sections, especially classification, in order to bring the information up to date and make it more accessible and easier to follow. Finally thanks must go to all concerned at Charles Letts & Co Ltd for their help, support and efficiency in making this fourth edition possible.

Susan and Glenn Toole 1988

Acknowledgements

During the preparation of this book we have been grateful for the constructive criticism of Morton Jenkins and in particular for his help with the evolution and ecology sections. We are also grateful to Gillian Frith for her work on co-ordination, to Brian Arnold for his assistance with the questions of the Scottish Examination Board, and Rebecca Thomas for help with the diagrams.

The following examining boards have made the publication of this book possible by giving permission to reproduce their questions: Associated Examining Board, University of London Entrance and School Examinations Council, Welsh Joint Education Committee, Southern Universities Joint Board for School Examinations, Oxford Local Examinations, Scottish Certificate of Education Examination Board, and Northern Ireland General Certificate of Education Examination Board. We should like to thank Philip Harris Biological Ltd for permission to use, and help in reproducing, the photographs.

Susan and Glenn Toole 1988

iv

Contents

Introduction and guide to using the book

The GCSE study aid *Revise Biology* is one of a series written to give advice on what GCSE syllabuses require, how and what candidates should learn, how to assess progress and an outline of how to answer examination questions. The emphasis in the GCSE book is on covering the factual information required by the main examination groups in a way that makes learning easy.

This book follows on from the GCSE study aid, but with one major difference. The range and depth of knowledge at A level make a comprehensive review of all the factual information impossible in a book of this size. The emphasis has therefore been shifted from providing factual information to giving practical advice on how to answer examination questions. It is apparent to any experienced teacher or A-level examiner that many candidates perform badly in examinations despite having a sound knowledge of their subject. One criticism levelled by students is that while there are a number of good, informative A-level textbooks on the market, little if any advice is available from books on how to put this information to maximum effect in answering A-level examination questions. One reason for this lack of information is that there is rarely one way of answering a question, especially an essay, at this level. A variety of different answers may bring equal marks. Each candidate must maximize his or her own strengths and only candidates themselves, or teachers, know what these are. At the same time there are many ways in which people experienced in the marking of A-level essays and examination scripts can help the candidate to succeed. They can explain the exact meaning of a question, something candidates frequently fail to understand. They can pinpoint the key words in a question and guide the candidate in how to marshal the information he has in such a way that the question is answered clearly, unambiguously and concisely. They can advise on planning answers, the use of techniques such as drawing diagrams, making graphs, the use of tables and a host of other technical points. Finally they can indicate the likely errors that candidates will make and show how these are best avoided. This book aims to provide this information. In order to gain maximum benefit from the book it should be used throughout the A-level course and not merely for revision.

The book is divided into 32 topic areas based on an analysis of the nine A-level syllabuses, four AS-level syllabuses and the Scottish Higher Grade syllabus. The factual content is not intended to cover the complete content of all 14 syllabuses, but rather the core material which is common to most of them. No attempt has been made to analyse the content of the A-level Nuffield biology course but Nuffield students will still find the book useful. The syllabus analysis chart indicates which topics are covered by each examination board.

For each topic area there are 7 main divisions:

Assumed previous knowledge

The background knowledge necessary to an understanding of the topic, but not covered by it, is listed. This information is either material covered by most GCSE syllabuses or another topic area within this book. In either case candidates are strongly advised to refer to this material before proceeding to the next section.

Underlying principles

This section deals with the fundamental ideas on which the topic is based. It contains no detailed knowledge but rather an outline of the basic biological concepts associated with the topic. It also provides a means of linking the topic with others in a general way and so helps to integrate them.

Points of perspective

In addition to the basic factual information any biological topic has historical, medical, social or practical implications. A knowledge of such material will broaden candidates' perception of the topic and amplify their knowledge in a way that may bring 'bonus' marks in examinations. The section outlines some of these points and candidates wishing to do well are strongly advised to explore them.

ESSENTIAL INFORMATION

This contains the basic factual information on a topic. It is not intended to be a detailed account and is in no way intended to act as a substitute for a textbook. Its purpose is to give candidates an outline of factual information on a topic sufficient for them to appreciate the relevant aspects and how they should be used in answering questions. It also serves as a summary of the information needed so that candidates can more easily find the relevant sections in their A-level textbooks, or in libraries and scientific journals.

QUESTION ANALYSIS

For each topic area a number of A-level questions are listed. These are either actual A-level questions from past examinations or good facsimiles. Their choice has been based on the following criteria:

1 The proportion of questions from each board reflects the number of candidates taking that board's examination over the past few years.
2 The proportion of essay, structured and multiple choice questions used reflects the proportion of these questions in A-level biology examinations as a whole.
3 As far as possible the questions are from recent papers. Where older ones are used they are of a style still commonly in use.
4 Many questions chosen are those which require interpretation or careful selection and application of facts and principles. At the same time some factual recall questions are included, especially where the topic is mostly examined by that type of question.

For each question a detailed analysis is made of what the question means and which elements of the candidate's knowledge should be used in answering it. An account is given of the sort of information the question requires and the form in which it is best expressed. Common mistakes and misunderstandings are mentioned and advice is given on how best to distribute the time allowed. The mark allocation and time allowance is given for each question to enable candidates to appreciate the level of detail required. For essay-type questions the factual information is summarized in an answer plan. Although each question has been set by a particular board the answers are intended to cover what might be expected by any of the boards. For this reason the answers provided are not necessarily precisely what a particular examination board would require. However the authors are themselves experienced A-level examiners.

SUGGESTED FURTHER READING

A list of books, usually monographs, is given, which will provide the candidate with more detailed information on the topic. They are mostly books in the excellent *Studies in Biology* series or the *Oxford/Carolina Biology Reader* series. This is because these books deal with specific topics, are relatively inexpensive and are commonly found in school and college libraries. The possible lack of availability to students of certain books has meant that only those the authors consider widely available have been listed. Only very rarely has a textbook been listed. It is felt that most candidates will possess one and so continual reference to relevant chapters would be unnecessary. The excellent *Biology – A Functional Approach* by M. B. V. Roberts (4th ed., Nelson 1986) is highly recommended. Other useful texts are *Biology* by Helena Curtis (2nd ed., Butterworth 1975), *Biological Science* by Green, Stout and Taylor (Cambridge University Press 1984) and *Understanding Biology for Advanced Level* by Glenn and Susan Toole (Hutchinson Education 1988).

LINK TOPICS

Examination questions often require information from more than one topic and so related ones have been listed in this section.

A comparison of GCSE and A-level study

While there are differences in the syllabus content it would be fair to say that A level is largely an extension of GCSE except that the topics are studied in greater depth. Most candidates expect such a difference and have few problems adjusting to A-level work in this respect. It is in differences in the methods of study and in the skills required that students find most difficulties. Some take many months to fully appreciate these differences and a few never fully succeed in making the change.

In biology one difference is that a practical examination is set by many examining boards – something which is rare, although not unknown, at GCSE. Assessment of practical work performed by students throughout the course may be made either by the subject teacher or external examiners. Some examining boards require the practical work performed during the course or in the examination to meet an overall minimum standard before an overall pass grade at A level can be given.

As with GCSE there is an emphasis on understanding and application of knowledge at A level rather than learning and the recall of factual information. An A-level candidate is expected to appreciate the underlying principles of each topic and appreciate the concepts and ideas behind them, as well as the factual content associated with them. As with GCSE an emphasis is placed on the ability to apply general biological knowledge and principles to novel situations rather than simply relate a previously learned set of facts. The ability to interpret data in a variety of forms is required and this is often deliberately obscure data which candidates will never have previously encountered. In this way examiners are able to test understanding as opposed to rote learning. Compared to GCSE the A-level student is expected to know a much wider range of species and be able to use these to illustrate specific points. More open-ended questions appear at A level and candidates may not be required to give specific answers, but rather argue a case for and/or against a particular view. At A level the ability to analyse experiments, data and information critically and to evaluate the accuracy of results and theories is an extension of the basic training given in these skills at GCSE. Such skills require the candidate to reason objectively, maintain a balanced view and argue logically. In order to develop these attributes candidates need to read widely on all aspects of biology. It is this, above all, that is essential if a student wishes to obtain a high grade at A level where a greater range of sources of information must be used. Scientific journals, for example, are invaluable as a means of keeping candidates up to date in a subject that is developing as rapidly as biology. Such reading should be done continuously throughout the course and all A-level students should be familiar with the biology section of their school, college, local and central libraries. At A level essays are not so much a means of testing a student's knowledge of a topic as a means of getting him to expand his knowledge and incorporate information from other sources. In this way a student has at least three different sources of information from which to revise – teacher notes, his own essays and a textbook. If the essays are solely compiled from the other two sources the candidate's range and depth of knowledge is restricted. Essays should be carefully researched over a long period rather than jotted down from memory.

One major difference from GCSE which the new A-level student will encounter is the large body of factual information which is required to be learnt. In order to do so the A-level student will need to learn quickly the skill of taking notes during class lectures, from films, videos, television programmes or books. The ability to isolate the main points from what is being communicated is essential. These skills are not evident at GCSE where lecturing and note-taking are less used.

There is a much greater onus on an A-level student to organize his own work, use his own initiative and develop his own pattern of study. There is greater freedom to pursue and develop one's own interests. However all this brings the need for greater self-discipline and personal sacrifice in order to work diligently and industriously over at least two years and not just in the final run-up to the examination. Probably the major cause of failing to attain the desired grades at A level is deluding oneself that provided adequate revision is done in the weeks preceding the examination, success is assured. A level can be stimulating and rewarding but it is always demanding and time-consuming. Above all else success requires a keen interest in the subject, the motivation to do well, the determination to succeed and the self-discipline to sustain one's efforts throughout the course.

A recent educational development has been the introduction of Advanced Supplementary (AS) courses in many subjects, including biology. Like A levels, the AS syllabuses are based upon the inter-board common core in biology agreed by the GCE boards in 1983.

The AS biology courses have been designed with 3 groups of students in mind:

1 Those doing A-level subjects unrelated to biology. e.g. arts subjects, in order to broaden their curriculum and give them some grounding in a science subject.

2 Those studying complementary A levels, e.g. physics, chemistry and maths, in order to act as a supporting subject.

3 Those not wishing to follow a full A-level course, but who have a personal or career interest in a biological qualification beyond GCSE, e.g. those hoping to do nursing.

AS syllabuses have been designed to occupy half the teaching and study time of an A-level subject. While usually extending over two years, candidates may be able to take AS examinations after a shorter period depending on the resources and policy of each school or college. A levels and AS levels may be taken in any combination, but typically students will study three A and one AS levels, two A and two AS levels, or two A and one AS levels. The same general calibre of work will be expected at AS level as is expected at A level; it is not a half-way house between GCSE and A level. Instead the amount, rather than the standard, of the work will be reduced to take account of the shorter teaching and study time available. The grading system will be the same as for A levels, and the grading standards will likewise correspond.

In biology, the AS syllabuses often consist of core material, dealing with basic cell structure, biochemistry, genetics and some aspects of physiology. In addition many boards offer a range of options from which candidates can choose a topic to study in greater depth. Popular topics include biotechnology, human physiology, disease and man's impact on the environment. On the whole the syllabuses have an emphasis on applied biology and consider problems of a personal, social, environmental, economic and technological nature.

The examination

THEORY PAPERS

Theory papers are generally designed to test, by a variety of means, the following abilities:

1 Recall of learned facts and principles.
2 Application of learned facts and principles.
3 Understanding and interpretation of data presented in a variety of forms, e.g. graphs or tables.
4 Construction of hypotheses and design of experiments.
5 Expression of information by means of prose, graphs, tables, drawings and diagrams.
6 Evaluation of scientific information and presentation of logical, written arguments.

Where there is no practical examination a number of questions will be set on the written papers based on problems, practical work and experimental procedure; these will often be compulsory. Other compulsory questions include multiple choice and highly structured ones. There is normally a choice of essay-type questions.

For any theory examination it is important to follow these basic rules:

1 Read all instructions carefully. Note especially the number of questions to be answered, any questions which are compulsory and instructions for the use of diagrams.
2 Read all questions carefully, preferably twice, and select possible questions to be attempted.
3 Always attempt the maximum number of questions required by the examiners.
4 Plan your time for answering according to the marks allocated. Do not spend more than the allocated time on a question.
5 For each question attempted underline the most significant words and plan your answer. It is important to refer back to the question during the writing of your answer.
6 Never include irrelevant material. This will not gain marks and will waste valuable time.
7 Diagrams should be included where they make a point clearly and save time. Do not repeat in prose information shown on a diagram. All diagrams should be large and clearly labelled or annotated. Do not waste time shading large areas.
8 Marks are often awarded for style on essay papers. Make sure your writing is legible, answers are well ordered and spellings, especially of words specific to biology, are correct.
9 When all the required answers have been attempted read through them and check that all instructions have been complied with.
10 On multiple choice papers make sure that you have answered every question. Normally no marks are deducted for incorrect guesses. If you have marked more than one response for a question neither will be marked as correct.
11 On highly structured papers the space allocated in the answer book may guide you in the length of your answer. Answer concisely and clearly.

PRACTICAL PAPERS

Although a practical examination is not set by all boards the skills tested are expected to have been acquired and are often assessed within the school or college as well as on theory papers. The most common skills required are:

1 To use external features of organisms as a basis for their identification into main group and sub-group.
2 To make accurate written or diagrammatic observations of living or preserved material.
3 To dissect, display, draw and label a small mammal. To follow detailed instructions for the dissection of other organisms.
4 To stain and make temporary mounts of specimens.
5 To carry out simple biochemical tests, e.g. food tests.
6 To perform simple physiological experiments recording the methods and results and accurately interpreting the data obtained.

7 To interpret and comment on photographs, e.g. stages of nuclear division, a variety of limb bones, electron micrographs of cell organelles.

8 To draw and accurately interpret permanent preparations of specimens.

9 To comment on apparatus and experiments set up as demonstrations, e.g. kymographs, klinostats.

10 To construct simple keys to distinguish between several related specimens.

The following points may be helpful when sitting a practical examination:

1 Read all instructions carefully. Normally all questions are compulsory.

2 Read through the whole paper carefully, preferably twice.

3 Speed and accuracy are required; allocate your time carefully.

4 Questions may be answered in any order but two main factors will probably determine the order chosen:

(a) Any physiological experiment requiring a series of readings over a period of time must be started immediately.

(b) Some specimens may have to be shared with other candidates and the supervisor will have drawn up a rota. Make sure you know when it will be your turn and use the specimens as soon as you receive them; there may be no second chance.

5 Identification of specimens: Your answer should normally be in the form of a table. Marks are not usually awarded for the correct sub-group if the main group given is incorrect. Where reasons for your classification are required visible features should be used.

6 Dissection: The required region must be clearly displayed. Use as many pins as necessary and cut away only tissue which obscures the part to be displayed. Keep your dissection tidy. Always use a dissecting lens for the dissection of small structures. Your drawing must be clear, in the correct proportion and be an accurate representation of **your** dissection. All labels should be accurate and detailed; do not allow labelling lines to cross. It is vital to complete your diagram in the time allowed; few marks can be awarded for a dissection, however good, if there is no labelled drawing of it.

7 Microscopic examination: Check that the microscope is properly adjusted so that as much detail as possible can be viewed on the specimen. Low-power plans should be drawn on a large scale and show the position of all the major types of tissue present; no individual cells should be drawn. With high-power drawings the number of cells to be drawn will normally be specified; do not exceed this. Draw each cell in its correct position and ensure that the proportions are correct. All visible cell contents must be included but all lines must be clear-cut and not sketchy.

8 Experiments: Plan your time carefully. Record all details of the method used, e.g. exact volumes used, exact times, exact temperatures, etc. Your own results must be clearly stated, usually in the form of a table and/or a graph. A high proportion of the marks available will be for interpretation of **your** results.

MATHEMATICAL AND PHYSICAL SCIENCE BACKGROUND

Although no questions will be set directly on these topics candidates are usually expected to have a basic knowledge of the following:

1 The electromagnetic spectrum.

2 Energy concepts (laws of thermodynamics, activation energy, chemical bond energy, potential energy).

3 Ions, molecules, acids, bases, pH and buffers.

4 Isotopes: stable and radioactive.

5 The colloidal state.

6 Oxidation, reduction, electron and hydrogen transfer.

7 Hydrolysis, condensation.

8 Arithmetic mean, mode and median.

9 Standard deviation.

10 Histograms, frequency diagrams and normal distribution curves.

11 Bimodal distribution.

12 Preparation of graphs using linear and logarithmic scales.

13 Meaning of integral positive and negative indices.

Analysis of examination syllabuses 1988 **A Level** **AS Level** (where available)

Key for syllabus analysis ● Probably studied at GCSE but required in detail at A level ○ Required in detail in an option	London	JMB	AEB	Cambridge	Oxford	Oxford and Cambridge	Welsh	Southern	SEB	NIGCEEB	London	AEB	COSSEC	Oxford
Theory papers (hours) c=one or more compulsory questions (% total mark)	3c 33⅓; 3 33⅓	3c 40; 3c 40	2½c 38½; 3c 46; 1c 15½	2½ 33⅓; 1c 16⅔; 1½c 25	3c 33⅓; 3 33⅓	2¼c 32.7; 3 40	2c 30; 3c 45	2½c 50; 3 50	1½ 37½; 2½ 62½	1½ 30; 3 40	2½c 50; 1½ 30	2½c 80	1½c 40; 1½c 20	2c 48; 2c 32
Practical paper (hours) c=one or more compulsory questions (% total mark)	2¼c 20			3c 25	2½c 33⅓	2½c 16.4	3c 25			2c 20				
Assessment (% total mark)	13⅓	20	15½	25		10.9					20	20	20	20
Optional Project			□	□		□		□		□ Not opt. 10	□ Not opt.		□ Not opt. 20	
1.1 Principles of Classification	●	●	●	●	●	●		●	●	●			○	
1.2 Viruses, Monera, Protista and Fungi	●	○	●	●	●	●	○	●	●	●	○	●	○	○
1.3 Plants	●	○	●	●	●	●	○	●	●	●	○		○	
1.4 Animals	●	○	●	●	●	●	○	●	●	●			○	
2.1 The Cell	●	●	●	●	●	●	●	●	●	●	●	●	●	●
2.2 Plant and Animal Tissues	●	●	●	●	●	●	●	●	●	●	○			
3.1 Carbohydrates, Lipids and Proteins	●	●	●	●	●	●	●	●	○	●	●	●	●	●
3.2 Enzymes (including Bio-chemical Techniques)	●	●	●	●	●	●	●	●	●	●	●	●	●	●
3.3 The Genetic Code	●	●	●	●	●	●	●	●	●	●	●			
4.1 Mitosis and Meiosis	●	●	●	●	●	●	●	●	●	●	●		●	
4.2 Heredity and Genetics	●	●	●	●	●	●	●	●	●	●	●		●	○
4.3 Genetic Variation and Evolution	●	●	●	●	●	●	●	●	●	●	●		●	○
5.1 Types of Reproduction and Life Cycles	●	●	●	●	●	●	○	●	●	●				●
5.2 Reproduction in Mammals	●	●	●	●	●	●	●	●	●	●		●		
5.3 Reproduction in Flowering Plants	●	●	●	●	●	●	●	●	●	●	○	●		
5.4 Growth and Development	●	●	●	●	●	●	●	●	●	●	○	●	○	
5.5 Control of Growth	●	●	●	●	●	●	●	●	●	●	○		○	○
6.1 Autotrophic Nutrition (Photosynthesis)	●	●	●	●	●	●	●	●	●	●	○	●	●	●
6.2 Heterotrophic Nutrition (Holozoic)	●	●	●	●	●	●	●	●	●	●				
6.3 Heterotrophic Nutrition (Saprotrophs and Parasites)	●	●	●	●	●	●	●	●	●					●
7.1 Cellular Respiration	●	●	●	●	●	●	●	●	●	●	●	●		●
7.2 Gaseous Exchange	●	●	●	●	●	●	●	●	●	●	○	●	○	
8.1 Properties and Importance of Water	●	●	●	●	●	●	●	●	●	●				
8.2 Blood and Circulation	●	●	●	●	●	●	●	●	●	●	○	●	○	
8.3 Uptake, Transport and Loss in Plants	●	●	●	●	●	●	●	●	●	●	○	●	○	
8.4 Osmoregulation and Excretion	●	●	●	●	●	●	●	●	●	●				
9.1 Locomotion and Support in Plants and Animals	●		●	●	●	●	●	●	●	●	○			
10.1 Hormones and Homeostasis	●	●	●	●	●	●	●	●	●	●			○	○
10.2 Nervous System Behaviour	●	●	●	●	●	●	●	●	●	●	○	●	○	○
10.3 Sense Organs	●	●	●	●	●	●	●	●	●	●				○
11.1 Ecology	●	●	●	●	●	●	●	●	●	●	●		●	
11.2 Man and his Environment	●	●	●		●	●	○	●	●	●	●		○	○

AEB	The Associated Examining Board Stag Hill House, Guildford, Surrey GU2 5XJ
Cambridge	University of Cambridge Local Examinations Syndicate Syndicate Buildings, 1 Hills Road, Cambridge CB1 2EU
COSSEC (AS only)	As for Cambridge, Oxford and Cambridge, or SUJB.
JMB	Joint Matriculation Board Devas Street, Manchester M15 6EU
London	University of London Schools Examination Board Stewart House, 32 Russell Square, London WC1B 5DN
NISEC	Northern Ireland Schools Examinations Council Beechill House, 42 Beechill Road, Belfast BT8 4RS
Oxford	Oxford Delegacy of Local Examinations Ewert Place, Summertown, Oxford OX2 7BZ
O and C	Oxford and Cambridge Schools Examinations Board 10 Trumpington Street, Cambridge and Elsfield Way, Oxford
SEB	Scottish Examinations Board Ironmills Road, Dalkeith, Midlothian EH22 1BR
SUJB	Southern Universities' Joint Board for School Examinations Cotham Road, Bristol BS6 6DD
WJEC	Welsh Joint Education Committee 245 Western Avenue, Cardiff CF5 2YX

1 Classification

1.1 PRINCIPLES OF CLASSIFICATION

Assumed previous knowledge None.

Underlying principles

Early life forms were simple prokaryotic cells (bacteria and blue-green algae) which, through the slow process of adaptation to new environmental conditions, evolved into millions of organisms, each sufficiently different to prevent them breeding with each other. It was inevitable that man should attempt to name these organisms and form them into an ordered hierarchy. Initially this was achieved on the basis of morphology and behaviour with the simplest at the bottom and man himself at the top (Aristotle's *Scala Naturae*). With the gradual acceptance that all species arose by adaptation of existing forms, the basis of this hierarchy became evolutionary, and simple and complex forms were placed in the same group because they shared a common ancestry. Species are groups that have diverged most recently, genera sometime earlier and so on up the taxonomic ranks. The difficult task of determining an organism's ancestry forms the basis of the science of classification.

Points of perspective

The historical background to taxonomy including the contributions made by Aristotle (384–322 BC), Ray (1628–1705), Linnaeus (1707–1778) and Darwin (1809–1882).

Essential information

The science of biological classification is called **taxonomy**.

Reasons for classification There are over two million species which must be organized into groups before any useful study of them can be made. A good universal system of classification aids communication between scientists and allows information about a particular organism to be found more readily (e.g. in libraries). Organisms form a continuum and any grouping of them is arbitrary and devised only for the convenience of man. Because evolution is a continuous process and new species are still being discovered, it is sometimes necessary to revise classifications.

What is a species? Species are groups of interbreeding populations which are reproductively isolated from other such groups. The biological definition of a species emphasizes its genetic isolation. Morphological distinctiveness provides only a general guide for delimiting species.

Classification by differences – artificial classification This takes two forms:

1 Arranging organisms in groups according to whether or not they possess a particular characteristic and then subdividing on the basis of presence or absence of another characteristic.
2 Removing from the group of organisms all those with an obvious peculiarity to one set and those with another to a second set and so on until a large number of mutually exclusive groups is formed.

These two methods rely at each division on the use of a single character and will not only group together unrelated forms but a situation may be reached where no obvious diagnoses can be made at all. However, the end-point of mutually exclusive groups makes these two methods useful for keys and the dichotomous key produced by the first system is most commonly used for identifying organisms.

Classification by similarities – natural classification The need for this was recognized as early as the eighteenth century by the taxonomist Linnaeus. Organisms are divided into groups and subgroups according to their basic similarities. Linnaeus worked before evolution was an accepted concept. In order to develop a natural classification today taxonomists use morphological, anatomical, biochemical, hybridization and chromosome studies to arrive at tentative conclusions about phylogenetic (evolutionary) relationships between species. Such relationships must be based on homologous not analogous characteristics.

Homologous characteristics An organ of one organism is said to be homologous with an organ of another if both have a fundamental similarity of origin, structure and position which is manifested especially during embryonic development, regardless of their functions in the adult, e.g. mammalian ear ossicles are homologous with the bone concerned with jaw attachment in a fish.

Analogous characteristics An organ of one species is said to be analogous to an organ of another when both organs have the same function and when they are not homologous, e.g. wings of birds and butterflies.

Rank It is convenient to distinguish large groups of organisms from smaller ones; a series of rank-names is used which allows one to estimate the position of any particular group in the natural hierarchy. Linnaeus was the first to devise a comprehensive scheme and the rank-names in use today derive mainly from him. Organisms are divided into large groups, phyla, each showing a radically different body plan; diversity within each allows the phylum to be divided into classes. Each class is divided into orders of organisms which have additional features in common. Orders are subdivided into families and at this level differences are less obvious. Each family is divided into genera and each genus into species.

Binomial nomenclature Every organism is given a scientific name according to an internationally accepted system of nomenclature, first devised by Linnaeus. The name is always in Latin and is in two parts. The first name indicates the genus and is written with an initial capital letter; the second name indicates the species and is written with a small initial letter. These names are always distinguished in text by the use of italics or by underlining.

Rank	Man	Sweet Pea
Phylum	Chordata	Spermatophyta
Class	Mammalia	Angiospermae
Order	Primates	Rosales
Family	Hominidae	Leguminosae
Genus	*Homo*	*Lathyrus*
Species	*sapiens*	*odoratus*

Listed above are the obligate ranks of classification to which every specimen must be assigned but a taxonomist may use any number of additional categories within this scheme. The principle rank-names in use today are listed below in hierarchical order.

Kingdom, Subkingdom, Grade, **Phylum,** Subphylum, Superclass, **Class,** Subclass, Infraclass, Superorder, **Order,** Suborder, Infraorder, Superfamily, **Family,** Subfamily, Tribe, **Genus,** Subgenus, **Species,** Subspecies, Variety.

Link topics

Section 1.2 Viruses, monera, protista and fungi   Section 1.3 Plants   Section 1.4 Animals Section 4.3 Genetic variation and evolution

Suggested further reading

Savory, T., *Animal Taxonomy* (Heinemann 1970)

QUESTION ANALYSIS

1 (a) (i) Arrange the following in the correct sequence with the largest group first:
order, species, phylum, class, genus
(ii) Define the term 'species'.
(iii) Describe TWO factors which favour the evolution of new species.
(b) The following four animals belong to the same taxonomic group: frog, mouse, fish, turtle
(i) Name the taxonomic group to which they all belong.
(ii) Give one feature which they all have in common.
Mark allocation 6/100   Time allowed 7 minutes   Associated Examining Board 1984, Paper I, No. 1

This structured short answer question predominantly tests knowledge of taxonomy and classification, but also includes evolution. It is surprising how many candidates fail to obtain credit in part (a)(i) simply because they reverse the order of the sequence. With the LARGEST group first, the order is phylum, class, order, genus and species.

A species is a group of interbreeding individuals which are reproductively and genetically isolated. Genetic isolation is one factor which favours evolution of a new species, another being environmental change.

The animals listed in part (b) are all chordates, the characteristics of which are listed in Table 1.7, page 24. Any one of these features could be given in answer to part (b)(ii).

2 The wing of a housefly and the wing of a bat are

A both homologous and analogous

B analogous not homologous

C homologous not analogous

D neither homologous nor analogous

Mark allocation 1/40   Time allowed 1½ minutes   *In the style of the Cambridge Board*

This is a multiple choice question testing the candidate's understanding of 'homologous' and 'analogous'. Although functionally similar the wings of the two organisms were derived as the structural solution to the problem of flight, at different times, and from different structures, during evolution. The two structures are therefore analogous but not homologous.

3 (a) Place the following list of taxonomic categories in their correct sequence beginning with the largest unit of classification and ending with the smallest: genus; phylum; family; species; order; class.   (5)

(b) Name a genus of flowering plant or conifer.   (1)

Mark allocation 6/80   Time allowed 7 minutes   *Oxford Specimen paper 1987, Paper 1, No. 1*

This is structured question requiring knowledge of taxonomic ranks and testing factual recall.

In (a) the order is important, not the actual position, as a mark is normally given for any two consecutive ranks correctly stated, i.e. 'genus' followed by 'species' gains a mark wherever it appears in the list. Note 'beginning with the largest . . .' which means start with phylum and work towards species.

The genus is the first of the two names given to an organism under the binomial system, i.e. the one shared with other closely related organisms. The Scots Pine has the genus *Pinus* and the species *sylvestris*.

1.2   Viruses, monera, protista and fungi

Assumed previous knowledge The basic structure of a protozoan.

Underlying principles

As living organisms evolved from non-living material, it is not surprising that the dividing line between some of them is thin. Viruses, for example, possess features of both living and non-living material. The other acellular organisms are far from being the simple structures they are sometimes thought to be. Within the confines of a single cell they show remarkable specialization of structures to suit a widely divergent range of environments. They are the oldest organisms in the world (3000 million years old); indeed more than three-quarters of the time since life first evolved on earth was occupied by acellular organisms alone. They are also the most abundant; one gram of fertile soil may contain as many as 2500 million bacteria.

Points of perspective

Differences between prokaryotic and eukaryotic cells and the theory of evolution of eukaryotic cell organelles from prokaryotic cells. Importance of early organisms in changing the composition of the earth's atmosphere, in particular in the production of free oxygen. Historical aspects of the control of disease caused by pathogenic protista, in particular the work of Jenner (1748–1823), Pasteur (1822–1895), Koch (1843–1910), Lister (1827–1912) and Fleming (1881–1955).

Essential information

In the past all organisms have been classified as plants or animals according to whether they possess chlorophyll and photosynthesize, as do plants, or lack chlorophyll and obtain their food from other organisms, as do animals. Many acellular organisms are capable of both modes of life. It is, therefore, convenient to distinguish other kingdoms of organisms, distinct from both plants and animals.

For the purposes of this book Viruses, Monera, Protista and Fungi will be considered as separate kingdoms.

Viruses exist outside living cells where they may be considered non-living chemicals. They enter living cells and once inside, multiply with the assistance of the host cells. Viruses contain very few enzymes and so, as intracellular parasites, they use the host's enzymes for their own metabolism. The presence of viruses was first postulated by the Russian Iwanowski at the end of the nineteenth century and they were isolated in crystalline form in 1935 by the American biochemist Stanley.

Viruses range in size from 20 nm–300 nm, i.e. below the resolving power of the light microscope. They are highly specific to both host and a particular tissue within a host. They are generally classified according to the host they infect, i.e. plant viruses, animal viruses and bacterial viruses (bacteriophages).

Essentially viruses are composed of a core of nucleic acid surrounded by a protein sheath. In some, e.g. fowl plague virus, fat and carbohydrate are also present. The nucleic acid in bacteriophages is always DNA; plant viruses all contain RNA and animal viruses may contain DNA or RNA.

There are many shapes of virus but they are frequently rod-shaped or polyhedral.

Life cycle of a virulent phage, e.g. T-phage of *Escherichia coli*

1 The virus becomes adsorbed onto the surface of *E.coli* by fibrils.
2 The needle is pushed into the bacterium and the DNA moves into the host cell.
3 Immediately replication of bacterial DNA (and therefore enzyme production) ceases.
4 All systems of the host cell are taken over by viral DNA and used to produce more strands of viral DNA.
5 The host enzyme and synthetic systems are used to produce protein coats etc. for this DNA. More than 100 new viruses are produced in this way.
6 Viral DNA causes the host systems to produce enzymes which cause lysis of the host cell wall, thus releasing the new viruses. The whole process generally takes about one hour, but may vary depending on the virus and environmental conditions.

Life cycle of a temperate phage

The process is much less rapid and the host and phage may exist together for many generations. Host DNA may become incorporated in the viral DNA, and this bacterial DNA is carried to the next host, thereby resulting in new characteristics. This process of **transduction** is an important method by which antibiotic-resistance spreads throughout a population of bacteria.

Diseases caused by viruses

Plant viruses cause mosaic diseases, e.g. of tobacco, potato, bean and sugar cane. Animal viruses cause fowl plague, mumps, cold sores, influenza, smallpox and poliomyelitis.

Monera. Their group includes bacteria and blue-green algae.

Bacteria

Structure

Bacteria range in size from $0.5\mu m - 4.0\mu m$.

Their shape may be spherical (coccus), rod-like (bacillus) or spiral (spirillum). The spherical forms may occur in chains (streptococcus) or in clusters (staphylococcus).

They are the smallest form of life organized on a cellular basis but the cells differ in a number of respects from those of other organisms. They are known as prokaryotic cells and have the following characteristics:

1 no distinct nucleus
2 DNA not incorporated in chromosomes but comprising a single, circular strand
3 no spindle forms at cell division
4 membrane-bounded organelles (e.g. Golgi, endoplasmic reticulum, mitochondria) are absent
5 no large vacuoles
6 cell wall of protein and polysaccharide (composition varies with genera).

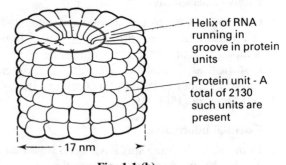

Helix of RNA running in groove in protein units

Protein unit - A total of 2130 such units are present

- 17 nm -

Fig. 1.1 (b)

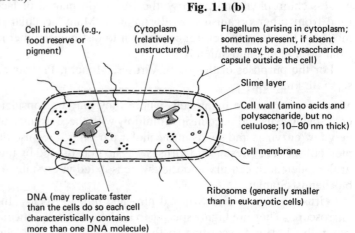

Head (hexagonal)

Protein coat

DNA

Sheath (protein)

Core

Tail

Fibre

Cell inclusion (e.g., food reserve or pigment)

Cytoplasm (relatively unstructured)

Flagellum (arising in cytoplasm; sometimes present, if absent there may be a polysaccharide capsule outside the cell)

Slime layer

Cell wall (amino acids and polysaccharide, but no cellulose; 10–80 nm thick)

Cell membrane

DNA (may replicate faster than the cells do so each cell characteristically contains more than one DNA molecule)

Ribosome (generally smaller than in eukaryotic cells)

Fig. 1.1 (a) Structure of a bacteriophage

Fig. 1.1 (c) Generalized bacterial cell

Mode of life

Bacteria are often classified on the basis of their method of obtaining energy:

1 Most obtain energy from the oxidation or breakdown of living or non-living organic matter, i.e. are heterotrophic. They are parasitic, saprophytic or symbiotic.

2 A few obtain energy from the oxidation of inorganic materials and use this energy to synthesize their own foods, i.e. are chemo-autotrophic; for example, iron bacteria oxidize ferrous compounds to ferric hydroxide and release energy; nitrifying bacteria oxidize ammonia to nitrate and release energy; colourless sulphur bacteria oxidize hydrogen sulphide to sulphur and release energy.

3 Some bacteria can use the energy of sunlight for the manufacture of food, i.e. are photo-autotrophic; for example, green and purple sulphur bacteria contain bacteriochlorophyll and photosynthesize using hydrogen sulphide (not water) as a source of hydrogen; sulphur, not oxygen, is a by-product.

Heterotrophic bacteria may be aerobic or anaerobic and some may live in either environment, for example, *Clostridium tetani* is a parasite in the absence of oxygen and a saprophyte in its presence. Autotrophs are anaerobic.

The importance of bacteria

1 The breakdown and recycling of plant and animal remains, in particular the recycling of essential elements such as carbon, nitrogen and phosphorus. The same processes account for the bacterial decomposition of sewage.

2 Symbiotic relationships with other organisms; for example, bacteria in the human gut synthesize some of the vitamin B complex and others break down cellulose in the guts of herbivores.

3 Food production, e.g. yoghurt, some cheeses, vinegar, coffee and tea.

4 Manufacturing processes such as tanning leather, retting flax to make linen, making soap powders.

5 A source of antibiotics, e.g. streptomycin.

6 They are easily cultured and therefore used for research.

7 Pathogenic bacteria are intercellular parasites and the symptoms of a disease are often caused by the toxins they produce. Only five genera of bacteria are thought to infect plants but they infect a wide range; for example *Xanthomonas phaseolus* causes common blight of beans. Human diseases caused by bacteria include whooping cough (*Bordetella pertussis*), some forms of pneumonia, leprosy, syphilis, tuberculosis, diphtheria, typhoid (*Salmonella typhi*), cholera and scarlet fever.

Protista. These consist of single-celled eukaryotic organisms. There are two main groups: the Protozoa and the Euglenophyta.

Protozoa

Table 1.1 Classes of protozoa

Phylum		Characteristics
Protozoa (first animals)		Acellular (unicellular) Mostly microscopic (2μm – 10μm long) Specialized organelles but no tissues or organs

Class	Characteristics	Examples
Rhizopoda (Sarcodina)	Move and feed by means of pseudopodia	*Amoeba*
Flagellata (Mastigophora)	Move using flagella; asexual reproduction by longitudinal binary fission	*Trypanosoma; Euglena*
Ciliata	Move using cilia; meganucleus and micronucleus; asexual reproduction by transverse binary fission	*Paramecium*

Estimates of the number of species vary from 1500–50 000. There are free-living, symbiotic and parasitic species which are widely distributed but only active in aqueous media. They are highly specialized with division of labour and specialization of organelles which has resulted in a wide variation of form and physiology. The flagellates include holophytic and holozoic forms. Protozoa form a significant part of plankton. Two orders of rhizopods, the Radiolaria and the

Foraminifera, have hard tests (shells) and over millions of years these have fallen to the bottom of the oceans to form deposits thousands of feet thick. Changes in sea level have exposed some of these as chalk and limestone. Most animals have one or more protozoan parasites, some of which cause disease in man including: *Trypanosoma* causing sleeping sickness and *Plasmodium* causing malaria.

Fungi. About 80 000 species of fungi have been recognized, including 30 000 Ascomycetes and 25 000 Basidiomycetes. They help maintain soil fertility by recycling many important minerals and decomposing the organic matter of both plants and animals. Industrial uses of fungi include the extraction of enzymes (e.g. invertases) and drugs (e.g. steroids and ergotamine); cheese making; baking and brewing. Fungi cause diseases in plants, e.g.
Puccinia graminis – rust of wheat
Phytophthora infestans – late blight of potatoes
A few fungi cause diseases in man, e.g.
Microsporum audouini – ringworm
Trichophyta interdigitale – athlete's foot
Stored food can be damaged by moulds; dry rots attack wooden construction materials and mildews affect cotton, wool and manufactured goods.

Table 1.2 Groups of fungi

Kingdom		Characteristics
Fungi		No chlorophyll; normally have hyphae, often grouped as a mycelium; hyphae are coenocytic, although they may have septa; cell wall mostly hemicellulose, chitin, lipid and protein

Division	Characteristics	Examples
Phycomycetes	Aseptate mycelium; asexual spore is a zoospore or sporangiospore; sexual stage is a zygospore or oospore	*Phytophthora* – causes late blight of potatoes; *Mucor* – large genus, mainly saprophytic
Ascomycetes	Septate hyphae; asexual reproduction by conidia or budding; sexual reproduction by ascospores	*Saccharomyces* (yeast) – simple, consisting of a single cell; *Penicillium*
Basidiomycetes	Septate hyphae; hyphae often massed into extensive three-dimensional structures (puffballs; toadstools; bracket fungi); sexual reproduction by basidiospores	*Agaricus* – mushroom

Link topics

Section 2.1 The cell
Section 6.3 Saprophytes, parasites and symbionts

Suggested further reading

Noble, W. C. and Naidoo, J., *Micro-organisms and Man,* Studies in Biology No. 111 (Arnold 1979)
Sleign, M. A., *The Biology of Protozoa,* 2nd ed. (Arnold 1987)
Vickerman, K. and Cox, F. E. G., *The Protozoa* (Murray 1967)

QUESTION ANALYSIS

4 (a) In the space below draw a diagram to show the general structure of a bacteriophage and label clearly two important components. (2)
(b) Which component of the bacteriophage enters the host cell during the process of infection? (1)
(c) What subsequently happens to this component? (1)
(d) Name one disease caused by a virus. (1)
(e) List two ways by which viruses are transmitted from one host to another. (2)
Mark allocation 7/80   Time allowed 8 minutes   *Oxford Local 1979, Paper 1A, No. 6*

The diagram in part (a) should be similar to Fig. 1.1a in the essential information with 'protein coat' and 'DNA strand' the important components.

The answers to (b), (c) and (d) are covered in the essential information. The role of DNA in producing mRNA that instructs the host to produce viral components should be mentioned in (c). The disease in (d) could be of a plant or animal.

In (e) it is necessary to look at the ways in which the diseases listed in the text are transmitted, e.g. influenza – through the air by droplet infection. Other viral diseases can be spread through water, contaminated articles, contact, or by vectors, etc.

5 Write an essay on viruses.
Mark allocation 25/100   Time allowed 45 minutes   In the style of the Joint Matriculation Board

The answer to this question should present a general account of the viruses, incorporating all aspects rather than being restricted to one aspect only, such as structure or reproduction.

The introduction could include a brief mention of their discovery and isolation, as outlined in the essential information, and continue with details of their size and structure emphasizing their simple organization and unique position between living and non-living forms.

A consideration of their life cycle should stress dependence on the host cell and follow the example given in the text above. Temperate and virulent types should be mentioned and specific examples used throughout (see text).

The economic importance of the group in causing disease in plants and animals should be covered with appropriate examples; but their beneficial features, such as their use in biological control of pests and in genetic studies, should not be overlooked.

In conclusion, mention should be made of the evolutionary significance of viruses in transmitting genetic material between organisms.

Answer plan

Introduction: historical aspects
Structure: size; diagram of specific example, e.g. T-phage
Reproduction: life cycle, e.g. T-phage
Economic importance: harmful and beneficial
Conclusion: evolutionary significance of transduction.

6 (a) What size are viruses and how can they be studied?
    (b) Describe the structure of two different types of virus.
    (c) Describe how one of these viruses infects living cells and reproduces.
    (d) With the aid of named examples compare viruses and bacteria as agents of disease.
Mark allocation 28/140   Time allowed 35 minutes   Oxford Local 1983, Paper II, No. 1

In the absence of information on the distribution of marks it is best to devote an equal amount of time to each part, although part (a) could possibly be completed more quickly.

The sizes of viruses required for part (a) are in the range 20-300nm. As such they are well below the resolving power of the light microscope and therefore must be studied using an electron microscope. Some detail of the basic principles of this instrument should be included (see Section 2.1 'Essential information'). Many candidates fail to appreciate that as viruses can be crystallized they can also be studied by X-ray crystallography.

In answering part (b) it is important to choose two examples which are distinctly different in structure. In making your choice, bear in mind that you are required in part (c) to show how one of your examples infects living cells and reproduces. This underlines the importance of always reading the whole of a question before attempting to answer any part of it. Your choice will clearly depend upon which virus you have studied in detail. A typical choice might be a bacteriophage, such as one of the seven strains of the T-phage virus which infects the bacterium *Escherischia coli*, the structure of which is shown in Fig. 1.1(a). A contrasting example could be the tobacco mosaic virus, whose structure is given in Fig. 1.1(b). As structure is much more clearly shown by diagrams, both diagrams should be included and should be well annotated.

Provided a sensible choice was made in part (b), the answer to part (c) should be straightforward. To take the above example of the T-phage virus, a slightly more detailed version of the six stages outlined in the essential information should be given. If a series of diagrams is used, these should be well annotated.

A general comparison which could be made to introduce part (d) is that while viruses are always parasitic, and therefore usually cause disease symptoms in their hosts, only a few bacteria are agents of disease. Viruses are generally transmitted by droplet infection, e.g. the myxovirus causing influenza or the rubella virus causing german measles. Bacteria also may be transmitted in this way, e.g. *Corynobacterium diptheriae,* although a much greater range of methods is involved, e.g. through water (*Vibrio cholerae* – cholera) or contaminated food (*Salmonella typhi* – typhoid*)*. Both may be transmitted by direct contact, e.g. the bacterium *Treponema pallidum* which causes syphilis is transmitted by sexual contact. Both types may be transferred by vectors e.g. the *Rickettsia* bacterium causing typhus has the rat flea as a vector, and the yellow fever virus is transmitted by mosquito. Many animal diseases are caused by viruses, e.g. foot and mouth disease, fowl pest and myxomatosis. Others are bacterial in origin, e.g. tuberculosis in cattle. The economically damaging diseases of plants are more often viral, e.g. tobacco mosaic virus and spotted wilt of tomato.

Whereas bacteria can be controlled by antibiotics such as a penicillin, viruses are only controllable by a substance called interferon. Diseases of both types may be controlled through immunization. Other minor comparisons could be included, but the emphasis throughout should be to include similarities and differences, and above all, specific examples.

7 Give a general account of fungi.
Mark allocation 20/100   Time allowed 35 minutes   *London June 1980, Paper II, No. 9*

This is a very common style of essay question on classification. As the basic outline for such questions has already been discussed, only a summary is given here.

Write a balanced essay. Do not restrict information to one aspect only, e.g. classification. Use specific examples to illustrate points. Use well annotated diagrams where applicable, but do not repeat the information in the written part of the essay. Remember that there is so much information that not all of it can be included so choose the important points. Work on the basis of 1 mark (possibly ½ mark) for every relevant detailed biological fact supported with an example.

Answer plan

1 **Classification/taxonomy:** major characteristics of the group and the three main classes (see essential information)

2 **Structure:** basic plan of fungal hyphae; variations between the classes

3 **Physiology:** some detail, with examples, on methods of: sexual and asexual reproduction; nutrition (extracellular, substrates, storage materials, saprophytic, parasitic and symbiotic examples); respiration and excretion (simple diffusion)

4 **Ecology:** distribution; relationships to other organisms, e.g. saprophyte, parasite, symbiont

5 **Economic importance:** beneficial effects, harmful effects. (For details see essential information.)

Emphasis should be on economic importance, but all five aspects should be fully covered.

1.3   PLANTS

Assumed previous knowledge Principles of classification, Section 1.1.

Underlying principles

Attempts to classify plants have been made since Theophrastus (c. 300 BC), but prior to Linnaeus (1707–1778) there was little, if any, uniformity in the nomenclature of plants; it was not until after Darwin (1809–1882) that attempts were made to classify plants according to their evolutionary relationships. Plants are grouped together according to their similarities of morphology, anatomy, embryology, biochemistry and reproductive behaviour. As the groups vary slightly depending on those aspects which the taxonomist considers most important, there are a number of different schemes of classification.

Points of perspective

A background knowledge of other plant classification schemes is useful. So is a knowledge of the geological periods during which the groups arose and were dominant, and a knowledge of some fossil representatives of the groups.

Essential information

Algae

Table 1.3 Characteristics of algae

Phylum	Characteristics
Algae	Plant body is a thallus, i.e. not differentiated into stem, root and leaves
	All or most cells contain chlorophyll
	Food-conducting tissue is rare

This is an artificial but convenient group which includes the aquatic, autotrophic, non-vascular plants. Sub-divisions are based mainly on structural and biochemical differences associated with photosynthesis.

There are over 7000 species of Chlorophyceae and 1100 species of Phaeophyceae. The algae are found almost everywhere except in sandy deserts and permanent ice and snow. As primary producers they are a major component of phytoplankton.

Derivatives of alginic acid, which is extracted from Phaeophyceae such as *Laminaria* and *Ascophyllum*, are used in the following commercial processes: as thickeners in food; in cosmetics; in latex formation; as gelling agents in confectionery, meat jellies and dental impression powders; as emulsifiers in ice cream, polishes, processed cheese and synthetic cream; as surface films on glazed tiles; as medical gauzes; and in some types of sausage casings.

Table 1.4 Classes of algae

Class	Characteristics	Examples
Chlorophyceae (green algae)	Chlorophyll is the main photosynthetic pigment; chiefly found in fresh water	*Spirogyra* – filamentous, chain of identical cells containing spiral chloroplasts; *Ulva* (sea lettuce) – flat thallus
Phaeophyceae (brown algae)	Multicellular; contain chlorophyll + fucoxanthin (a brown pigment); store the carbohydrate laminarin	*Ectocarpus* – filamentous, prostrate anchoring filaments and upright photosynthetic reproductive filaments; *Fucus* (common intertidal wrack) – holdfast, stipe and flattened frond

Carrageenin is used in much the same way as alginic acid and is extracted from red seaweeds (Rhodophyceae) such as *Gigartina* and *Chondrus crispus*. These algae are also the richest source of agar, which is used as a medium for algal, bacterial and fungal cultures. The siliceous walls of diatoms form large deposits of diatomite which are mined and used as an insulation material and as a filtration aid (e.g. sugar refining; brewing). In coastal areas seaweeds may be used as a fertilizer and for animal fodder. Algae aid the oxidation of sewage and thus promote the growth of aerobic bacteria. They are used in percolation beds in water purification plants but in reservoirs they cause problems by blocking filters and giving the water an unpleasant taste. They prevent erosion by binding the surface of the soil. They may be used as a measure of water pollution.

Bryophytes

Table 1.5 Classification of bryophytes

Phylum		Characteristics
Bryophyta		No filamentous forms (except one stage in mosses) Clear alternation of generations, with gametophyte independent, sporophyte dependent Gamete-producing cells surrounded by jacket of sterile vegetative cells No asexual spores produced
Class	**Characteristics**	**Examples**
Hepaticae (liverworts)	Thalloid gametophyte is flat ribbon or leafy shoot; found in moist, shady places	*Riccia; Marchantia; Pellia*
Musci (mosses)	More conspicuous; better able to withstand drought; gametophyte has two growth stages: filamentous protonema and upright plant with spirally arranged leaves; more complex capsule	*Polytrichum; Funaria; Mnium*

There are over 23 000 species of bryophytes of which about 9000 are liverworts and 14 500 are mosses. They are mostly terrestrial, but water is essential for fertilization. They occur on soil and the trunks of trees; many are fully aquatic.

Pteridophytes

Table 1.6 Classification of pteridophytes

Phylum		Characteristics
Pteridophyta		Clear alternation of generations Sporophyte is most prominent phase Gametophyte reduced to simple prothallus Sporophyte has roots, stems, leaves and simple vascular tissues
Class	**Characteristics**	**Examples**
Lycopodiales (club mosses)	Densely packed small leaves borne on branched stems	*Selaginella* – heterosporous *Lycopodium* – homosporous
Equisitales (horse-tails)	Whorls of small leaves on upright stems; strobili (cones) at apex	*Equisetum*
Filicales (true ferns)	Prominent frond-like leaves; often sporangia on lower surface; underground rhizomes	*Dryopteris; Polypodium*

There are 11 000 species of Filicales and 100 species of Lycopodiales; the class Equisitales has one genus with two dozen living species. Most Pteridophyta today are relatively small but of the numerous fossil forms many were forest trees.

Table 1.7 Classification of spermatophytes

Phylum		Characteristics
Spermatophyta (Spermaphyta) (seed plants)		Heterosporous: male spore is pollen grain, female spore is embryo sac Embryo sac enclosed in ovule Seeds formed after fertilization Vascular tissues complex Gametophyte very reduced
Class	**Characteristics**	**Examples**
Gymnospermae (naked seeds)	No protective case (e.g. ovary wall) around ovule; most have cones (not *Ginkgo* or *Taxus*); no vessels, only tracheids in xylem; many have needle-like leaves which are often evergreen	*Pinus* – pine; *Picea* – spruce; *Larix* – larch

Of the 6500 species of Gymnosperms, most are xerophytic. They are common in temperate regions where the soil water may be frozen. Conifers grow quickly to produce the soft wood often used in building and for inexpensive furniture. However, their greatest importance is in pulping and paper manufacture; about 40% of the world's timber harvest is used for paper. Tannins obtained from the bark of some conifers (e.g. spruce) are used to prevent the growth of bacteria on leather, to prevent scaling in boilers and for clarifying wines and beers. By-products of the timber trade are resins, which are used in paints, varnishes and printing inks.

Table 1.8 Classification of angiosperms

Class		Characteristics
Angiospermae		Flowers produced; ovules develop inside ovary; ovary wall develops into fruit; vessels in xylem
Sub group	**Characteristics**	**Examples**
Monocotyledons	Embryo has a single cotyledon; leaves usually show parallel venation; flower parts typically in multiples of 3; stem contains scattered vascular bundles; no secondary growth	*Endymion non-scriptus* (bluebell) – family Liliaceae; *Triticum* (wheat) – family Graminae
Dicotyledons	Embryo has two cotyledons; leaves have a network of veins; flowers often in multiples of 4 or 5; stem contains ring of vascular bundles; secondary growth occurs	*Ulex* (gorse) – family Leguminosae; *Helianthus* (sunflower) – family Compositae; *Prunus domestica* (plum) – family Rosaceae

These are the dominant plants of the world today with over 250 000 species, of which 60 000 are monocotyledons and 190 000 are dicotyledons. They vary in form from simple duckweed (*Lemna*) through herbaceous and shrubby plants to trees such as chestnut and oak. They are the most important terrestrial primary producers. Angiosperms are man's most important food plants: they include the cereals, vegetables, succulent fruits and sugar cane. As well as providing food for man's domesticated animals (pasture grasses and clover), they are a source of oils, drugs and insecticides. The Leguminosae are very important for the nitrogen-fixing bacteria present in their root nodules. The angiosperms provide a variety of hardwoods.

Link topics

Section 1.1 Principles of classification
Section 4.3 Genetic variation and evolution

Suggested further reading

Heywood, V. H., *Plant Taxonomy*, Studies in Biology No. 5, 3rd ed. (Arnold 1987)
Holmes, S. *Outline of Plant Classification* (Longman 1983)

Miller, R., *Plant Types I* (Hutchinson 1983)
Miller, R., *Plant Types II* (Hutchinson 1985)

QUESTION ANALYSIS

8 (a) For what purposes are organisms classified?
(b) Using the Crustacea, Nematoda, Algae and Bryophyta describe the main features that allow organisms to be placed into each of these groups.
Mark allocation 20/80   Time allowed 35 minutes   *In the style of the Cambridge Board*

This is an essay question testing knowledge of the taxonomic characters of four groups and the candidates' understanding of why organisms are classified.

In part (a) the key word is 'purposes' indicating that the reasons for classification are required, not an account of the methods of classification as some candidates would be tempted to give. For details see 'Reasons for classification' in the essential information for Section 1.1.

In part (b) the key words are 'describe' and 'main'. 'Describe' indicates that a mere list is inadequate; a little more detail on each feature, explaining what it is, its appearance and its importance in taxonomy, is required. 'Main' indicates that only the most important features should be described.

The question requires description of only those features which enable organisms to be placed into their groups; the candidate must therefore use only those of taxonomic value. Lists of the features required are given in this section for Algae and Bryophyta and in Section 1.4 for Nematoda and Crustacea, under the heading *Characteristics* in the essential information. Take care with Crustacea to keep to the group's own diagnostic features and not to include those of the Arthropoda.

Answer plan

(a) Convenience, easier access to information, communication without ambiguity, means of indicating similarities and evolutionary relationships.
(b) A description of the following features including their taxonomic value, i.e.

Crustacea: exoskeleton impregnated with calcite; cephalothorax; gills; 2 pairs of antennae.

Nematoda: Unsegmented; worm-like; pseudocoel; mouth and anus; circular muscles absent.

Algae: thalloid; contain chlorophyll; food-conducting tissue is rare.

Bryophyta: alternation of generations with dominant gametophyte upon which sporophyte depends; sterile jacket around gametangia; no asexual spores.

9 (a) State four similarities between angiosperms and gymnosperms that allow them to be classified as spermatophyta.
(b) State four differences between angiosperms and gymnosperms which mean they belong to separate sub-groups.
(c) List three features which distinguish dicotyledons from monocotyledons.
Mark allocation 11/60   Time allocation 15 minutes.   *In the style of the Cambridge Board*

In effect (a) is asking for the major features of the spermatophyta, so choose any four of the following:
(1) Embryo sac enclosed in the ovule
(2) Sporophyte is dominant and the gametophyte is much reduced
(3) Separate male and female spores
(4) Pollen tube present
(5) Fertilized embryo sac develops into a seed
(6) Xylem and phloem present
In (b) the answers should be features which clearly distinguish the two groups. Choose from:

Angiosperms	Gymnosperms
Ovule protected in ovary	Ovule unprotected
Stigma and style present	Stigma and style absent
Cones absent	Cones present
Fruits formed after fertilization	No fruits formed
Companion cells present in phloem	No companion cells
Xylem has tracheids and vessels	Only tracheids in xylem, no vessels

Again, in (c) be sure that the features chosen clearly distinguish the two groups. Give both parts of the answer, not just one, in each case. Select any three from the following:

Dicotyledons	Monocotyledons
Two cotyledons	One cotyledon
Net (reticulate) veined leaves	Parallel veined leaves.
Floral parts in groups of four or five (or multiples thereof)	Floral parts in threes (or multiples thereof)
Vascular bundles in stem arranged cylindrically near periphery	Vascular bundles in stem scattered throughout – some central
Vascular cambium in all types showing secondary growth	Little, if any, vascular cambium

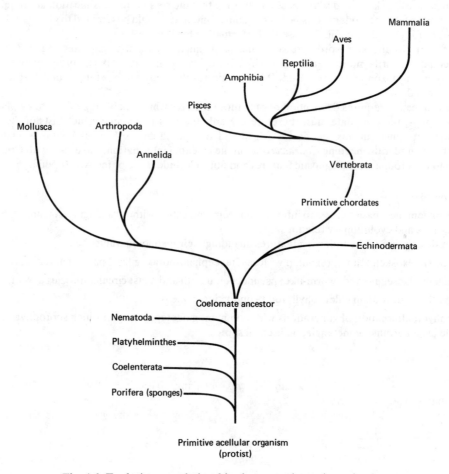

Fig. 1.2 Evolutionary relationships between the major animal groups

1.4   ANIMALS

Assumed previous knowledge Principles of classification, Section 1.1

Underlying principles

If one accepts the Darwinian theory of evolution that each group has evolved from a pre-existing one, the problem is to determine the order in which they arose and which groups are most closely related (phylogeny). The original ancestors of each group are now extinct, as are many of the intermediate groups. Many soft-bodied forms have left no fossil record and others that have remain to be discovered. The picture of animal evolution is hence fragmentary and open to a number of interpretations. Fig. 1.2 shows one 'evolutionary tree' incorporating a number of possible theories. These theories have been based on the embryological and morphological relationships of the groups.

Points of perspective

Theories of the origin of the Metazoa. The geological periods during which the groups arose and were dominant and a knowledge of a fossil example of each group.

Essential information

Coelenterata

Table 1.9 Classification of coelenterates

Phylum	Characteristics
Coelenterata (hollow-intestine)	Diploblastic, i.e. all cells derived from ectoderm or endoderm Radially symmetrical Single body cavity (enteron; gastrovascular cavity) with only one opening to exterior Nematoblasts (stinging cells) Polymorphism common (hydroid or medusoid forms)

Class	Characteristics	Examples
Hydrozoa	Typically hydroid and medusoid forms	*Hydra; Obelia; Physalia* – Portuguese man-of-war
Anthozoa	No medusoid stage	*Actinia* – sea-anemone; corals
Scyphozoa	Reduced hydroid stage	*Aurelia* – jellyfish

There are over 9000 species of coelenterates which are entirely aquatic and mostly marine. They feed only on living prey. Coral reefs form and protect certain islands and are useful for determining the age of rock deposits.

Platyhelminthes

Table 1.10 Classification of platyhelminthes

Phylum	Characteristics
Platyhelminthes (flatworms)	Triploblastic, i.e. cells derived from ectoderm, endoderm and mesoderm Bilaterally symmetrical (prerequisite for efficient locomotion) Dorso-ventrally flattened Acoelomate Gut, where present, branched with one opening

Class	Characteristics	Examples
Turbellaria	Free-living; ciliated epidermis	*Planaria*
Trematoda	Parasitic with one vertebrate and one invertebrate host	*Fasciola* – liver fluke; *Schistosoma* – blood fluke
Cestoda	Parasitic with two vertebrate hosts; body divided into proglottids	*Taenia* – tapeworm

The platyhelminthes include aquatic, free-living and parasitic forms. Many of them cause disease in man and/or in domesticated animals; for example, bilharzia in man and liver rot in sheep are caused by *Schistosoma* and *Fasciola*, respectively.

Nematoda

Table 1.11 Classification of nematodes

Phylum	Characteristics
Nematoda (thread worms)	Unsegmented Cylindrical, worm-like shape Body cavity an unlined pseudocoel Mouth and anus present No circular muscles (movement restricted to dorso-ventral plane)

There are over 10 000 species of free-living and parasitic nematodes found in every conceivable niche from the poles to the tropics. Many cause disease in man, animals and crops, e.g. hookworm in man – *Ascaris*; pin-worm in man – *Enterobius*; elephantiasis – *Wuchereria (Filaria) bancroftii*; *Heterodera* is a parasite in the roots of tomatoes, cucumbers and beet.

Annelida

Table 1.12 Classification of annelids

Phylum		Characteristics
Annelida (ringed worms)		Coelom well developed (see below)
		Metamerically segmented (see below)
		Chaetae usually present
		Thin cuticle of collagen

Class	Characteristics	Examples
Oligochaeta	Obvious clitellum; few chaetae	*Lumbricus* – earthworm; *Allolobophora* – earthworm
Polychaeta	Many chaetae on parapodia	*Nereis* – ragworm; *Arenicola* – lugworm
Hirudinea	Ectoparasitic; suckers	*Hirudo* – leech

There are about 9000 species of freshwater, marine and terrestrial annelids which are mainly free-living. Earthworms improve soil fertility by increasing aeration, drainage and availability of nutrients. Leeches cause wounds through which pathogens may enter the body.

Importance of the coelom In coelomates the mesoderm is split into two layers divided by a fluid-filled cavity, the coelom. The development of the coelom allows the gut muscles to be separated from those of the body wall so that:

1 locomotion and digestion can take place independently and at the same time; this leads to the development of specialized locomotory and digestive organs.

2 a transport system (usually a blood system) must develop because the body cannot be flattened to produce a large surface area/volume ratio.

3 the presence of a transport system permits the development of specialized respiratory surfaces.

Metameric segmentation This is present in most coelomates. Basically the body of a metamerically segmented animal is made up of a series of identical segments of the same age. Each segment has a similar pattern of excretory organs, blood vessels etc., but they act independently (unlike a proglottis). Often metamerism is difficult to detect in the adult, especially at the anterior end (due to cephalization) and near the limbs. The repetition of structures inherent in metamerism lends itself readily to the adaptation of various parts of the body.

Arthropoda

Table 1.13 Classification of arthropods

Phylum		Characteristics
Arthropoda (jointed limbs)		Metamerically segmented
		Coelomate (haemocoel)
		Chitinous exoskeleton
		Jointed appendages
		Compound eyes often present (unique to arthropods)

Class	Characteristics	Examples
Crustacea	Exoskeleton hardened by calcite; Cephalothorax; 2 pairs antennae; Gills	*Daphnia* – water flea (subclass Branchiopoda) freshwater, parthenogenetic, filter feeder; *Oniscus* – woodlouse (subclass Malacostraca, order Isopoda) terrestrial; *Astacus* – crayfish ⎫ (subclass Malacostraca, order *Carcinus* – crab ⎬ Decapoda) 5 pairs walking legs, *Crangon* – shrimp ⎭ stalked compound eyes.
Chilopoda	Flattened body; 15–20 pairs long legs	*Lithobius* – centipede
Diplopoda	Cylindrical body; Up to 200 pairs short legs	*Julus* – millipede
Arachnida	Prosoma and opisthosoma (2 main body regions); 1 pair chelicerae; 1 pair pedipalps; 4 pairs walking legs; Lung books for respiration	*Araneus* – spider

Table 1.13 continued

Insecta	Head, thorax and abdomen; Exoskeleton strengthened by chitin and protein; 1 pair antennae; 3 pairs walking legs; Respiration by tracheal system	*Locusta* – locust (order Orthoptera) usually large herbivores, hind legs modified for jumping, stridulatory (sound producing) organs in male; *Libellula* – dragonfly (order Odonata) strong fliers, adult and nymph predatory carnivores; *Pieris brassica* – cabbage white butterfly (order Lepidoptera), scales on large wings, adults feed on liquids, larva is a caterpillar; *Musca domestica* – house-fly (order Diptera), hind wings modified to form halteres to control equilibrium in flight, legless larvae (maggots); *Apis mellifica* – honey-bee (order Hymenoptera) wide variety of mouthparts, many social forms

The Arthropoda is the largest and most successful animal phylum; over 800 000 species have been described. There are over 30 000 species of Crustacea, which are typically marine and form an important component of zooplankton; some are eaten by man. The 3000 or so species of Chilopods are all fast-moving carnivores and the 9000 species of Diplopods are slow-moving herbivorous scavengers. The Chilopoda and Diplopoda were formerly placed in the Class Myriapoda. There are over 35 000 species of arachnids which are parasitic or free-living carnivores. The early forms were mostly aquatic but present forms are mainly terrestrial. Many ticks carry disease. By far the most significant class of arthropods is the Insecta with more than 750 000 species described. Insects are mainly terrestrial and, although there are some freshwater forms, none are truly marine. The insect body plan can become specialized for many different modes of life and therefore the group shows great adaptive radiation and is widely distributed because of the ability to fly. Insects can be beneficial or harmful to man.

Table 1.14 Beneficial and harmful insects

Beneficial	Harmful
1 Use of insect products: silk; honey; bees' wax; cochineal; shellac	1 Destroy leaves and fruit of plants, e.g. locusts, boll weevils, fruit flies.
2 Commercial production of fruits dependent on insect pollination	2 Transmit disease, e.g. mosquitoes (malaria and yellow fever); house-flies (typhoid and dysentery); fleas (bubonic plague and typhoid); tsetse flies (sleeping sickness)
3 Biological control of harmful organisms, e.g. ladybirds eat aphids	3 Annoy or harm domestic animals, e.g. lice
4 Recycling of dead and decaying plant and animal remains, e.g. by beetles and flies	4 Cause destruction of materials in and around houses, e.g. cockroaches (spoil food); moth larvae (feed on carpets and clothing); termites (bite into wooden furnishings, or gnaw through the wooden supports of houses)
5 Food source of many economically important animals, e.g. trout	

Mollusca

Table 1.15 Classification of molluscs

Phylum		Characteristics
Mollusca (soft bodied)		Unsegmented Head, muscular foot and visceral mass Mantle secretes shell
Class	**Characteristics**	**Examples**
Gastropoda	Asymmetrical due to torsion; single shell, often coiled; feed using radula	*Littorina* – periwinkle; *Helix* – snail
Bivalvia	Shell in two halves; filter feeders	*Mytilus* – mussel; *Cardium* – cockle

There are over 80 000 living species of molluscs, which are typically marine but which have invaded freshwater habitats and land. Molluscs are valued as food for man, for the purple dye that can be obtained from *Murex,* and for pearls from oysters. Deleterious effects include damage done by the shipworm *Teredo* to wooden jetties; furthermore slugs and snails are harmful to crops, and snails are intermediate hosts in the life cycles of some parasites of economic importance, e.g. *Schistosoma.*

Echinodermata

Table 1.16 Characteristics of echinoderms

Phylum	Characteristics
Echinodermata (spiny skinned)	Acquired radial symmetry (often pentamerous) No head Endoskeleton of calcareous ossicles with movable spines Watervascular system with tube-feet and madreporite

Examples

Asterias (Class Asteroidea) flattened, star-shaped flexible body;
Echinus – sea urchin
There are over 5000 living species of echinoderms, which are exclusively marine. The 'Star of Thorns' starfish destroys coral.

Chordata

Table 1.17 Classification of chordates

Phylum	Characteristics
Chordata	Notochord – at least in embryo CNS tubular and dorsal Ventral heart Paired visceral clefts at some stage in life history Post-anal tail (usually)

Subphylum	Characteristics
Vertebrate (Craniata)	Vertebral column replaces notochord in adult Cranium Paired limbs) except in lampreys Jaws } and hagfish

Class	Characteristics	Examples
Chondrichthyes (cartilagenous fishes)	Cartilagenous skeleton; no operculum; no swimbladder	*Scyliorhinus* – dogfish
Osteichthyes (bony fishes)	Bony skeleton and scales; operculum; swimbladder	*Clupea* – herring
Amphibia	Dependent on freshwater for development; smooth, glandular skin	*Rana* – frog; *Triturus* – newt
Reptilia	Dry skin with epidermal scales; amniote eggs with thick leathery shells	*Lacerta* – lizard; *Natrix natrix* – grass-snake
Aves (birds)	Endothermic; epidermal feathers; beak; pectoral limbs modified to form wings	*Troglodytes troglodytes* – wren
Mammalia	Endothermic; hairy, glandular skin; diaphragm; secondary palate; mammary glands (except in monotremes); heterodont dentition	*Talpa* – mole (order Insectivora) feeds on insects, usually small and nocturnal, numerous pointed teeth; *Plecotus* – long-eared bat (order Chiroptera), ability to fly, usually nocturnal, echolocation; *Rattus* – rat (order Rodentia), gnaws, forelimbs often manipulate food; *Canis* – dog (order Carnivora), large canines, only vertical movements of jaw possible, carnivorous hunters; *Delphinus* – dolphin (order Cetacea), marine, carnivorous, blubber, flippers, streamlined shape.

Link topics

Section 1.1 Principles of classification
Section 4.3 Genetic variation and evolution

Suggested further reading

Barnes, R. D., *Invertebrate Biology* (Saunders 1963). Particularly Chapter 21
Cratchley, K., *Handbook of Animal Types* (Longman 1980)
Kershaw, D. R., *Animal Diversity* (University Tutorial Press 1983)
Goto, H. E., *Animal Taxonomy* Studies in Biology No. 143 (Arnold 1982)
Robinson, M. and Wiggins, J., *Animal Types – Invertebrates* (Hutchinson 1971)
Robinson, M. and Wiggins, J., *Animal Types – Vertebrates* (Hutchinson 1973)

QUESTION ANALYSIS

10 (a) Give *three* reasons why a frog and a mammal are classified in the same phylum.
   (b) Give *three* ways in which a frog and a lizard differ structurally from each other.
   (c) State *three* features of birds that have contributed to the success of this group.
   Mark allocation 9/200   Time allowed 7 minutes   *London June 1982 Paper 1, No. 1*

This structured question tests straightforward recall of learned facts and should therefore present few problems to a well-prepared candidate. The difficulty, however, is in selecting the best facts and in giving adequate detail. The answers must be short, and the words used must therefore be chosen with the utmost care.

In part (a) for example, a frog and a mammal both have a notochord at some stage of their life history and this answer would doubtlessly be accepted. To use the term 'vertebral column' instead of 'notochord' would probably be acceptable, but it is unlikely that 'backbone' would be an adequate alternative at this level. Candidates should always use A-level terminology and detail and not expect O-level standard answers to bring the same credit. Keep to obvious similarities that are of sufficient biological importance, e.g. the presence of a post-anal tail or pharyngeal gill clefts at some stage in the life history. The presence of a cranium, a dorsal hollow nerve cord or a ventral heart are other suitable alternatives. Avoid trite or superficial similarities such as 'they both have limbs, brains and internal skeletons'. Keep the answers strictly at an A-level standard.

In part (b) the two key words are 'differ structurally'. Any features that are even remotely common to both should be ignored as should any functional differences. Some typical differences are listed in the plan below.

Answers to part (c) must only include features that have clearly made birds as a group successful. Migration is commonly associated with birds but in fact only a minority of the bird species exhibit this behaviour and it consequently cannot be responsible for the success of birds as a whole. Again keep the terminology at an A-level standard. 'Endothermy (Homoiothermy)' would be acceptable, 'warm-blooded' would not.

Answer plan

(a) Choose any 3 from:

1 dorsal hollow nerve cord
2 cranium
3 notochord (vertebral column)*
4 ventral heart pumping blood posteriorly in the dorsal vessel and anteriorly in the ventral vessel
5 metamerically segmented post-anal tail*
6 pharyngeal gill slits (clefts)*
7 jaws present
*at some stage in the life history

(b) Choose any 3 from:

Frog	Lizard
Scales absent	Scales present
4 digits on forelimb	5 digits on forelimb
Webbed feet	No webbed feet
Tongue attached anteriorly	Tongue attached posteriorly
10 pairs of cranial nerves	12 pairs of cranial nerves
Ventricle of heart not divided at all	Ventricle of heart partly divided
Hind limbs considerably longer and more powerful than forelimbs	Hind and forelimbs roughly comparable in size
Jacobson's organ absent	Jacobson's organ present

(c) Choose any 3 from (the earlier features are of greater importance):

   1 feathers for flight/insulation
   2 endothermic (homoiothermic) allowing the maintenance of a high metabolic rate essential for flight
   3 chalky, waterproof egg
   4 parental care/nest building well developed
   5 forelimbs modified to form wings

6 excretion of uric acid to conserve water
7 internal fertilization
8 hollow bones, reducing weight for flight
9 air sacs
10 good eyesight (rapid accommodation).

11 Write an essay on either the arthropods or the angiosperms.
      Mark allocation 20/100   Time allowed 35 minutes   *London June 1981, Paper II, No. 6*

This style of question is a common way of examining knowledge of biological classification and a choice is usually given, often between one plant and one animal group. Regardless of the invitation to 'write an essay', 'write a general account of', 'make notes on' or 'discuss' the method of approaching the question is similar. Look for 'either/or' and 'one of the following' and do not exceed the number of groups required.

Although the question requires information to be restricted to one major group, the scope of the essay beyond that point is quite open. Examiners expect a broad essay incorporating all aspects of the group and it should be approached under the headings of:

1 classification/taxonomy
2 morphology/anatomy
3 physiology
4 ecology/distribution
5 economic importance/relevance to man.

Information on these groups is so vast that is is impossible to say exactly what should be included. It is best to consider that there will be a mark (possibly half a mark) for each clear, relevant, detailed biological fact, supported by an example. In other words there would be a large number of possible answers all warranting maximum marks, unlike many other questions where the candidate is working towards one specific answer. To maintain the balance of the essay there will be a maximum number of marks possible on any one topic, i.e. on a 25-mark question on arthropods there may be a maximum of five marks for each of the five topics listed above, but the distribution of marks will vary from group to group. In fungi and angiosperms for instance, there would probably be more credit for economic importance than for classification, whereas bryophytes and pteridophytes there would be few, if any, marks for economic importance. As plant and animal groups are frequently taught under a general heading of classification (often in conjunction with practical identification), it is a common failing of candidates to restrict the answer to taxonomy and classification which, however well answered, would bring a maximum of five or six marks.

At the same time as ensuring that the answer incorporates all aspects of the group, it is essential that clear, detailed, biological facts are included, supported where possible with specific examples and the use of appropriate biological terms.

Diagrams are an excellent way of conveying information and should be used wherever relevant but only where they are quicker and/or clearer than a written account, or add something new. Never repeat in words the information shown on a diagram; this unnecessary duplication is time-consuming and marks for a particular point can only be given once. Use annotations to supplement the drawing.

Answer plan (for essay on arthropods)

1 **Classification/taxonomy:** characteristics of the phylum. Classification to include the 5 classes and if possible some subclasses and orders of Insecta and Crustacea. If possible include, or at least refer to, fossil members of the group.

2 **Morphology/anatomy:** details of typical members of the group, e.g. basic body plan of an insect, crustacean and spider (possibly also a diplopod or chilopod); reference to diversity of form.

3 **Physiology:** some detail with examples on methods of:
   (a) reproduction – hemi/holometabolous; parthenogenesis
   (b) respiration – tracheae; gills; lungbooks
   (c) nutrition – carni/herbi/omnivore; parasitic
   (d) excretion – ureo/uricotelic
   (e) sensitivity – range of sense organs
   (f) locomotion – walking; swimming; flying

4 **Ecology:**
   (a) habitats – terrestrial; freshwater; marine
   (b) distribution
   (c) behaviour – e.g. social types
   (d) relationships to other organisms – e.g. symbiosis
   (e) position in food chain – trophic level

5 **Economic importance:**
   (a) beneficial types
      (i) directly, e.g. food source
      (ii) indirectly, e.g. pollination

(b) harmful types
(i) directly – stings; bites
(ii) indirectly – disease vectors

While insects are the most numerous group in terms of number of species, deserving slightly more attention than any of the other classes, the essay should not be confined to them.

12 (a) What are the distinguishing features of chordate animals?   (5)
(b) Using a named example of (i) a fish, (ii) an amphibian, (iii) a bird, describe the ways in which their external features show adaptation to their environments.   (5, 5, 5)
Mark allocation 20/100   Time allowed 35 minutes   *London June 1979, Paper I, No. 4*

This is a structured essay requiring a knowledge of the chordates and their features, especially external features. Part (a) is strictly recall of learned facts and needs only a list of features, whereas a similar list for the groups mentioned in (b) would be useless if given in isolation. Instead, in (b), the examiner is testing the ability of a candidate to relate the learned features to the animals' environment. The common failing is to omit or quickly forget this aspect and to describe the features and the environment separately without linking the two together. A greater depth of understanding and an appreciation of the relationship between structure and function are both essential to a good answer.

The important word in part (a) is 'distinguishing' and indicates that the features chosen should be ones that separate chordates from other groups, i.e. while bilateral symmetry and segmentation may be features of chordates they are shared by many other groups and thus are not distinguishing. On the other hand the five features listed in the essential information are unique to chordates and are those required by the question. Care should be taken not to include features that occur only in one class of chordates, e.g. hair, mammary glands, feathers, etc.

In part (b) important words are 'a named example'. Note that such an example is required for all three groups and that these examples should not be chosen randomly but with the greatest care, since the choice may well determine the quality of your answer. For a comprehensive answer, the examples should be ones whose external features you know and that show a high degree of specialization of external form to suit their environment. Having chosen wisely, keep strictly to these examples and do not inadvertently widen the discussion to others or to the group as a whole. List for each example as many external features as you can, and then add a note of how each helps the organism survive in its own environment. Some features will not readily lend themselves to this and will need to be rejected. Remember that the environment encompasses not only the medium (air/water) but also the habitat in which the organism lives, together with temperature and climate. As a guide it is useful to think of nutritional, locomotory, respiratory, nervous, reproductive, excretory, thermoregulatory and protective adaptations, remembering to keep strictly to external features.

The essential point throughout is to clearly 'describe the ways' the chosen features show 'adaptation to their environment'. It is not enough to merely describe the features: they must be related to the organism's mode of life. For example, 'The feet of passerine birds with opposing digits and claws are adapted to perching'.

Answer plan

(a) Characters as listed under chordates in the essential information.

(b) Adaptations of each of the features listed under each example
(i) **fish, e.g. mackerel:** fins; operculum; terminal mouth; overlapping scales; nostril; laterally flattened body and overall body shape.
(ii) **amphibian, e.g. frog:** streamlined body; dorsally prominent eyes; dorsal nares; ear-drum; short, stout forelimbs; long, powerful hindlimbs; webbed hind feet; loosely fitting, moist, slimy skin; mouth with wide gape. Mention of gills, tail and cement gland in the tadpole.
(iii) **bird, e.g. chaffinch:** short, strong beak; contour, flight, down and tail feathers; clawed, opposing digits; eye with eyelid and nictitating membrane; streamlined shape; wings.

Note the mark distribution and allocate a quarter of the time to (a) and three quarters to (b), dividing this equally between the three examples.

13 An animal is free-living, acoelomate, triploblastic, bilaterally symmetrical and possesses flame cells.
(a) Name
(i) the phylum to which this animal belongs.
(ii) the class to which this animal belongs.
(b) Suggest why most members of this class:
(i) are flattened and leaf-like.
(ii) live in water.
(c)
(i) Name a parasitic member of this phylum.
(ii) Name one body system which is reduced or absent in the named parasite and which is functional in the free-living members of the phylum.
(iii) Suggest why the system named in (ii) is not required by the parasite.
Mark allocation 6/100   Time allowed 7 minutes
          Associated Examining Board, November 1979, Paper I, No. 2

This is a structured question requiring knowledge of the classification and mode of life of the Platyhelminthes. Part (a) requires factual recall. In the initial list of features, the key ones for answering (i) are 'acoelomate' which limits the animal to a lower invertebrate and 'flame cells' which restrict it to the Platyhelminthes. For (ii) the key feature is 'free-living'; Turbellaria is the only class in this phylum with free-living members.

In (b) a deeper understanding of the mode of life of turbellarians is required, especially that the absence of a specialized respiratory system demands a large surface area to volume ratio and flattening achieves this at the same time as allowing a large surface area of ciliated epidermis to remain in contact with the substrate for locomotion. The large surface area must be thin to allow gaseous diffusion, and this would lead to desiccation if the animal lived out of water.

In (c) it is essential to read to the end of the question to determine what is required. Then decide on the named parasite in (i) as the choice of a suitable example will greatly facilitate the answering of the remaining parts of the question. An individual named organism is required and not the name of a parasitic group. In (ii) the reduced system will depend on the example chosen: if a trematode is chosen, locomotory and sensory systems are good answers; if a cestode, the digestive system is a further alternative.

Reasons for them not being required (iii), should be related to their reduced importance in a parasitic mode of life compared with a free-living one; often the host performs these functions on the parasite's behalf.

2 Cytology and Histology

2.1   THE CELL

Assumed previous knowledge The appearance of a plant and an animal cell as seen through a light microscope.

Underlying principles

Modern cell theory states that:

1 all living matter is composed of cells.
2 all cells arise from other cells.
3 all metabolic reactions of an organism take place in cells.
4 cells contain the hereditary information of the organisms of which they are a part, and this is passed from parent to daughter cell.

All cells are similar in comprising a self-contained and more or less self-sufficient unit surrounded by a cell membrane and having a nucleus at some stage of their existence. At the same time cells show remarkable diversity of structure and function. In shape cells are basically spherical, although modification to suit function leads to a degree of diversity. In size they mostly range from $10-30\mu m$ in diameter. Their size is restricted by:

1 the surface area to volume ratio, which must be large to allow exchange of metabolic substances
2 the capacity of the nucleus to exercise control over the rest of the cell.

Points of perspective

Historical background including the discovery of cells by Robert Hooke in 1665 and the cell theory by Theodore Schwann and Matthias Schleiden in 1839.

Electron microscopy (EM), including transmission EM and scanning EM, should be understood. Basic details of section-cutting, fixing and staining for EM preparations would be helpful.

A wide range of photoelectron micrographs of both plant and animal cells should be studied and practice obtained in identifying the parts and in calculating magnifications.

Essential information

Cells were first seen by Hooke in 1665. Schleiden and Schwann in 1839 stated that all living systems are composed of cells and cell products (Cell Theory).

Cells are studied using light or optical microscopes which have a resolution of approximately

0.25µm, imposed by the wavelength of light. The electron microscope uses an electron beam instead of light and electromagnets instead of glass lenses. Electrons have a wavelength of 0.005 nm and in theory can resolve objects of 0.0025 nm in diameter, although in practice this is limited to 1 nm, i.e. an electron microscope has a resolving power approximately 400× as great as a light microscope.

Disadvantages of the electron microscope:
1 preliminary treatment of material may cause distortion
2 specimens must be mounted in a vacuum and therefore be dead
3 expensive to make and run
4 only provides black and white image
5 preparation is lengthy and difficult
6 large, and needs a special room
7 affected by magnetic objects.

Differences between plant and animal cells:

Table 2.1 Plant and animal cell differences

Plant	Animal
Cellulose cell wall as well as membrane	No cellulose cell wall, only membrane
Pits in cell wall	No pits
Plasmodesmata present	No plasmodesmata
Large vacuole filled with cell sap	Some vacuoles but usually small and numerous
Cytoplasm peripheral	Cytoplasm throughout the cell
Nucleus usually peripheral	Nucleus anywhere in cytoplasm but often central
2 cytoplasmic membranes: outer plasmalemma, inner tonoplast	Only 1 cytoplasmic membrane
Variety of plastids, e.g. chloroplast; leucoplast	Not normally plastids
Cilia and flagella absent in higher plants	Cilia common in higher animals
Centrioles absent in higher plants	Centrioles present

Cell organelles

Plasma membrane This separates the cell from its surroundings and has the same basic structure – known as the unit membrane – as the membranes of mitochondria, endoplasmic reticulum and other organelles. There are a number of proposed models for the arrangement of protein and lipid in the unit membrane but most are variants of the Davson-Danielli hypothesis of the 1930s. See Fig 2.1(b). In the early 1970s Singer and Nicholson proposed the **fluid-mosaic** model of the cell membrane based on chemical analyses, freeze-etching and electron microscopy. Their work confirms the bimolecular structure of the lipid part of the membrane but indicates that the protein molecules form irregular globules and not a continuous layer. See Fig. 2.1 (d).

Large particles enter the cell by phagocytosis: the cell membrane invaginates to form a depression containing the particles, and the invagination then becomes sealed off as a phagocytic vacuole. The process is selective, e.g. *Amoeba* engulfing food and phagocytes engulfing bacteria.

Pinocytosis is essentially similar to phagocytosis but involves the ingestion of liquids. In *Amoeba* pinocytic channels provide a means by which liquids are brought into the cell. Vacuoles form from the channels but the plasma membrane remains intact and liquids still have to cross it in order to enter the body of the cell. Smaller, micropinocytic vacuoles have been seen through the electron microscope, for example at the base of microvilli in the mammalian small intestine.

Nucleus This is bounded by a double unit membrane containing pores of 40 – 100 nm, allowing exchange of materials between the nucleus and the cytoplasm. The outer membrane is granulated and the inner one is smooth. Inside the membrane is a meshwork of nucleoplasm interspersed with nuclear ribosomes and DNA. There may also be one or two small, round bodies, the nucleoli, which are rich in RNA and are not surrounded by a membrane.

Endoplasmic reticulum Throughout the matrix of the cell is a series of interconnected cavities bounded by unit membranes which are continuous with the nuclear membrane. This system of parallel cavities is known as the endoplasmic reticulum, or ER. Most of the ER has granules called ribosomes attached to the matrix-side of the membranes. This is rough ER. Smooth ER lacks ribosomes and is thought to have a different function. Rough ER provides a means by

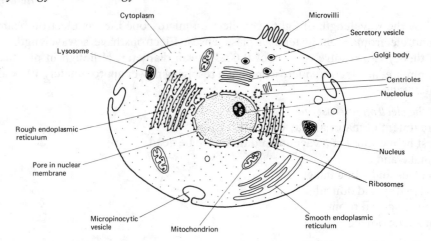

Fig.2.1 (a) Ultrastructure of a generalized animal cell

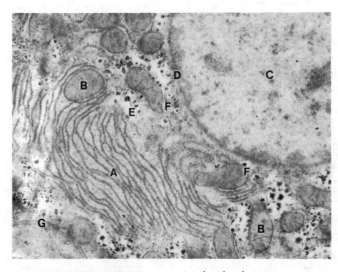

Key: A Rough endoplasmic reticulum
B Mitochondrion
C Nucleoplasm
D Nuclear envelope
E Glycogen granules
F Mitochondrial envelope
G Golgi apparatus

Fig. 2.1 (b) Rat liver cells, magnification 25 000 ×

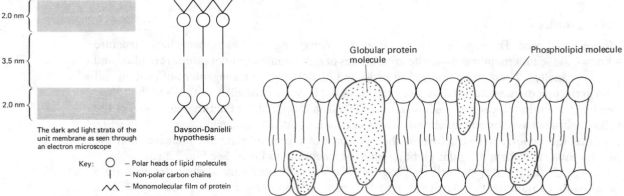

Fig. 2.1 (c) The organization of the unit membrane

Fig 2.1(d) Fluid-mosaic model of membrane structure

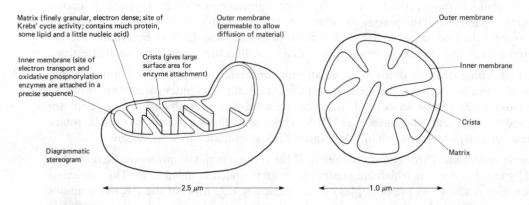

Fig 2.1 (e) Structure of a mitochondrion

Fig.2.1 (f) TS of a mitochondrion

which the polypeptides synthesized by the ribosomes (see Section 3.3) are transported within the cell. Smooth ER is not continuous with rough ER and is thought to be concerned with the transport of lipids and steroids.

Golgi body This consists of stacks of flattened cavities, similar to smooth ER, but always associated with secretory granules. It is thought that newly synthesized proteins move from the rough ER to the Golgi body where carbohydrate is added to them to form the glycoprotein secretions common in cells. Part of the Golgi membrane surrounds these glycoproteins to form vesicles which move to the surface of the cell and discharge their contents.

Mitochondria These vary in shape and size but are usually rod-shaped with a diameter of approximately 1.0 μm and length about 2.5 μm. The wall consists of two thin membranes, the inner one of which is folded to form cristae. Cell respiration takes place in mitochondria, most of the necessary enzymes being attached to the inner membrane and cristae, or occurring in the matrix. Thus cells which expend a lot of energy have numerous mitochondria each with many cristae, e.g. muscle cells.

Lysosomes Sometimes called 'suicide-bags', these are spherical organelles containing enzymes capable of splitting complex substances into smaller units. Their function is to destroy unwanted organelles (e.g. mitochondria), which are surrounded by a membrane into which the lysosomes discharge their enzymes. Lysosomes may also destroy entire 'worn-out' cells by liberating their enzymes within the cell.

Microvilli These finger-like projections, approximately 1.0 μm long and 0.08 μm in diameter, on the surface of the plasma membrane increase the surface area of the cell, thereby aiding absorption, e.g. in convoluted tubules of the kidney and in the lining of the small intestine.

Cilia and flagella Both may be concerned with locomotion in acellular organisms but cilia are also widespread in higher animal groups. Cilia are about 5–10 μm long and flagella 100 μm; both have an average diameter of less than 0.3 μm. They have the same basic structure. At the base of each cilium is a basal body composed of nine peripheral fibres but not the central two. Movement of cilia and flagella requires ATP and the bending process is initiated at the base and transmitted towards the tip.

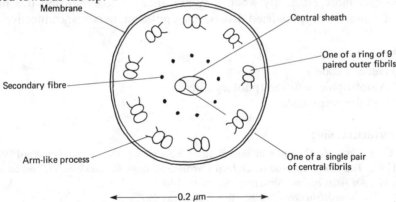

Fig. 2.2 TS of a cilium, based on an electron micrograph

Centriole In animal cells two are found at right angles to each other. Centrioles have the same basic structure as the basal bodies of cilia. They function in the formation of the spindle at cell division and also give rise to cilia and flagella.

Microtubules These are made of protein and are widespread in the cytoplasm of the cell, occurring either singly or in bundles. They are thought to be concerned with cellular movements and transport in cells. They are easily broken down and reassembled. The spindle fibres formed at cell division are microtubules.

Chloroplasts These are found only in plant cells. Their shape and size vary with species but there is always a double unit membrane surrounding a matrix (stroma) in which are stacks of lamellae (grana). The lamellae hold the chlorophyll in the most suitable position for photosynthesis. The stroma contains numerous starch granules and enzymes for the reduction of carbon dioxide.

Cell wall This is a non-living layer found only in plant cells. A young plant cell has a primary cell wall composed of calcium pectate and some cellulose. As the cell ages a secondary cell wall is laid down; this is usually composed of layers of cellulose possibly impregnated with other substances, chiefly lignin. In places the secondary cell wall is absent, giving rise to pits and the endoplasmic reticulum of adjacent cells may be continuous through the plasmodesmata.

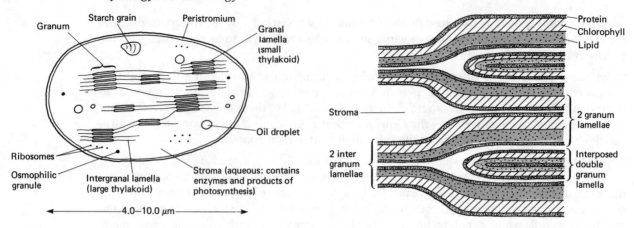

Fig. 2.3 (a) Structure of a chloroplast

Fig. 2.3 (b) Section through grana showing arrangement of chlorophyll

Differentiation of cells Acellular organisms carry out all vital metabolic processes and have, within the cell, special organelles to carry out various functions. However, there is a limit to the degree of complexity possible within a single cell. To allow for increasing complexity the multicellular condition arose. At first colonies of cells simply stick together, each cell being independent of the others. In the next stage a ball of cells is produced, but only those on the outside have flagella for movement. This marks the first stage of differentiation, i.e. a cell changing its normal structure so that it can become specialized to a particular function, in this case locomotion. With differentiation comes loss of other functions, a process carried to extremes in higher plants and animals, e.g. nerve cells are so highly specialized to one function that they are incapable of dividing. Some functions must be carried out by every cell, e.g. respiration, but others can be 'delegated', e.g. digestion of food is left to cells lining the alimentary canal; muscle cells are responsible for locomotion. Cells which carry out a particular function operate more efficiently when they are grouped together and this has led to the formation of a number of specialized tissues, e.g. epithelial; nervous; connective.

Link topics

Section 3.3 Genetic code
Section 6.1 Autotrophic nutrition (photosynthesis)
Section 7.1 Cellular respiration

Suggested further reading

Allison, A. C., *Lysosomes,* Oxford/Carolina Biology Reader No. 58, 2nd ed. (Packard 1977)
Cook, G. M. W., *The Golgi Apparatus,* Oxford/Carolina Biology Reader No. 77, 2nd ed. (Packard 1980)
Fawcett, D. W., *An Atlas of Fine Structure* (Saunders 1966)
Hurry, S. W., *The Microstructure of cells,* 4th ed. (Murray 1972)
Jordan, E. G., *The Nucleolus,* Oxford/Carolina Biology Reader No. 16, 2nd ed. (Packard 1978)
Lockwood, A. P. M., *The Membranes of Animal Cells,* Studies in Biology No. 27, 3rd ed. (Arnold 1984)
Lucy, J. A., *The Plasma Membrane,* Oxford/Carolina Biology Reader No 81, 2nd Ed. (Packard 1978)
Satir, P., *Cilia and Related Organelles,* Oxford/Carolina Biology Reader No. 123 (Packard 1983)
Tribe, M. and Whitaker, P. A., *Chloroplasts and Mitochondria;* Studies in Biology No. 31, 2nd ed. (Arnold 1982)

QUESTION ANALYSIS

1 (a) Describe the structure and function of cilia. (10)

(b) Compare the advantages and disadvantages of the electron microscope with the light microscope. (7)

(c) Describe three differences between a plant palisade cell and an animal smooth muscle cell as seen under the electron microscope. (3)

Mark allocation 20/100   Time allowed 35 minutes   *In the style of the London Board*

This is an essay question covering knowledge of cytology and microscopy. In (a) the structure should include a diagram similar to Figure 2.2 and a separate written account of its overall appearance and function. Full details of functions are needed including how a cilium moves (see Section 9.1 for outline) and the uses to which this movement is put, e.g. locomotion; movement of materials within an organism; feeding, etc.

In part (b) the key words are 'compare', and 'advantages and disadvantages'. Make certain the answer is comparative, i.e. an advantage of the light microscope is that it can magnify living specimens whereas the electron microscope can only be used to view dead material. A table is the most expedient way of presenting the answer. State clearly whether the point being made is an advantage or a disadvantage for each of the microscopes; on the whole, an advantage of one will be a disadvantage of the other.

Part (c) is straightforward, but all too often candidates read the question as 'differences between plants and animals' ignoring key words such as cells and 'as seen under the electron microscope' and thus giving answers such as 'plants contain chlorophyll whereas animals do not'. If the word 'chloroplast' is substituted for 'chlorophyll' the answer becomes acceptable because the former is visible under an electron microscope whereas the latter is not. Similarly differences in feeding, locomotion, overall shape, etc. cannot be seen under an electron microscope and should not be included.

Answer plan

(a) structure: see essential information, including Figure 2.2
    functions: method of action of cilium (Section 9.1)
    locomotion, e.g. *Paramecium*
    feeding, e.g. *Mytilus*
    movement within organism, e.g. mucus in mammalian trachea; ovum along oviduct; circulation of cerebrospinal fluid.

(b)

Electron microscope	Light microscope
Advantages	**Disadvantages**
Greater magnification up to 500 000×	Magnification only up to 1500×
Greater depth of field	Lesser depth of field
No complex lighting	May involve phase contrast or other complex lighting
Higher resolving power (1 nm)	Lower resolving power (0.5 μm)
Disadvantages	**Advantages**
Expensive to buy (£1 million +)	Cheaper (£100)
Expensive to run	Cheaper
Black and white image	Coloured image
Specimen dead	Specimen living or dead
Vacuum required	No vacuum required
Lengthy, expert preparation needed	Preparation is short and relatively simple
Large, requiring special room	Small and portable
Affected by magnetic fields	Unaffected by magnetic fields

(c) Choose from:

Structure	Plant	Animal
Chloroplasts	Present	Absent
Cellulose micelles (cell wall)	Present	Absent
Centrioles	Absent	Present
Microvilli	Absent	Present

2 Give an account of the distribution and functions of the membranes of cells.
   Mark allocation 20/100   Time allowed 35 minutes   *London June 1981, Paper II, No. 7*

This is a broad essay question requiring considerable knowledge, not only of membranes, but of a wide variety of structures in which membranes occur.

The words 'membranes of cells' do not mean 'cell membrane', i.e the candidate is expected to include not only the limiting plasma membrane that surrounds every cell but all other internal organelles and structures that have membranes. In addition to the basic cell organelles such as the nucleus, mitochondria and endoplasmic reticulum found in all cells, specialist organelles such as chloroplasts, contractile vacuoles and pinocytotic vesicles which do not occur universally must not be forgotten. The variety of plasma membrane adaptations also needs mentioning, e.g. microvilli in the ileum; root hairs of plants (both to increase the surface area for absorption).

The other keys words are 'distribution' and 'function'. By 'distribution' the question means where the membrane is found, e.g. forming the lamellae in the chloroplast as well as the limiting membrane around it, or limiting the mitochondrion and forming the cristae within it. 'Functions' means the role played by the membrane in the structure being discussed. Here a degree of detail is necessary, for instance it would not be adequate to say the nuclear membrane 'surrounds the nucleus' or 'keeps it together'; more specific detail is needed such as 'isolates the genetic material DNA and allows mRNA to diffuse outwardly only'. Similarly the functions of mitochondrial membranes could be stated as 'allowing diffusion of pyruvic acid and other glycolytic products into the organelle and the regular arrangement of the Krebs' cycle and electron carrier system enzymes on the cristal membranes'.

Under no circumstances should the answer be concerned primarily with the structure of the basic membrane although an initial simple annotated diagram of a unit membrane or a short written account would be a valuable introduction and would clarify the description of membrane function.

A final point to be made is that the plasma membrane, apart from its specialist functions, serves to limit a small volume of cytoplasm over which a nucleus exerts control, i.e. it simply acts to divide one cell from another to form a series of fundamental units. The overriding point for the candidate to remember is that an examiner is looking for a wide range of distribution and functions with brief details on each, rather than a long, detailed account of the plasma membrane alone. Bearing this in mind the final answer plan below gives some idea of the range required, although it is unlikely that any candidate would include all those listed. Those at the top of the list are most important and proportionately more time should be devoted to them.

Answer plan

Distribution	Function
Plasma membrane	Differential permeability; allows osmosis, diffusion, active transport; limits cells
Nuclear membrane	Limits DNA; allows mRNA out
Mitochondria: outer	Allows glycolytic products in
inner	Attachment of respiratory enzymes
Endoplasmic reticulum	Cellular transport; attachment of ribosomes
Chloroplast: outer	Allows photosynthetic products out and substrates in;
lamellae	Reservoir of chlorophyll, carotenes etc.;
Golgi	Storage of glycoprotein; synthesis of polysaccharides (e.g. cellulose in plants)
Lysosomes	Limit autolytic enzymes
Tonoplast	Limits cell sap
Pinocytotic/phagocytotic vesicles	Uptake of materials
Other specialized membranes:	
Root hairs	Increase surface area
Microvilli (e.g. kidney tubule)	Increase surface area
Myelin sheath membrane	Insulation of nerve fibre
Neurilemma	Diffusion of Na^+ and K^+ allowing depolarization

3 In the space below make a diagram to show the structure of a mitochondrion as revealed by the electron microscope.

(a) Label four of its component parts (2)

(b) Add a scale to your diagram which shows the approximate size of the organelle. (1)

(c) How would you proceed to isolate mitochondria from the cells of a living tissue such as the brain or the liver? (2)

(d) Having obtained a suitable sample of mitochondria in (c) above, briefly describe how you would demonstrate any one of their functions experimentally. (4)

Mark allocation 9/140   Time allowed 10 minutes   *Oxford Local, 1982, Paper I, No. 3*

This is a structured question and as such requires concise and yet detailed answers. With an overall time allowance of ten minutes the drawing should be a simple LS or TS (see Fig 2.1(f)). The actual structure as seen under the electron microscope can be seen in Fig 2.1 (b). The labels needed for (a) are shown in Fig 2.1 (f); the time available does not warrant any annotations. The scale required in (b) is shown in the same diagram.

The method of isolating mitochondria asked for in (c) is centrifugation. A short account of the process could be included; such an account is given under 'Biochemical techniques' in Section 3.2.

The major funtion of mitochondria is the conversion of pyruvic acid into carbon dioxide and water with the liberation of energy in a process called the Krebs' Cycle. One method of demonstrating this function is to add pyruvic acid to the isolated mitochondria. The liberation of carbon dioxide would indicate that the mitochondria were capable of carrying out the Krebs' Cycle. A further confirmation would be the addition of cyanide to the mitochondria and pyruvic acid. As an inhibitor of cytochrome oxidation, the cyanide should prevent any carbon dioxide being produced. This indicates that the mitochondria must possess these enzymes and therefore be the site of the electron transport pathway as well as of the Krebs' Cycle.

4 (a) Make a fully labelled drawing of a chloroplast to show its structure as seen under an electron microscope.
(b) Using diagrams, give an account of the structure and function of the endoplasmic reticulum and Golgi apparatus.
(c) Explain what lysosomes are and why they are sometimes called 'suicide bags'. Why are they more frequent in phagocytic cells than other cells?
Mark allocation 20/100   Time allowed 40 minutes   In the style of the Joint Matriculation Board

This is a structured essay examining a wide range of knowledge on cell organelles. The candidate is required to present information on both structure and function, to draw clear diagrams and to answer precisely. Each part should be answered in isolation and be clearly labelled with the appropriate letter. In these types of question some credit could be lost for putting material in the wrong section, e.g. if the Golgi apparatus were drawn under the heading (b) it could be considered as a poor representation of the endoplasmic reticulum and all or some of the marks could be forfeited.

In (a) the diagram should be large and clearly labelled down to the detail revealed by an electron microscope, but not including the arrangement of the pigment molecules as these are beyond the resolving power of an electron microscope. No separate written account is needed so do not waste time with details of photosynthesis or how the chloroplast functions.

In (b) the key words are 'diagrams', 'structure' and 'function'. Diagrams of the ER and Golgi apparatus should be enhanced with annotations or a separate written account given. The organelles' functions should be related to the structures.

Part (c) is straightforward recall of facts included in the essential information. 'Suicide bags' refers to their ability to destroy certain organelles within the cell. Their high frequency in phagocytic cells is because of these cells' function of ingesting unwanted material for the purpose of breaking it down enzymatically in order to render it harmless.

Answer plan

The highly structured nature of this essay makes a full plan unnecessary. An outline of the information needed for the answers is covered in the essential information. Figures 2.1a and 2.3 give guidance for the diagrams.

2.2   PLANT AND ANIMAL TISSUES

Assumed previous knowledge The cell, Section 2.1

Underlying principles

Fertilized egg cells divide to give a mass of similar (undifferentiated) cells. Although single cells can carry out all necessary functions, there are times when these conflict and it then becomes more efficient for each cell to take on a different function (differentiation) acquiring special characteristics to suit its function (specialization). This inevitably leads to the loss of some functions and so to the dependence of one cell upon another. For further efficiency cells performing similar functions are grouped together to form a tissue. Some tissues contain very similar cells, e.g. epithelial tissue, and others are more diverse, e.g. connective tissue.

Points of perspective

Embryological derivation of tissues from germ layers in animals. The derivation of plant tissues from meristems.

It is important to be able to correlate the structure of a tissue with its function.

The ability to interpret the three-dimensional nature of a tissue from longitudinal and transverse sections would be a considerable advantage. Students should be able to prepare temporary microscope slides of tissues, e.g. blood; leaf epidermis.

Essential information

A tissue This is a region consisting mainly of cells of a similar type and function bound together by cell walls (plants) or by intercellular material (animals).

Table 2.2 Functions of plant and animal tissues

Function	Plant	Animal
Protection	Epidermis	Skin
Mechanical	Collenchyma Sclerenchyma Cell sap of parenchyma Xylem fibres	Blood Bone Cartilage Connective tissue proper
Secretion	Gland cells	Gland cells
Conduction of water or food	Xylem vessels Tracheids Sieve tubes	—
Conduction of impulses	—	Nerve
Contraction	—	Muscle

Animal tissues The principal types and their derivations are:
   Epithelial – from all germ layers
   Connective – from mesoderm
   Muscular – from mesoderm
   Nervous – from ectoderm.

Epithelial tissue Associated with its role as a covering tissue subject to mechanical damage, epithelial tissue has the following characteristics:
1 cells frequently attached to basement membane
2 adjacent cells joined together by intercellular cement
3 may be interconnecting bridges of cytoplasm between cells
4 may build up large number of layers to resist abrasion, e.g. epidermis of skin
5 rapid division of epithelial layer to resist abrasion and maintain only a thin layer of cells, e.g. intestines
6 resistance to abrasion may be increased by presence of keratin, e.g. epidermis of skin

Types of epithelia

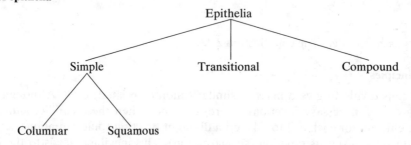

Any of the various types may be ciliated and/or secretory.

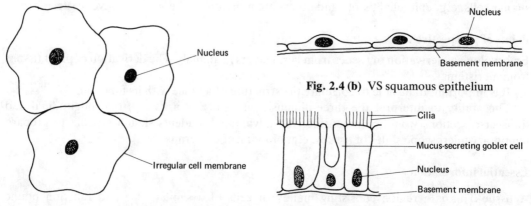

Fig. 2.4 (a) Squamous epithelium, surface view

Fig. 2.4 (b) VS squamous epithelium

Fig. 2.4 (c) VS ciliated columnar epithelium

Columnar epithelium: cilia beat in a definite rhythm, moving along mucus and other particles. Found in, for example, nasal cavities; trachea; oviducts; ventricles of brain.

Squamous epithelium: sometimes called 'pavement' epithelium. Flattened, single layer of thin nucleated cells. Found where rapid diffusion is essential, e.g. Bowman's capsule; alveoli; endothelium.

Transitional epithelium: found where protection and distention are both required, e.g. bladder and urinary tract.

Compound (stratified) epithelium: many layers of cells; the ones nearest the basement membrane are usually columnar and those farthest from it are often flattened and dead, possibly impregnated with keratin. Found where there is· considerable mechanical stress, e.g. epidermis of skin; oesophagus; anal canal; vagina.

Connective tissue

Characteristics: variable appearance; includes a cellular element and the non-cellular product of these cells (the intercellular substance, or matrix).

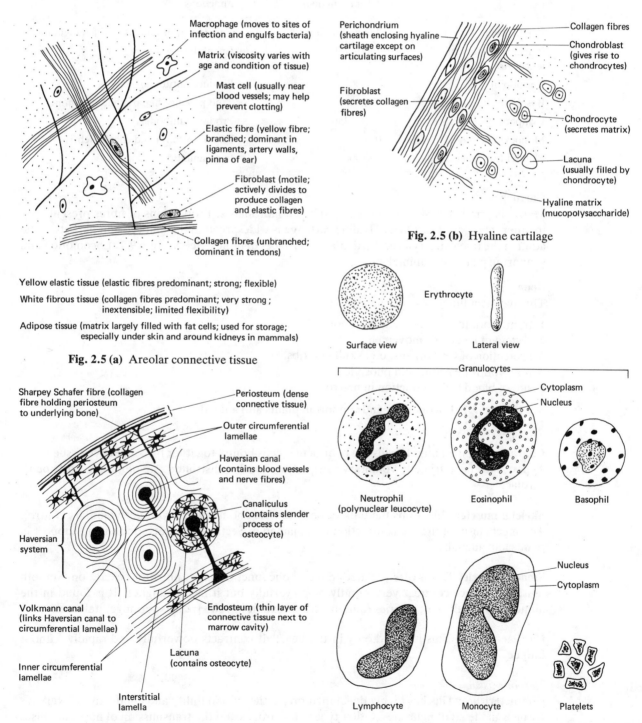

Macrophage (moves to sites of infection and engulfs bacteria)

Matrix (viscosity varies with age and condition of tissue)

Mast cell (usually near blood vessels; may help prevent clotting)

Elastic fibre (yellow fibre; branched; dominant in ligaments, artery walls, pinna of ear)

Fibroblast (motile; actively divides to produce collagen and elastic fibres)

Collagen fibres (unbranched; dominant in tendons)

Yellow elastic tissue (elastic fibres predominant; strong; flexible)

White fibrous tissue (collagen fibres predominant; very strong; inextensible; limited flexibility)

Adipose tissue (matrix largely filled with fat cells; used for storage; especially under skin and around kidneys in mammals)

Fig. 2.5 (a) Areolar connective tissue

Perichondrium (sheath enclosing hyaline cartilage except on articulating surfaces)

Fibroblast (secretes collagen fibres)

Collagen fibres

Chondroblast (gives rise to chondrocytes)

Chondrocyte (secretes matrix)

Lacuna (usually filled by chondrocyte)

Hyaline matrix (mucopolysaccharide)

Fig. 2.5 (b) Hyaline cartilage

Erythrocyte

Surface view   Lateral view

Granulocytes

Cytoplasm

Nucleus

Neutrophil (polynuclear leucocyte)   Eosinophil   Basophil

Sharpey Schafer fibre (collagen fibre holding periosteum to underlying bone)

Periosteum (dense connective tissue)

Outer circumferential lamellae

Haversian canal (contains blood vessels and nerve fibres)

Canaliculus (contains slender process of osteocyte)

Haversian system

Volkmann canal (links Haversian canal to circumferential lamellae)

Endosteum (thin layer of connective tissue next to marrow cavity)

Inner circumferential lamellae

Lacuna (contains osteocyte)

Interstitial lamella

Fig. 2.5 (c) TS compact bone

Nucleus

Cytoplasm

Lymphocyte   Monocyte   Platelets

Fig. 2.5 (d) Types of blood cells

Types of connective tissue

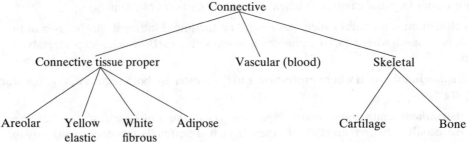

Blood
Plasma is the matrix and the cells present include those shown in the following table.

Table 2.3 Frequency and functions of blood cells

Cell	No/mm^3 in man	Functions
Erythrocyte	4.2–6.4 million	Carriage of oxygen as oxyhaemoglobin
Polymorphonuclear leucocyte	1500–7500	Engulf bacteria
Eosinophil	0–400	Produce antitoxins
Basophil	0–200	? but produce heparin and histamine
Lymphocyte	1000–4500	Produce antibodies
Monocyte	0–800	Engulf bacteria and coarse debris
Platelet (not truly cellular)	300 000	In coagulation

Cartilage
Matrix is firm and resilient and secreted by chondroblasts. Proportions of fibres present vary in different types of cartilage. Hyaline cartilage is widespread in the embryo and persists in the adult in the trachea, larynx and articulating surfaces of bones. See Section 9.1, Locomotion and support in plants and animals.

Bone
The five main functions of bone are:

1 framework for support, e.g. backbone
2 system of levers for movement
3 protection of vital organs, e.g. skull and ribs
4 reservoir for calcium and phosphorus
5 site of blood cell production in marrow.

See Section 9.1, Locomotion and support in plants and animals

Muscle
Characteristics: all muscle cells are contractile; cells bound together by connective tissue.
Types: skeletal (striated; voluntary; striped); smooth (unstriated; involuntary; unstriped); cardiac.

Skeletal muscle: This forms the bulk of body muscle and is generally under voluntary control. The mechanism of muscle contraction is explained in Section 9.1, Locomotion and support in plants and animals.

Smooth muscle: This is often related to metabolic functions, e.g. nutrition; excretion. Smooth muscle does not contract very rapidly or powerfully but it never fatigues. It is found in the alimentary canal; skin; arteries (and to a lesser extent in veins); ducts; urinogenital tract.

Cardiac muscle: This is found only in the heart. It contracts powerfully and rapidly without fatigue.

Nervous tissue
Characteristics: This has highly developed properties of irritability and conductivity. Nervous tissue is made up of neurones. Other types of neurones and the transmission of nerve impulses are dealt with in Section 10.2, Nervous system and behaviour.

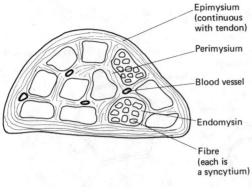

Fig. 2.6 (a) (i) TS whole skeletal muscle

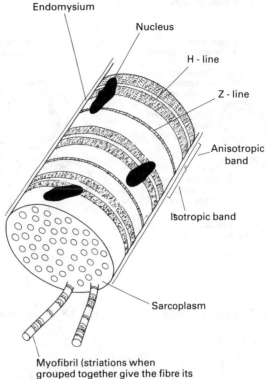

Fig. 2.6 (a) (iii) LS skeletal muscle fibre

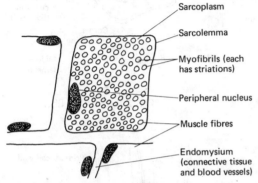

Fig. 2.6 (a) (ii) TS skeletal muscle fibres

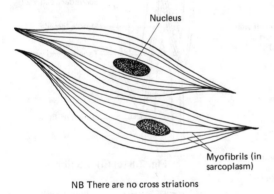

NB There are no cross striations

Fig. 2.6 (b) LS smooth muscle fibres

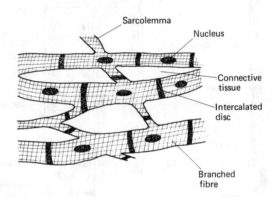

Fig. 2.6 (c) Cardiac muscle

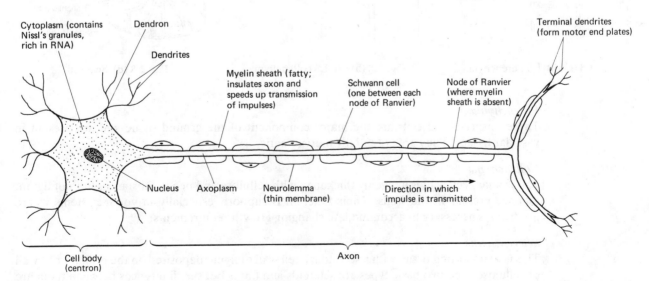

Fig.2.7 Typical motor neurone

Plant tissues

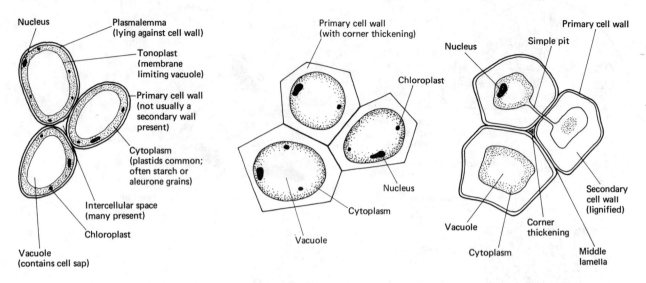

Fig. 2.8 (a) (i) TS parenchyma   **Fig. 2.8 (b) (i)** TS collenchyma   **Fig. 2.8 (c) (i)** TS fibre

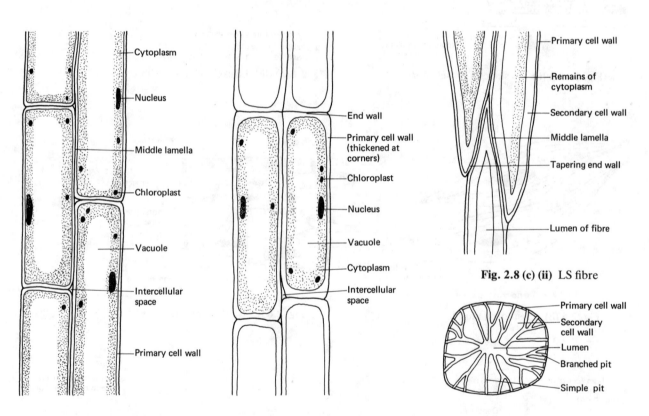

Fig. 2.8 (a) (ii) LS parenchyma   **Fig. 2.8 (b) (ii)** LS collenchyma   **Fig. 2.8 (d)** Sclereid

Parenchyma

These unspecialized cells are the major component of the ground tissue and are present in vascular tissues. They are potentially meristematic.

Collenchyma

The primary cell wall is unevenly thickened with cellulose and pectic substances but not lignin. Pits are present in the walls. Their function is support, especially in younger stems where plasticity is necessary to accommodate changing growth requirements.

Sclerenchyma

This is a supporting tissue with a secondary cell wall of lignin deposited on the primary cell wall of cellulose. The two basic types are sclereids and fibres but the differences between them are not always clear-cut. The pits in the cell walls may be simple or bordered.

Fibres

These are often found in the vascular bundles of dicotyledons or around the vascular bundles of monocotyledons. They are often grouped in strands. Jute, hemp and flax are economically important fibres.

Sclereids

These are widely distributed, heavily lignified cells with numerous pits.

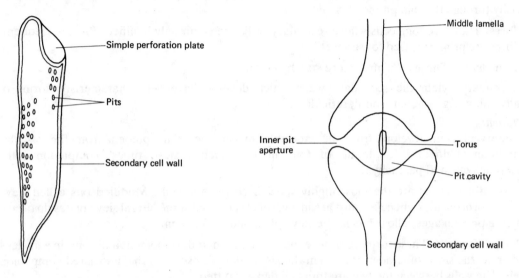

Fig. 2.9 (a) Vessel   **Fig. 2.9 (b)** LS bordered pit (from tracheid)

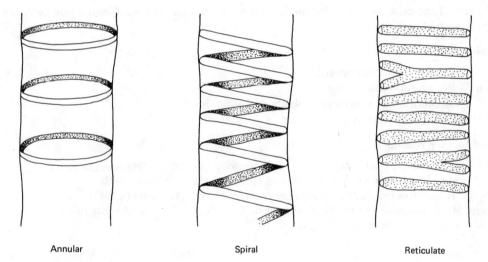

Annular   Spiral   Reticulate

Fig. 2.9 (c) Thickenings of secondary cell walls in primary xylem

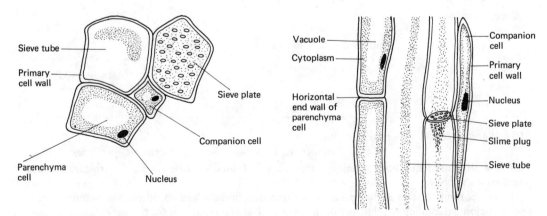

Fig. 2.9 (d) (i) TS phloem   **Fig. 2.9 (d) (ii)** LS phloem

Xylem

Primary xylem is formed from the embryo and the resultant meristems. Secondary xylem develops later, during secondary thickening. Xylem is made up of four main elements: vessels, tracheids, fibres and parenchyma.

Vessels: These cells have one or more perforations at each end so that water can move easily from cell to cell. Cell walls may be simple or perforated by bordered pits.

Tracheids: These are imperforate cells but the pits occur in pairs so that the water can pass easily through the thin pit membrane.

Fibres: These are long cells whose secondary walls are commonly lignified. Pits are frequent. Fibres are primarily used for support.

Parenchyma: These cells often store starch and oils.

In primary xylem the secondary walls which develop have very characteristic forms of thickening, e.g. annular; spiral; reticulate.

Phloem

Primary phloem develops from the procambium and secondary phloem from the vascular cambium. Phloem is made up of four main elements: sieve tubes, companion cells, sclerenchyma and parenchyma.

Sieve tubes: These are the most highly specialized phloem cells. Modified pits called sieve plates form on adjacent cells; most are in a vertical series but some lateral sieve plates do occur. As the phloem ages callose blocks the sieve plate and the element dies.

Companion cells: These form from sieve tubes early in their development and contain a nucleus whereas the sieve tube does not; when the sieve tube dies so does the associated companion cell. The walls between the two are thin and densely pitted.

Sclerenchyma: Sclereids and fibres are common components of the primary and secondary phloem.

Parenchyma: These cells contain substances such as starch and tannins. Companion cells are specialized parenchyma cells.

Link topics

Section 5.4 Growth and development
Section 8 Water relations and transport
Section 9.1 Locomotion and support in plants and animals
Section 10.2 Nervous system and behaviour

Suggested further reading

Bracegirdle, B. and Miles, P. H., *An Atlas of Plant Structure* (2 volumes) (Heinemann 1971)
Clegg C. J. and Cox G., *Anatomy and Activities of Flowering Plants* (Murray 1978)
Freeman, W. H. and Bracegirdle, B. *An Atlas of History,* 2nd ed. (Heinemann 1967)
Richardson, M., *Translocation in Plants,* Studies in Biology No. 10, 2nd ed. (Arnold 1975)
Rudall, P., *Anatomy of Flowering Plants* (Arnold 1987)

QUESTION ANALYSIS

5 Complete the following table relating to four different tissues.

Name of tissue	Special property	Location
Cardiac muscle		Heart
	Controlled by involuntary part of nervous system	
		Attached to bones
	Elastic and tough	At joints

Mark allocation 6/100   Time allowed 7 minutes

Associated Examining Board November 1979, Paper I, No. 11

This is a structured question examining knowledge of tissues, their names, characteristics and positions in the body.

It is important to read all the parts of the question initially and so be certain exactly what is needed before attempting to fill in any of the blanks. The first column is for tissues – make sure your answers are not organs but are confined to those tissues listed in the essential information in this section. From the examples given in the second column it is clear that what is required is a feature that is of particular importance in the functioning of the tissue. The third column requires a specific place in the body where the tissue is typically found.

Cardiac muscle's 'special property' is the fact that it is not fatigued. In the second row the tissue that is controlled 'by the involuntary part of the nervous system' is smooth muscle, which is found in a number of locations, including the wall of the alimentary canal. The only information available on the next tissue is 'attached to bones'. Two structures frequently confused by candidates are ligaments and tendons, both of which are attached to bones. Ligaments join bone to bone at joints; in order to permit movement they are made up largely of yellow elastic tissue. Tendons on the other hand attach muscles to bones. If these were elastic, muscle contraction would simply stretch the tendon rather than effect movement of the bone. Tendons must therefore be strong and inelastic, and for this reason comprise mainly white fibrous connective tissue.

Since the fourth tissue is 'elastic and tough' and occurs at joints this must be the yellow elastic connective tissue of ligaments. The question states that the tissues are different so the third answer must be the white fibrous connective tissue of a tendon. Its special property is that it is 'strong and inelastic'.

6 What is a tissue? Explain your answer by reference to suitable examples from both animals
and plants.   (25)
Mark allocation 25/100   Time allowed 45 minutes   *London January 1979, Paper II, No. 5*

This is an essay question testing the candidates' understanding of the term 'tissue' and their ability to select suitable examples to illustrate that they comprehend its meaning.

The definition of a tissue is given at the beginning of the essential information in this section and would serve as the starting point for the essay. It is important to relate the rest of the answer to this definition, in particular the similarity of the cells in both structure and function. The examples should be carefully chosen with this in mind. For instance blood is considered a tissue but is a poor example since it has a variety of cells and functions and does not clearly fit the usual definition of a tissue. In the same way striated muscle with its syncytial arrangement (has many nuclei within the plasma membrane), while being a tissue, does not illustrate the definition well enough. In this respect the key word is 'suitable', doubtlessly added by the examiner as a caution to candidates to choose examples with care. In choosing the examples look for simple tissues, comprising similar cells, grouped to form one main function. In plants, parenchyma, collenchyma and sclerenchyma are good choices, performing the functions of support in varying degrees, but since parenchyma may have other functions, such as storage and photosynthesis, the much more specialized sclerenchyma is certainly the best of the three.

Divide your time equally between plants and animals and give a range of examples of tissues in each. In plants use sclerenchyma, epidermis and phloem for instance and in animals one type of epithelial tissue, smooth muscle and cartilage or nervous tissue. Avoid striated muscle and blood for reasons given above; xylem (which consists of dead cells) and bone (composed largely of secreted material) are other poor examples.

Use drawings to illustrate the similar nature of the chosen tissue's cells and bear this in mind when choosing examples: sclerenchyma cells (Fig. 2.8c(i)) are easier and quicker to draw than xylem vessels (Fig. 2.9a); smooth muscle (Fig. 2.6b) is easier than bone (Fig. 2.5c).

Do not draw too many cells, three or four of one type are adequate and more would be a waste of valuable time. Avoid simply describing various tissues in turn; rather relate them to the definition given: sclerenchyma comprises groups of similar cells (diagram) grouped together, e.g. around vascular bundles in herbaceous stems, to perform the one main function of mechanical support.

A logical approach to the answer would be to deal with a specific function such as support and give one suitable plant example (sclerenchyma) and one animal one (cartilage). Doing this with other functions would ensure a correct balance between plants and animals.

Answer plan

Definition of tissue (see essential information). Relate each of the tissues in turn to that definition.

Function	Plant	Animal
Support	Sclerenchyma Fig. 2.8c(i), 2.8c(ii)	Cartilage Fig 2.5b
Protection/lining	Epidermis	Columnar epithelium Fig. 2.4c
Transport/movement	Phloem Fig. 2.9d(i), 2.9d(ii)	Smooth muscle Fig. 2.6b
Storage	Parenchyma Fig. 2.8a(i), 2.8a(ii)	Adipose

7 Structure and function are closely related. By reference to (a) striated (skeletal) muscular tissue and (b) parenchyma in plants, discuss how far this statement is true.

Mark allocation 20/100   Time allowed 40 minutes   *London June 1983, Paper 2, No. 10*

This type of essay question in section B of paper 2 carries marks for the way in which the question is answered as well as for the subject content. To obtain marks candidates' answers should be completely relevant, all facts should be clear and the ideas and arguments should be well marshalled. The candidate must show a reasonable command of language and write lucidly. Spelling should be accurate and diagrams must be neatly drawn, well-labelled and make a useful contribution to the answer.

It is essential throughout the answer to relate structure to function. Candidates should not describe the structure and simply state the function, but rather point out clearly how the stucture enables the stated function to be carried out efficiently. In part (a), for instance, it would be inadequate to state that skeletal muscle comprises fibres that contract. More completely the candidate should make the point that the fibres are 'long' or 'elongated' in order to allow considerable contraction. Other points to be made include the parallel arrangement of fibres which mean that they all shorten along the same axis and so produce a more powerful contraction of the muscle as a whole. The ends of the fibres are tapered and interwoven, giving additional strength to the muscle.

The answer should not be restricted to detail visible under the light microscope but include structures visible using an electron microscope. The numerous mitochondria in the sarcoplasm should be related to the need for considerable amounts of ATP to effect contraction. The arrangement of the actin and myosin filaments of a sarcomere should be related to the way in which the filaments slide over one another during contraction and the ratchet mechanism involved.

Although not strictly muscle tissue it would be worth mentioning associated tissues which contribute to muscle contraction. For example, the presence of a rich blood supply could be related to the muscles' need for an adequate supply of oxygen and glucose. Similarly the presence of myoglobin could be associated with the storing of oxygen in the muscle for use when the partial pressure of it in the blood is low. Motor end plates and their role in stimulating muscle contraction could be included as could the arrangement of fibres in motor units in order to permit various degrees of contraction according to the number of units stimulated.

Some illustrations could be used to elucidate the structure of muscle, possibly one diagram of the muscle fibres as seen under a light microscope and another showing the electron microscope detail. The diagrams should be included with the part of the text they relate to rather than be drawn together at the end of the section.

Part (b) should be approached in a similar way to part (a). Care should be taken to restrict the points to basic parenchyma and to avoid including xylem or phloem parenchyma or collenchyma. A good starting point would be the unspecialized appearance of parenchyma cells; this could be related to the wide variety of functions they perform. The isodiametric shape of the cells should be related to their packing function, their thin walls to the need for material to pass in and out and the transparent walls to the need to allow light to enter for photosynthesis. The permeability of the walls is associated with the cells' ability to take up water and so provide turgidity to herbaceous parts of a plant.

As well as individual cells the tissue as a whole should be considered. In this respect the presence of intercellular air spaces should be related to the need to allow gases to diffuse rapidly through the tissue. These spaces are especially large in the aerenchyma tissues of submerged parts of aquatic plants where they are particularly important in allowing oxygen to diffuse to the roots and in providing buoyancy. The contents of individual cells should also be discussed. In particular the presence of various plastids and their functions should be considered. For example chloroplasts and their role in photosynthesis, chromoplasts and their role in colouring petals in order to attract insects for pollination and leucoplasts for the storage of starch.

As with muscle some illustrations would be helpful, for instance one diagram showing a few cells arranged together and a second giving detail of the internal structure of a single parenchyma cell.

Answer plan

In each instance it is essential to show clearly how the structure permits a specific function to be carried out efficiently.

(a) Striated (skeletal) muscle

Structure	**Function**
Elongated fibres	Allow considerable contraction
Parallel fibres	Give maximum contractile effect
Fibre ends tapered and interwoven	Provide strength
Large number of mitochondria	Provide large amount of ATP
Actin and myosin arrangement in sarcomere	Allows contraction by filaments sliding over each other
Rich supply of blood vessels	Provide adequate supply of oxygen and glucose
Myoglobin present	To store oxgyen for release when blood oxygen levels are low
Motor end plates	Allow stimulation of muscle
Fibres arranged in motor units	To permit variable degrees of contraction

Diagrams of skeletal muscle as seen under light and electron microscopes.

(b) Parenchyma cells

Structure	Function
Unspecialized tissue	Variety of functions
Many intercellular spaces	Diffusion of gases
Isodiametric cells	Packing material
Thin cellulose cell walls	Permit passage of materials
Transparent cell wall	Permits entry of light for photosynthesis
Permeable walls	Allow water entry for turgidity
Large cells/large vacuoles	Provide storage space
Chloroplasts present	Allow photosynthesis
Chromoplasts present	In petals provide colour to attract insects for pollination
Leucoplasts present	To store starch

Diagrams of parenchyma tissue and an individual cell showing internal detail.

3 Molecular Biology and Biochemistry

3.1 CARBOHYDRATES, LIPIDS AND PROTEINS

Assumed previous knowledge The elements present in carbohydrates, lipids and proteins.

Underlying principles

Hydrolysis: the splitting of a molecule by the addition of water.

Condensation: the combination of two simple molecules to form a complex one with the release of water.

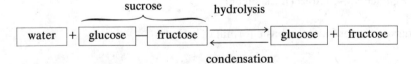

Polymers: long chains of similar units joined together to form a complex molecule.

Isomers: chemical compounds having the same number and kind of atoms, but differing in the arrangement of these atoms and hence in their properties.

Points of perspective

The structure and function of carbohydrate derivatives and related substances, e.g. lignin, suberin, chitin and pectin, could be useful. It may help to have some knowledge of the nature of the changes that occur in each food test:

Table 3.1 Food tests

Test for	Reagent(s)	Active constituent of reagent	Nature of the colour change	Additional notes
Reducing and non-reducing sugars	Benedicts	Copper (II) sulphate	Copper (II) sulphate (blue) → copper (I) oxide (orange/red)	Reaction only occurs in alkaline conditions
Starch	Iodine dissolved in potassium iodide solution	Iodine	Formation of a complex with the iodine trapped in the centre of the helical starch molecule	Heating above 70°C causes unwinding of the starch helix and loss of colour. Test must be carried out at room temperature or below
Lipids	Alcohol water	Alcohol	Alcohol dissolves the lipid. On mixing this with water the lipids become finely dispersed and refract light, giving a milky appearance	Thorough mixing of water and lipid-laden alcohol gives the best result. Solution goes clear in time due to settling out of the lipids.
Protein	Millons	Mercuric nitrate (in nitric acid)	Reacts with certain phenol rings to give a red coloration	Of the amino acids present in proteins, only tyrosine gives a positive result
	Biuret	Copper (II) sulphate	A purple complex is formed with adjacent pairs of $-CONH_2$ groups	If excess copper sulphate is added a blue precipitate of copper hydroxide is formed

Essential information

Carbohydrates The general formula for carbohydrates is $(CH_2O)_n$

where n = 3 it is called a triose sugar (glyceraldehyde is $C_3H_6O_3$)
where n = 5 it is called a pentose sugar (ribose is $C_5H_{10}O_5$)
where n = 6 it is called a hexose sugar (glucose is $C_6H_{12}O_6$)

These are all examples of sugars (saccharides). Where there is only one sugar it is called a monosaccharide; two monosaccharides can join together to form a disaccharide, and many monosaccharide units form polysaccharides.

Monosaccharides 'single sugars'
e.g. glucose; fructose; galactose
All three have the same basic formula, $C_6H_{12}O_6$, but differ in the arrangement of the atoms, i.e. are isomers; for example, glucose is an aldose sugar whereas fructose is a ketose sugar. They all reduce Benedict's (Fehlings) Reagent, are sweet, soluble and easily transported, and are the main respiratory substrates.

Disaccharides 'double sugars'
These are formed by the condensation of any two monosaccharides
e.g. glucose + glucose = maltose (malt sugar)
glucose + fructose = sucrose (cane sugar)
glucose + galactose = lactose (milk sugar)

These three disaccharides have the basic formula $C_{12}H_{22}O_{11}$.

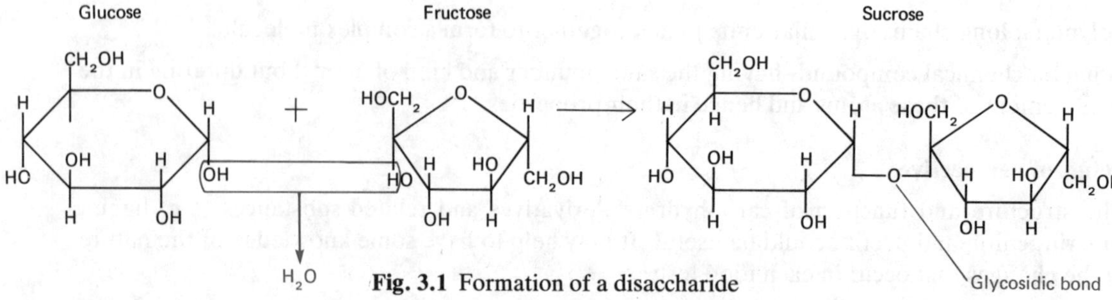

Fig. 3.1 Formation of a disaccharide

Some disaccharides, e.g. maltose, will reduce Benedict's (Fehlings) Reagent but others, e.g. sucrose, are non-reducing sugars. All disaccharides are sweet and soluble in water. One important disaccharide, sucrose, is transported in the sieve elements of the phloem and occasionally stored in parenchyma cells, as in the scale leaves of onion. Disaccharides are readily converted into respiratory substrates.

Polysaccharides 'many sugars'

These are formed by the condensation of numerous monosaccharides.

e.g. starch; glycogen; cellulose

They have the general formula $(C_6H_{10}O_5)_n$

They will not reduce Benedict's (Fehlings) Reagent, are not sweet, are insoluble and are used for storage (starch; glycogen) or for structural support (cellulose).

Tests for carbohydrates

Reducing sugars Using a little of the carbohydrate (if necessary by grinding with a little water in a mortar and pestle), add an equal volume of Benedict's (Fehlings) Reagent and boil for a few minutes. Presence of a reducing sugar is indicated by a green/yellow/brown/red precipitate depending on the concentration of the sugar.

Non-reducing sugars A non-reducing sugar, e.g. sucrose, will not give a precipitate when heated directly with Benedict's (Fehlings) Reagent unless it is first hydrolysed into its component reducing sugars. This hydrolysis can be carried out by boiling with a weak acid but since Benedict's (Fehlings) Reagent works only in alkaline conditions this acid must be neutralized before boiling with the reagent.

1 Boil equal volumes of test solution and dilute hydrochloric acid for about five minutes.
2 Add 10% sodium hydroxide solution until the solution is neutral (test with pH paper).
3 Add sufficient Benedict's Reagent to give a definite blue colour.
4 Boil for a few minutes.
5 Green/yellow/brown/red precipitate forms depending on the concentration of the sugar.

If the above procedure is carried out on a reducing sugar a positive result will also be obtained. A non-reducing sugar is therefore indicated by a negative result for a reducing sugar followed by a positive one for a non-reducing sugar. See Table 3.2.

Table 3.2 Tests for sugars

	glucose (reducing sugar)	sucrose (non-reducing sugar)
Reducing sugar test	+ve result	−ve result
Non-reducing sugar test	+ve result	+ve result

Where both sucrose and glucose are present in a solution the following procedure should be used:

The reducing sugar test and the non-reducing sugar test are carried out separately on two equivalent samples and, provided equal volumes of Benedict's Reagent are used, the amount of precipitate after the non-reducing sugar test will be greater than after the reducing sugar test.

Starch Add iodine in potassium iodide solution to the substance to be tested. Do not heat. The presence of starch is indicated by a blue-black coloration.

Lipids (Fats) Lipids are formed as shown in Figure 3.2.

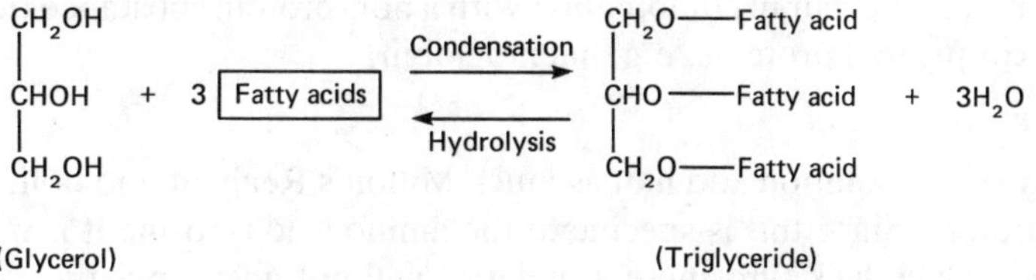

Fig. 3.2 Formation of a fat

Some lipids have one of the fatty acids replaced by phosphoric acid and are called phospholipids.

Lipids usually contain glycerol but the component fatty acids may vary considerably. The resulting diversity of lipids is reflected in the wide range of lipid functions. For example,

1 **energy source:** 38 kJ/g.
2 **storage:** since lipids store over twice as much energy for the same mass as carbohydrates, they are used when it is important to keep weight to a minimum, e.g. in fruits and seeds.
3 **protection:** stored around delicate organs, e.g. kidneys.
4 **insulation:** especially useful where hair/feathers etc. are of little use, i.e. under water (e.g. blubber in whales).
5 **waterproofing:** leaf cuticles are made of waxes; secretion from sebaceous glands.

Tests for lipids

1 If a few drops of lipid are dissolved in ethanol and an equal volume of water added, a cloudy white suspension is produced.
2 Lipids take up the red stain Sudan III readily. If a lipid and water are shaken with Sudan III, only the lipid stains red.
3 Greasemark test. If a drop of lipid and water are placed on filter paper and left to dry, only the lipid leaves a residual mark when held up to the light.

Proteins These consist of long chains of amino acids, the basic formula of which is shown in Figure 3.3a.

There are over twenty naturally occurring amino acids which differ in the composition of the R group.

Two amino acids may be linked together by condensation (Fig. 3.3b).

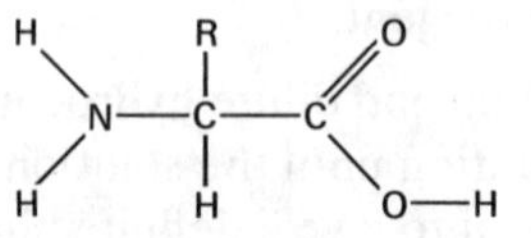

Fig. 3.3 (a) Typical amino-acid

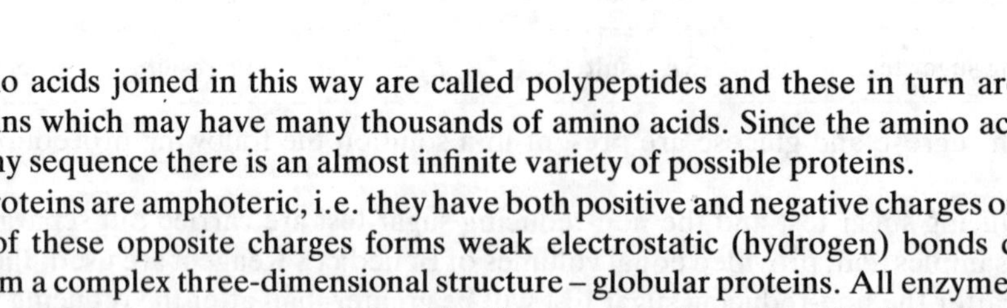

Fig. 3.3 (b) Formation of a dipeptide

Many amino acids joined in this way are called polypeptides and these in turn are linked to form proteins which may have many thousands of amino acids. Since the amino acids may be joined in any sequence there is an almost infinite variety of possible proteins.

Since proteins are amphoteric, i.e. they have both positive and negative charges on them, the attraction of these opposite charges forms weak electrostatic (hydrogen) bonds causing the chain to form a complex three-dimensional structure – globular proteins. All enzymes and some hormones are globular proteins and they also form antitoxins and antibodies. Sometimes the protein consists of long parallel chains with cross links – fibrous proteins. These are insoluble and have structural functions, e.g collagen in cartilage; keratin in hooves, feathers, hair; actin and myosin in muscle; in cell membranes. If a globular protein is heated or treated with a strong acid or alkali the hydrogen bonds are broken and it reverts to a more fibrous nature – denaturation. Proteins sometimes occur in combination with a non-protein substance (prosthetic group); these are called conjugated proteins, e.g. haemoglobin.

Tests for proteins

1 To a quantity of test solution add half as much Millon's Reagent and boil. A red coloration indicates a protein. Since this is specific to the amino acid tyrosine it is possible that a few proteins (ones which lack tyrosine e.g. gelatin) will not give a positive result.
2 Biuret Test. To the test solution add an equal volume of 10% potassium hydroxide solution. Add 0.5% copper sulphate solution a drop at a time. A purple coloration indicates a protein. Care is needed since the addition of excess copper sulphate solution results in the disappearance of the coloration.

Link topics

Section 3.2 Enzymes (including biochemical techniques)
Section 3.3 Genetic code

Suggested further reading

Phelps, C. F., *Polysaccharides,* Oxford/Carolina Biology Reader No. 27 (Packard 1972)
Phillips, D. C. and North, A. C. T., *Protein Structure,* Oxford/Carolina Biology Reader No. 34, 2nd ed.
(Packard 1978)

QUESTION ANALYSIS

1 (a) What is a protein? (3)
   (b) Discuss with suitable examples the variety of functions of proteins. (13)
   (c) Explain how their structure permits this wide variety of functions. (4)
   Mark allocation 20/100   Time allowed 30 minutes   In the style of the Oxford and Cambridge Board

This is a structured essay question examining candidates' knowledge of protein structure and function, and their ability to relate them to each other.

The simplest definition of a protein would be: a complex organic compound composed of a chain of amino acids linked by peptide bonds to form a three-dimensional structure of large molecular mass. For three marks however, a little more detail should be included, e.g. the nature of an amino acid (Fig. 3.3a) and a peptide link (Fig. 3.3b), globular and fibrous types and their amphoteric nature.

Part (b) forms the bulk of the question and the key word here is 'variety'. The answer should aim to cover as wide a range of functions as possible rather than concentrating on one aspect such as enzymes. There are two possible approaches: one is to look at the functions of fibrous proteins (usually structural roles such as cell membranes, keratin in skin and actin/myosin in muscles) and then discuss the functions of globular proteins (enzymes, some hormones, antibodies, blood pigments, etc.); however, a better approach, which is more likely to produce the required variety of functions, is as follows. As proteins are used in almost every aspect of an organism's life it is logical to consider their functions under the seven headings that constitute the basic characteristics of living organisms, namely: respiration, excretion, reproduction, movement, sensitivity, nutrition and growth. Add to these 'cell structure' and give a few examples under each heading, and a balanced, diverse answer is assured. Although the sections on reproduction, movement (including support) and nutrition will be longer than those on excretion or growth, try to maintain a reasonable balance.

Select examples from a range of plant and animal groups (see Sections 1.2, 1.3, 1.4). It may not be possible to use all groups but making a conscious effort to do so will avoid the classic error of limiting the examples to mammals in general and man in particular. Above all include some plant examples.

Part (c) requires the candidate to relate the wide range of functions given in (b) to the diversity of protein structure. The answer must show how this structural diversity is possible by the arrangement of twenty or so amino acids in any order to make up long chains. Diversity is enhanced by three-dimensional folding and hydrogen bonding of these chains in a number of different ways. The largely metabolic functions of globular proteins should be related to this diversity of three-dimensional structure, whereas the largely structural functions of fibrous proteins should be related to the mechanical strength achieved by the arrangement of parallel protein chains.

Answer plan

(a) Protein = long chain of amino acids (detail) linked by peptide bonds (detail); globular and fibrous types; amphoteric nature.

(b) Functions:

Heading	Protein	Function
Cells	Fibrous	Component of cell membranes/organelles
	Globular	Metabolic pathways
Respiration (including blood)	Haemoglobin⎫ Haemoerythrin⎬ Haemocyanin Chlorocruorin⎭	Transport of oxygen
	Fibrinogen⎫ Prothrombin⎬	Blood clotting
	Mucin	In mucus to keep respiratory surface moist
	Antibody	Defence
Excretion	Urease⎫ Arginase⎬	Enzymes in ornithine cycle

Heading	Protein	Function
Reproduction	Chromatin	Structural component of chromosomes
	Gluten	Storage protein in seeds
	Casein	Nutritive protein in milk
	Keratin	Horns/antlers for sexual display
Movement/ locomotion/ support	Ossein	Protein in bone – support
	Actin/myosin	Muscle contraction
	Keratin	Feathers, claws, nails, scales – protection
	Collagen	Strength in cartilage and tendons
	Elastin	Flexibility with strength in ligaments
	Mucin	Lubrication
	Folding and unfolding molecules	Pseudopodial motion
	With chitin	Exoskeleton for protection
Sensitivity (co-ordination)	Pigments	Pigment spot for orientation
	Rhodopsin	Vision
	Insulin	Regulates blood sugar level
	Prolactin	Milk production
	Vasopressin	Regulates blood pressure
Nutrition	Pepsin	Protein digestion
	Amylase	Starch digestion
	Protein in granal lamellae	Photosynthesis
	Mucin	Trapping food in filter feeders; prevents autolysis; lubrication
Growth	Thyroxin	Growth hormone

(c) Variety of function due to variety of structure. Variety of structure due to almost infinite arrangement of amino acids in protein chain and different means of folding this chain.
globular – three-dimensional structure – metabolic
fibrous – straight chains – structural.

2 Natural fats and oils are compounds formed from glycerol and fatty acids.
  (a) Write a general formula for a fatty acid.
  (b) (i) How many molecules of glycerol and fatty acid are used in the synthesis of a triglyceride?
   (ii) Outline the principal features of this reaction either diagrammatically with chemical formulae or by a concise statement of the events.
  (iii) What is the name given to this type of reaction?
  (iv) Give the name of the general group of enzymes that react with fats in the lumen of the small intestine.
  (c) Plant oils, such as sunflower oil, are important foodstuffs.
  (i) From which part of the plant are they extracted?
  (ii) If you were to make a similar extraction in a school laboratory what would you use as a solvent?
  (d) Fatty acids are also constituents of many plant waxes.
  (i) Where are these waxes found in the plant?
  (ii) State their function.

Mark allocation 4/40   Time allowed 6 minutes

Welsh Joint Education Committee June 1979, Paper A1, No. 1

This is a structured question requiring recall of factual information about all aspects of fats. In (a) the general formula required is R–COOH where R is a hydrocarbon chain of variable length. In (b) (i) the only likely problem is that some candidates may think that three molecules of each compound are required for a triglyceride, rather than three molecules of fatty acid and just one of glycerol (Fig. 3.2). The answers to (b) (ii) and (b) (iii) – condensation – are clearly set out in Figure 3.2.

The general name of the enzymes referred to in (b) (iv) is derived from the alternative term for fats – lipids. The addition of -ase to the substrate is the typical way of naming enzymes; hence lipids are acted upon by lipases.

The answer to (c) can be deduced from remembering the major role of fats in organisms. Owing to their relatively concentrated energy content (more than double the energy/unit mass of carbohydrate), they are used where energy is required but weight needs to be minimized. In plants these two require-ments appear in the seeds, especially wind dispersed ones. The extraction of fat in a school laboratory is normally carried out using ethanol, which readily dissolves them.

Part (d) concerns plant waxes and the answers can be derived by looking at the properties and functions of fats. Since they do not mix with water, they readily form waterproof barriers in organisms. In plants this is particularly necessary over the leaves, where the large surface area (for photosynthesis) is conducive to water evaporation. A waxy cuticle on the leaves reduces this water loss without affecting light transmission for photosynthesis. The answer to (d) is therefore (i) over the leaf surface and (ii) to reduce evaporative water loss.

3 (a) How would you detect the presence of 'soluble' starch and glucose in a solution of both substances?
   (b) How would you detect the presence of sucrose in a solution composed of sucrose and glucose?
   (c) Name a sugar other than sucrose or glucose and state precisely one place it is found naturally and the role it performs there.
Mark allocation 7/100   Time allowed 11 minutes   *In the style of the Joint Matriculation Board*

This is a structured question examining recall of experimental tests for carbohydrates and the application of these in the detection of the components of carbohydrate mixtures. The natural occurrence and function of one of the less common sugars are also required.

In (a) and (b) remember to state clearly the results of each test, e.g. 'a clear blue solution of Benedict's produces a red precipitate'.

For (a) the tests for starch and reducing sugars (glucose) should be described as in the essential information. Be as detailed as possible by giving precise volumes used, e.g. $0.5\,cm^3$ of iodine solution + $2\,cm^3$ of test solution and $2\,cm^3$ of Benedict's Reagent + $2\,cm^3$ of test solution. The 'solution' in the question should be divided into two samples, and the two tests carried out separately, since the results of one test could mask those of the other if both were performed on the same sample. Remember that the starch test must be carried out at room temperature whereas the Benedict's test requires heating.

In (b) the tests for reducing sugars (glucose) and non-reducing sugars (sucrose) are required. Since the non-reducing sugar test gives a positive result when glucose is present, the sucrose can only be detected by comparing the amounts of precipitate after each of the tests. Details are given towards the end of the tests for carbohydrates in the essential information, but remember the success of the method depends on the two samples being of exactly equal volume and their being heated to the same temperature for the same length of time (use a water bath). Some examples of the answer to (c) are given below:

Sugar	Location	Function
Fructose	Bee's honey; Many succulent fruits	Food store Attraction of animals leading to dispersal
Lactose	Mammalian milk	Nutrition of sucklings
Ribose	RNA of all cells, especially in nucleolus	Synthesis of proteins within the cell

Avoid maltose and galactose since these usually occur as the result of the hydrolysis of starch and lactose respectively, and only occur rarely in the natural state.

3.2   ENZYMES (INCLUDING BIOCHEMICAL TECHNIQUES)

Assumed previous knowledge The basic definition of enzymes, their protein nature and their universal occurrence in cells; that enzyme activity is affected by certain conditions.

Underlying principles

The Laws of Thermodynamics The First Law of Thermodynamics (Law of Conservation of Energy) states that energy may be transformed from one form into another but is neither created nor destroyed.

The Second Law of Thermodynamics states that processes involving energy transformations will not occur spontaneously unless there is a degradation of energy from a non-random to a random form.

Blackman's Law of Limiting Factors states that when the speed of a process is conditioned by a number of separate factors, the rate of the process is limited by the pace of the slowest factor.

Reaction Kinetics Chemical reactions are reversible, the direction being determined by the conditions in the reaction:

$$A \rightleftharpoons B$$

The reaction may move from left to right at a high pH and in the reverse direction at a low pH. The addition of more A or the removal of B will tend to move the reaction from left to right and the addition of B or removal of A tends to move it from right to left.

Points of perspective

Historical background should include an awareness of Buchner who named enzymes, the word meaning 'in yeast'.

Candidates should consider medical, industrial and agricultural uses of enzymes, e.g. enzyme inhibitors as a basis of sulphonamide drugs and certain pesticides and the use of enzymes in biological washing powders.

Essential information

Names of enzymes are normally derived from the addition of the suffix -ase to the name of the substrate on which they act, e.g. lactase acts on lactose; maltase on maltose, etc. Enzymes are also grouped according to the type of reaction they carry out.

Table 3.3 Enzyme group reactions

Enzyme group	Reaction catalysed	Example
1 Oxidoreductases	All reactions of the oxidation–reduction type	Dehydrogenases; oxidases
2 Transferases	The transfer of a group from one substrate to another	Transaminases; phosphorylases
3 Hydrolases	Hydrolytic reactions	Phosphatases; peptidases; amidases; lipases
4 Lyases	Additions to a double bond or removal of a group from a substrate without hydrolysis, often leaving a compound containing a double bond	Decarboxylases
5 Isomerases	Reactions where the net result is an intramolecular rearrangement	All-trans retinal $\rightarrow$ all-cis retinal
6 Ligases	The formation of bonds between two substrate molecules using energy derived from the cleavage of a pyrophosphate bond such as ATP	Acetate + CoA–SH + ATP $\rightarrow$ acetyl–CoA + AMP + P–P

Characteristics of enzymes

Specificity
The degree of specificity varies from those which will catalyse reactions of one isomer (Section 3.1) to those which catalyse reactions involving a particular type of linkage, e.g. peptide link, wherever it occurs.

Reversibility
Enzymes will catalyse reactions in either direction, e.g. carbonic anhydrase:
carbonic acid $\longrightarrow$ carbon dioxide and water (in lungs)
$H_2CO_3 \longrightarrow CO_2 + H_2O$
$CO_2 + H_2O \longrightarrow H_2CO_3$ (in tissues)
The direction depends on conditions such as pH and substrate concentration.

Effect of enzyme concentration
The rate of an enzyme-catalysed reaction increases with enzyme concentration provided there is excess substrate.

Effect of substrate concentration
The rate of an enzyme-catalysed reaction increases as substrate concentration increases provided there is excess enzyme.

Temperature
Up to about 40°C the rate of an enzyme-catalysed reaction doubles for each 10°C rise in temperature; it then tails off until at about 60°C it stops. Temperature limits vary from enzyme to enzyme.

pH
Every enzyme has its own optimum pH at a given temperature, concentration and substrate concentration.

e.g. pepsin   about pH 2.0
        salivary amylase   about pH 7.0
        arginase   about pH 10.0

Activation
Many enzymes require the presence of another substance before they will catalyse a reaction.

If such a substance is inorganic it is called a **co-factor,** e.g. calcium activates thrombokinase in the clotting of blood (see Section 8.2, Blood and circulation).

If it is a separate, non-protein organic molecule it is known as a **co-enzyme,** e.g. nicotinamide adenine dinucleotide (NAD) is a co-enzyme for dehydrogenases; the vitamin thiamine is a co-enzyme for pyruvic dehydrogenase.

If the non-protein, organic molecule forms an integral part of the enzyme it is called a **prosthetic group,** e.g. cytochrome oxidase has a prosthetic group containing iron.

Inhibition

Certain substances slow down or stop enzyme-controlled reactions. Where the inhibitor and substrate compete for the enzyme, neither forming a permanent combination, the inhibition is called competitive; the relative concentrations of substrate and inhibitor determine the degree of inhibition, e.g. malonic acid competes with succinic acid for succinic dehydrogenase.

In non-competitive inhibition, the inhibitor forms a permanent combination with the enzyme to the exclusion of the substrate; the degree of inhibition therefore depends on the concentration of inhibitor only, e.g. cyanide inhibits cytochrome oxidase thus arresting respiration.

Inhibition of enzymes is a normal part of the control of metabolic pathways (negative feedback).

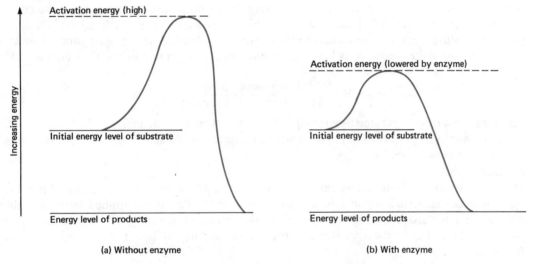

(a) Without enzyme   (b) With enzyme

Fig. 3.4 (a) How enzymes lower activation energy levels

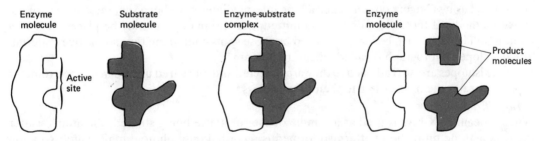

Fig. 3.4 (b) The lock and key hypothesis

Mechanism of enzyme action

In any spontaneous reaction the energy level of the products is lower than the energy level of the initial substrate. Before the reaction can occur, the substrate molecule must surmount an energy barrier known as the activation energy. The presence of an enzyme lowers the activation energy. The active site is a precise region of the enzyme molecule whose shape is complementary to that of the substrate and into which the substrate fits.

In molecular terms this is thought to be by enzyme and substrate forming a complex in which the two molecules fit neatly together – an explanation known as the lock and key hypothesis.

Enzyme specificity is explained by the need for enzyme and substrate to fit together precisely. The molecular configuration of enzymes is determined by weak electrostatic (hydrogen) bonds which are broken by heat, acids and alkalis; hence their sensitivity to temperature and pH.

Biochemical techniques An individual cell may carry out up to 1000 chemical reactions, which must be carefully controlled. To assist control the conversion of one chemical to another takes place in a series of small steps known as a metabolic pathway. In addition the energy is released gradually rather than in a violent manner.

To determine the order of intermediates, a number of laboratory techniques are used:

1 the addition of an intermediate which should lead to an increase in all intermediates between it and the product.
2 the addition of an enzyme inhibitor which should cause an accumulation of the enzyme's substrate and the substances formed prior to it.

Centrifugation

Cells are broken down mechanically to form an homogenate suspension which is spun at a high speed causing the heavy structural components to separate out and form a sediment below a supernatant liquid. Particles of a different mass may be separated by centrifuging at different speeds (differential centrifugation).

Dialysis

Small molecules such as inorganic ions may be separated from larger ones such as protein by placing them within a membrane permeable only to the ions. The membrane is then placed in water into which the ions diffuse.

Chromatography

This is a means of separating out mixtures of chemicals by using their different solubilities in certain solvents. A concentrated spot of the solution to be separated is placed at one corner of a strip of paper. This is dipped in a solvent which carries the chemicals dissolved in it up the paper depositing them, the least soluble first, the most soluble last. If necessary other chemicals may be used to make the spots visible., e.g. Ninhydrin colours amino acids. They are identified by their colour and R_f values:

$$R_f = \frac{\text{distance moved by spot}}{\text{distance moved by solvent}}$$

Spots may be further separated by running a different solvent at right angles to the first (two-way chromatography), or, instead of paper, a layer of suitable material, e.g. silica gel, may be used (thin layer chromatography).

Electrophoresis

This is similar to chromatography except that the spot is put in the centre of a strip of paper or the thin layer soaked in a suitable electrolyte. A potential difference is applied across it so that cations move towards the anode and anions to the cathode separating the substances, not on their solubilities, but on their relative electropositivity and negativity.

Isotopes

Since an isotope differs from the normal element its progress along a metabolic pathway can be traced in a number of ways. If radioactive, e.g. ^{14}C, its presence in intermediates can be determined using Geiger counters once all the intermediates have been separated by one of the above methods. If too little radiation is emitted, the chromatogram may be placed next to a photographic plate which becomes exposed by those spots containing the radioactive material. The developed photograph is called an autoradiogram.

If the isotopes are not radioactive they are detected and measured using a mass spectrometer which separates them according to their atomic mass.

X-ray diffraction

When a beam of X-rays is fired at a sample of the substance being studied, the atoms scatter the rays and the diffraction pattern so formed is recorded on a photographic plate. Rotation of the sample to allow patterns from different angles to be made gives a complete analysis. Computers are used to determine the precise positions of atoms from the patterns. This is used, for example, in the determining of DNA and protein structure.

Colorimetry

Many substances are detected by the production of a colour on the addition of an indicator or test reagent. The intensity of this colour is usually proportional to the amount of the substance present, e.g. the concentration of a starch solution may be determined by the intensity of the blue produced on the addition of iodine in potassium iodide solution. A colorimeter is an instrument which passes a beam of light through a sample of the solution under test (which is usually contained in a special flat-sided tube called a cuvette). The amount of light passing through alters the electrical resistance of a photocell, producing a deflection on the meter that measures the voltage across the cell. The scale shows either optical density and/or percentage transmission of light. The colorimeter has many applications such as the measurement of the rate of starch hydrolysis by amylase over a period of time. It is necessary to draw a calibration graph of colorimeter readings against known concentrations of starch, from which the actual starch concentration for any given colorimeter reading can be found.

Link topics

Section 6.1 Autotrophic nutrition (photosynthesis)
Section 6.2 Heterotrophic nutrition (holozoic)
Section 7.1 Cellular respiration
Section 8.4 Osmoregulation and excretion

Suggested further reading

Barker. G. R.. *Chemistry of the Cell*, Studies in Biology No. 13, 2nd ed. (Arnold 1982)
Rose, S., *The Chemistry of Life* (Pelican 1966)
Wynn, C. H., *The Structure and Function of Enzymes,* Studies in Biology No. 42, 2nd ed. (Arnold 1979)

QUESTION ANALYSIS

4 For each of the following:
 (i) chromatography
 (ii) electrophoresis
 (iii) dialysis
 (iv) centrifugation

 (a) State the principles upon which each procedure is based.

 (b) Outline, for one example of each, its application in biology.
 Mark allocation 20/100   Time allowed 40 minutes   In the style of the Joint Matriculation Board

This is an essay question testing knowledge of biochemical techniques; it is not so straightforward as it seems. Superficially it appears to be a question on the four techniques and the candidate who has learned what they are may be tempted to choose the question and give an account of each technique. The more discerning candidate will appreciate the key words '. . . principles on which each . . . is based', and realize that a mere factual account of them is inadequate. Interpretation of the question is all important, in this case the interpretation of the word 'principles' – which here means 'the underlying scientific theories and concepts'. The word must not be ignored as simply a convenient mode of expression but thought of as something the examiners included after much consideration. It indicates that they are not interested in the precise mechanism by which a technique is carried out but the scientific basis that led to the mechanism being developed.

For instance in '(i) chromatography' it is not the number of spots of solution used, how the spots are dried, what solvents are used or the time scale that matters, but such principles as:

1 Chromatography is a means of separating mixtures of chemical substances, usually from living tissues.

2 To extract the substances, a solvent in which they all dissolve is required.

3 The solution so formed must be concentrated, otherwise on separation the amount of each individual substance will be too small for identification.

4 Suitable solvent or solvents must be found in which the substances in the mixture have different solubilities.

5 This solvent must be drawn along a suitable material by capillarity (i.e. the material must comprise a matrix or meshwork of particles or fibres with spaces between).

6 As the solvent front moves along the material it carries with it the substances, depositing them successively according to their solubilities, least soluble first, most soluble last, thus separating them.

7 If not visible the substances may be 'developed' using a suitable chemical that colours them.

8 The distance moved by any substance relative to the distance moved by the solvent front will be constant for any given solvent, and therefore the substance can be identified by reference to a table of these values (R_f values).

The above account contains no details of technique, solvents, type of paper, etc.; it is purely the principles. The candidates who produce accounts, however accurate, of how they separated chlorophyll pigments during a practical lesson may leave the examination room confident but they will be disappointed when the results are published. A similar account is needed for the remaining three techniques (see answer plan).

In (b) note the emphasis on **'each'** and be certain to include four separate examples, and the words 'application in biology' and choose examples accordingly–those studying A-level chemistry may know good examples from their chemistry studies but these are useless here unless they are specifically biological.

Answer plan

(a) (i) **Chromatography:** this example is covered earlier in the question analysis.
 (ii) **Electrophoresis:** principles similar to (i) except separation is based on substances having different charges of varying magnitude. In an electric field these will separate according to their relative charges.

(iii) **Dialysis:** membrane used has differential permeability and separates according to moleculai size. Based on principle of diffusion.

(iv) **Centrifugation:** rapid rotation produces a centrifugal force, particles of greater mass separating out first, i.e. separating according to relative mass.

(b) (i) separation of photosynthetic pigments.

(ii) separation of amino acids.

(iii) separation of proteins from inorganic salts.

(iv) separation of cell organelles.

5 (a) Complete the graphs below to show the curves which would be obtained when experiments on enzyme activity are carried out under constant conditions and using

(i) excess substrate   (ii) a fixed quantity of enzyme

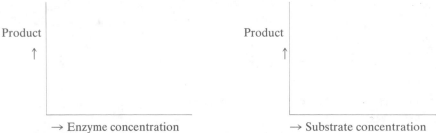

(b) Explain the shape of each of the curves drawn in (a).

(c) What is the 'active site' of an enzyme?

Mark allocation 6/100   Time allowed 7 minutes

Associated Examining Board November 1980, Paper 1, No. 7

This is a structured question testing knowledge of enzymes, in particular their properties. Although not difficult, any question involving graphs needs to be interpreted carefully, and while many candidates will draw the lines accurately a number will be unable to interpret them accurately enough to answer (b) effectively.

In (a) 'constant conditions' can be interpreted as meaning that no other factor, except those mentioned in the question, will affect the rate (or at least its effect will not vary throughout the experiment) and can therefore be discounted.

In the essential information you are told that 'the rate of an enzyme-catalysed reaction increases with enzyme concentration provided there is excess substrate'. The latter is true in (a) (i) and therefore the curve rises – but in what way? It is logical to suppose that in the absence of any other variable the rise would be linear, i.e. a straight line. Do not be put off by the word curve. Since no particular enzyme is specified and no scale is given, it is impossible to predict the angle of the line and therefore any reasonably accurate one (not vertical or horizontal) would be acceptable. The line should not tail off as the substrate is in excess.

In (a) (ii) the graph would be similar except that the limited amount of enzyme means a point is reached where it is all being used and therefore further increases in substrate concentration have no effect and at higher substrate concentrations the graph flattens to become horizontal.

The explanations for (b) have been partly covered above, in determining the shape of the graphs, but in addition the candidate should make reference to the lock and key theory of enzyme action (see essential information) explaining that in (a) twice as many enzyme molecules means twice as much substrate can be converted to twice as much product, hence the linear rise, whereas in (b) when all enzyme molecules are in use, further substrate can have no effect and the rate reaches a maximum which can only be further increased on the addition of more enzyme.

Part (c) is straightforward and covered in the essential information. Diagrams might help if time and space are available.

6 (a) What is meant by (i) an enzyme, and (ii) a metabolic pathway?   (6)

(b) Give some account of the factors that affect the rate of enzyme-catalysed reactions.   (10)

(c) Describe two of the techniques that are used in the study of metabolic pathways.   (4)

Mark allocation 20/100   Time allowed 35 minutes   *London June 1980, Paper 2, No. 1*

This is a structured essay that broadly covers much of the work on this topic. Though basically requiring recall of facts, the question goes a little deeper since definitions of 'enzyme' and 'metabolic pathway' are descriptive and the answers could take a number of forms.

In (a) (i) it is reasonable to assume that three marks are available and hence the 'definition' should be brief, referring to an enzyme's protein nature, catalytic properties, complex molecular shape, universal occurrence in cells and importance in cell chemistry.

In (a) (ii) the reasons for small-step sequences of chemical change as outlined under 'metabolic pathways' should be mentioned and the idea of a controlled pathway from substrate to product emphasized.

In part (b) the factors discussed under 'characteristics of enzymes' form a basis of the answer but the words 'factors that affect the rate' mean that 'specificity' and 'reversibility' are not applicable here. The other important words are 'some account', which can be interpreted as meaning that the examiner appreciates that the volume of information is such that a full or complete account is not possible in the time allowed (50% of the question time, say 15–20 minutes) and therefore only a moderate amount of detail is required, though more than the mere outline given in the essential information would be expected. For instance in 'effect of enzyme concentration' the account should include a graph showing rate of reaction against enzyme concentration, tailing off when substrate concentration limits the reaction.

Though part (c) says 'describe', the allocation of four marks for two techniques leaves little room for more than a very basic outline. The danger here is that the candidate will write on chromatography or electrophoresis, when these are methods of separation and not directly involved in the study of metabolic pathways. Instead the candidate should choose any two of the two techniques for determining intermediates (see essential information) and the use of tracer experiments using radioactive or other isotopes. The use of isotopes is of such importance that it warrants inclusion, leaving a choice of one of the remaining two.

Answer plan

(a) (i) Protein; biological catalyst; complex molecular shape
   (ii) Sequence of small steps from substrate to product.
(b) Basic details of pH, temperature, inhibitors, activators, enzyme concentration, substrate concentration.
(c) Use of isotopes and either addition of intermediates or use of enzyme inhibitors.

7 Exposed photographic film has black silver salts bonded to it by a thin layer of gelatin (a protein). In an investigation into the digestion of gelatin by the enzyme trypsin the end point is shown by the clearing of the film, as in the diagram.
Seven test tubes, each with a different buffered pH solution and 1 cm^3 of 0.5% trypsin solution, were placed in a water bath at 35°C for 5 minutes. Small pieces of exposed film were simultaneously placed into each test tube and the time taken for the film to clear was noted.

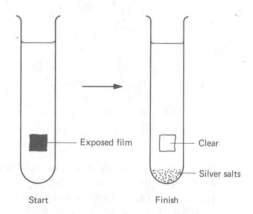

Fig. 3.5 Action of an enzyme on exposed film

pH	Time taken to clear in minutes
6.0	30
6.5	20
7.0	13
8.0	5
9.0	8
9.5	20
10.0	35

  (i) Plot a graph of 'time to clear' against pH. (2)
 (ii) What appears to be the optimum pH for trypsin? (1)
(iii) Explain briefly what is meant by a buffered solution. (1)
(iv) Give details of controls required for this experiment. (1)
 (v) Why are controls necessary? (1)
(vi) Explain the need for a waterbath in this experiment. (1)
(vii) Explain why it is necessary to place the seven test tubes in the waterbath for 5 minutes before inserting the film into each. (1)
(viii) Predict what would have happened to the 'time to clear' had the experiment been carried out at 65°C. (1)
Mark allocation 9/125   Time allowed 11 minutes   *Scottish Higher II, 1980, No. 10*

This structured question tests knowledge of experimental techniques and the ability to draw and interpret graphs. Candidates may have carried out a similar experiment in which case they will be at an advantage. Many will not have done, but provided sound principles of experimentation are applied this should not prevent them obtaining full marks.

In (i) the graph should be titled, have axes clearly labelled and the units accurately stated. The graph should be a smooth U shaped curve. Choose scales for the axes that are easy to use but make maximum use of the graph paper provided; be prepared to turn the graph paper if necessary to achieve this. In (ii) it is important to remember that the lowest point of the curve is the optimum pH because this is the shortest time taken for the trypsin to break down the gelatin, release the silver salts and so clear the film. The result should be around pH 8 but need not be exactly so. Read the figure from the graph rather than the table of data, it will most likely be slightly higher than pH 8.0, but the actual value will depend on your own interpretation of the points plotted.

For (iii) the candidate should state that a buffered solution is one in which the pH remains constant despite the addition of small amounts of solution of different pH. There are a number of possible controls for (iv). The enzyme could be excluded and replaced by an equal volume of distilled water. However, a control should be as near as possible the same as the experiment, except in respect of the factor under test. As it is the action of an enzyme that is under test it is preferable to use the same volume of the enzyme but to boil it for at least 5 minutes beforehand to ensure it is denatured. If the control did not clear the film it would show not only that trypsin was responsible for the change in the experimental tube, but also that its action was prevented by heating and it was acting enzymatically. The necessity for a control (v) was to show that the change was due to trypsin acting enzymatically and that the pH was therefore affecting trypsin and not the film directly. The waterbath mentioned in (vi) was to maintain a constant temperature. To be effective and allow conclusions to be drawn only the pH should vary, all other factors should remain constant. Water with its high thermal capacity effectively buffers external temperature changes. The point of placing the tubes in the waterbath for 5 minutes (vii) before adding the film is to bring them up to the experimental temperature, otherwise the temperature would be changing during the experiment. At 65°C (viii) the film would fail to clear as the heat energy would disrupt the molecular configuration of trypsin (denaturation) and it would fail to catalyse the breakdown of gelatin.

8 In an experiment ^{18}O was used to label water and carbon dioxide which was fed to green plants. As a result of the experiment the following information was obtained:
Only when $H_2^{18}O$ was fed to the plants was ^{18}O given off.
When 0.85% of the water supplied to the plant was $H_2^{18}O$ then 0.84% of the oxygen evolved from the plants was ^{18}O.

(a) (i) What is meant by the term 'labelled water'?   (1)
    (ii) How is the ^{18}O detected?   (1)
   (iii) From the information given, where does the oxygen given off from green plants appear
          to originate?   (1)

(b) In which part of the chloroplast is oxygen produced?   (1)

Mark allocation 4/125   Time allowed 5 minutes   *In the style of the Scottish Higher*

This short structured question tests recall of factual information on biochemical techniques. Candidates should read the information and try to understand what is occurring, even before reading the questions. In (a) (i) 'labelled water' refers to the fact that the oxygen atom in each water molecule is the radioactive isotope ^{18}O. The progress of this atom can be traced during the biochemical changes the water undergoes. For (a) (ii) the best way of detecting ^{18}O in the amounts likely to be encountered is using a mass spectrometer. Take care in (a) (iii) to use 'the information given'. Since ^{18}O is given off when the plant is fed only $H_2^{18}O$ it is reasonable to assume the ^{18}O originated in the water. The answer required in (b) is the grana.

3.3   GENETIC CODE

Assumed previous knowledge The chemical nature of amino acids and proteins (see Section 3.1). The properties of enzymes, in particular their protein nature, specificity and mode of action (see Section 3.2).
Basic cell ultrastructure, especially nucleus and ribosomes (see Section 2.1).

Underlying principles

The nature of an organism is determined by the arrangement of its cells, whose nature is in turn determined by the chemical processes within them. These processes are controlled by enzymes, all of which are proteins. Any cell produces a wide variety of enzymes, differing from cell to cell and organism to organism. As the catalytic properties of these enzymes depend on the precise sequence of their amino acids, the process by which they are made must be accurate and, as enzymes are only made when required, carefully regulated. The vast set of instructions for the production of these enzymes must be contained within the minute amount of genetic material which passes from generation to generation.

Points of perspective

Details of the elucidation of DNA structure by Watson and Crick using X-ray diffraction, etc.

The Jacob–Monod theory of gene action in controlling protein synthesis.

The Meselsohn–Stahl experiment using ^{15}N to demonstrate semi-conservative replication of DNA.

Hämmerling's experiment on *Acetabularia* to show that the nucleus is involved in heredity.

Essential information

Structure of nucleic acids Nucleic acids are large molecules of high molecular mass composed of chains of nucleotides which differ only in their organic bases.

Nucleotides
Each organic base is attached to a pentose sugar called ribose and phosphoric acid (H_3PO_4) to give a nucleotide.

Nucleic acids
Ribonucleic acid (RNA) comprises a long chain of nucleotides linked to form a single strand. It contains the organic bases adenine, guanine, cytosine and uracil (not thymine).

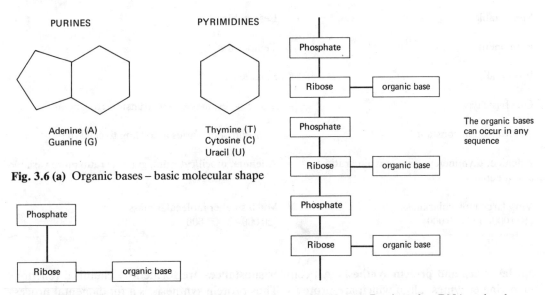

Fig. 3.6 (a) Organic bases – basic molecular shape

Fig. 3.6 (b) Basic structure of a nucleotide

Fig. 3.6 (c) Portion of an RNA molecule

Deoxyribonucleic acid (DNA) is similar to RNA except that it contains the organic bases adenine, guanine, cytosine and thymine (not uracil), has one less oxygen in the ribose sugar, and is composed of a double strand wound into a helix (Watson and Crick hypothesis). The structure may be likened to a twisted ladder in which the ribose and phosphate molecules form the uprights and the organic bases the rungs. The 'rungs' of this ladder must be of the same length so it follows that, as both purines and pyrimidines are found in DNA, each 'rung' must be made up of a purine linked with a pyrimidine. Two purines would make the 'rung' too long; two pyrimidines too short (see basic molecular shape). Analysis of DNA shows the quantity of adenine and thymine to be the same and the quantity of guanine and cytosine to be the same, indicating that they form the organic base pairs.

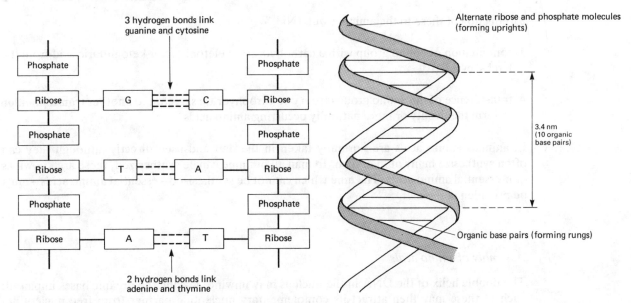

Fig. 3.7 (a) Basic DNA structure – helix unwound

Fig. 3.7 (b) Double helical structure of DNA

Molecular biology and biochemistry

Table 3.4 Differences in structure between DNA and RNA

DNA	RNA
Organic bases are adenine, thymine, guanine, cytosine	Organic bases are adenine, uracil, guanine, cytosine
Double helix	Single strand
Deoxyribose (H)	Ribose (OH)
Mostly nuclear	Throughout the cell
More stable	Less stable
Permanent	Temporary
Insoluble	Soluble
One basic type	Three types: messenger, transfer, ribosomal
Concentration constant	Concentration varies according to cell type
Adenine, thymine/cytosine, guanine ratio about equal	Adenine, uracil/cytosine, guanine ratio more variable
Very large molecular mass (100 000–120 000 000)	Much smaller molecular mass (20 000–2 000 000)

Nucleic acids and protein synthesis All cellular substances are formed by chemical reactions involving enzymes, all of which are proteins. Thus protein synthesis is a fundamental process of life. There are two stages:

1 synthesis of amino acids.

2 assembly of amino acids in the correct sequence to form a particular protein.

Synthesis of amino acids

In plants, synthesis occurs in the mitochondria and chloroplasts in the following stages:

1 abstraction of nitrates from the soil

2 reduction of these to the amino group (NH_2)

3 combination of amino group with a carbohydrate skeleton, e.g. α ketoglutaric acid from the Krebs' cycle.

4 transference of the amino groups from one carbohydrate skeleton to another **transamination** to form the twenty or more naturally occurring amino acids.

In animals amino acids are normally taken in the diet and used directly although they can often synthesize many of their own. In man all but nine can be synthesized; these are known as non-essential amino acids. The nine which cannot be synthesized – essential amino acids – must be provided in the diet.

Assembly of amino acids

The double helix of the DNA in the nucleus may unwind, leaving the organic bases unpaired; each of these may then attract its complementary nucleotide partner from free nucleotides within the nucleus to form two separate strands of DNA = **semi-conservative replication**

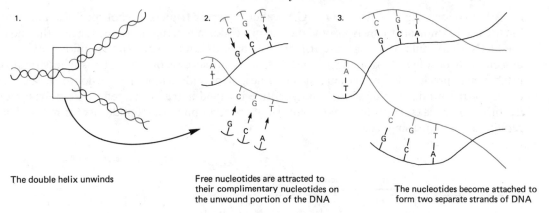

| The double helix unwinds | Free nucleotides are attracted to their complimentary nucleotides on the unwound portion of the DNA | The nucleotides become attached to form two separate strands of DNA |

Fig. 3.8(a) Semi-conservative replication of DNA

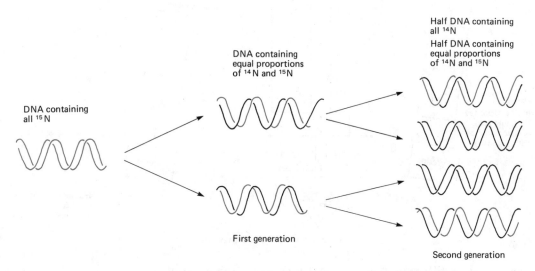

Fig. 3.8 (b) Evidence for semi-conservative replication of DNA

Evidence for semi-conservative DNA replication

This was provided by the work of Meselsohn and Stahl in the late 1950s. Many generations of *E. coli* cells were grown in cultures containing ^{15}N (heavy nitrogen) to ensure that their DNA contained ^{15}N. They were then transferred to a medium containing only ^{14}N (light nitrogen). The next two generations of *E. coli* each had their DNA separated by centrifugation and the presence of light and heavy nitrogen in the DNA was detected by its absorption of ultra violet light. The results showed the first generation of bacteria to have DNA containing equal proportions of ^{14}N and ^{15}N. In the second generation half the bacteria had DNA containing all ^{14}N, the remainder having DNA with equal proportions of ^{14}N and ^{15}N.

Sometimes the DNA only partially unwinds and the new strands produced become detached and leave the nucleus; these molecules are called messenger RNA (mRNA). The mRNA has a definite sequence of organic bases along it which are complementary to the DNA from which it came.

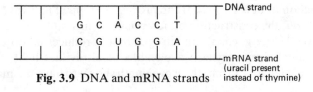

Fig. 3.9 DNA and mRNA strands

The mRNA moves out of the nucleus and wraps itself around groups of 5–50 ribosomes (polysomes) in the cytoplasm. In the cytoplasm are smaller strands of RNA called transfer RNA (tRNA) each of which links up with a specific amino acid at one end. At the other end are three organic bases known as **anticodons**. When the mRNA wraps itself around the polysome, one of the ribosomes moves along it; as it passes each set of three bases (codons), the appropriate anticodon of the tRNA attaches.

As each codon on the mRNA consists of three bases, it is called a triplet code. Any amino acids may have up to six different triplet codes, e.g. in the diagram arginine is coded by CGU but it may also be coded for by CGC, CGA, CGG, AGA and AGG.

A few codons, e.g. UAA; UAG; UGA, called **nonsense triplets**, do not specify amino acids but are used to indicate to the ribosome the end of a code. When the ribosome reaches the end of the mRNA molecule or a nonsense triplet, the chain of amino acids (polypeptides) becomes detached. Ribosomes pass along the mRNA in rapid succession so that a single strand of mRNA may produce large amounts of the same polypeptide chain. The region of DNA that codes for the production of a single polypeptide is called a **gene** (one gene–one polypeptide hypothesis). Because a protein may comprise more than one polypeptide, its formation may require more than a single gene.

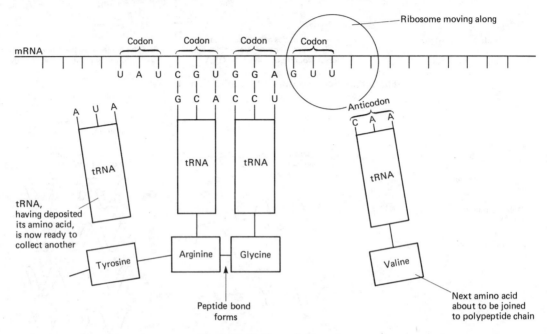

Fig. 3.10 Coding of a protein

Experimental evidence to show that DNA is the hereditary material
In *Pneumococcus,* the bacterium causing pneumonia, there are two easily distinguished strains: **avirulent**, which has no capsule (coat) and produces rough colonies (R) when grown on an agar plate; **virulent**, which has a polysaccharide capsule and produces smooth colonies (S). Smooth strains mutate into rough strains rarely but the reverse never occurs. In 1928 Griffith injected mice with dead virulent (S) cells and living avirulent (R) cells. The mice frequently died of pneumonia and living virulent (S) cells were isolated from them. Thus the ability to cause the disease is somehow passed from the dead virulent (S) types to the living avirulent (R) types. In 1933 Alloway showed that a cell-free extract was capable of bringing about this change; in 1944, by testing each component of the extract in turn, Avery, Macleod and McCarty showed that only a highly purified DNA extract was capable of bringing about this transformation. Since the virulence passed on by the DNA is inherited in successive generations, DNA must be the hereditary material.

In the 1950s Hershey and Chase, by radioactive labelling of the DNA in bacteriophages that attack *E.coli,* showed that only DNA entered the *E.coli*; as new phage particles were produced, it could only be the DNA which possessed the hereditary information necessary for the production of new viruses. If the process is stopped halfway, i.e. only half the DNA is passed into *E.coli*, the construction of new phages is incomplete.

Damage to DNA (e.g. change in nitrogen base) produces inherited changes (mutations).

When illuminated by varying wavelengths of ultraviolet light, the maximum absorption of DNA is of wavelength 254 nm, the wavelength shown to have the maximum mutagenic effect.

Link topics

Section 4.1 Mitosis and meiosis
Section 4.2 Heredity and genetics

Suggested further reading

Clark, B. F. C., *The Genetic Code and Protein Biosynthesis* Studies in Biology No. 83 2nd ed. (Arnold 1984)
Jackson, R., *Protein Biosynthesis,* Oxford/Carolina Biology Reader No. 86 (Packard 1978)
Kendrew, J., *The Thread of Life* (BBC 1966) Particularly the last five chapters.

Travers, A. A., *Transcription of DNA*, Oxford/Carolina Biology Reader No. 75, 2nd ed. (Packard 1978)
Warr, J. R., *Genetic Engineering in Higher Organisms*, Studies in Biology No. 162 (Arnold 1984)
Watson, J. D., *The Double Helix* (Weidenfeld and Nicolson 1976; Penguin 1970)

QUESTION ANALYSIS

9 (a) Biochemical analysis of a sample of DNA showed that 33 per cent of the nitrogenous bases was guanine. Calculate the percentage of the bases in the sample which would be adenine. Explain how you arrived at your answer.
(b) What name is given to the triplet of three bases which designates individual amino acids?
(c) If the triplet of mRNA bases which designates the amino acid lysine is AAG (where A = adenine and G = guanine), what is the complementary triplet of three bases on the tRNA molecule? Give a key for the letters that you use.
Mark allocation 6/80   Time allowed 9 minutes

Associated Examining Board, June 1984, Paper 1, No. 8

This short answer structured question requires application of knowledge on DNA structure and protein synthesis.

In part (a) the candidate needs to know that in the DNA molecule guanine always pairs with cytosine and thymine with adenine. Therefore the quantity of guanine is always the same as cytosine, and thymine the same as adenine. If 33 per cent of the nitrogenous bases are guanine, 33 per cent must be cytosine, giving a total of 66 per cent. The remaining 34 per cent must therefore comprise thymine and adenine. As these are found in equal amounts, each must account for 17 per cent of the total nitrogenous bases. The triplet of bases which designates individual amino acids is called a codon. The complementary triplet of bases on tRNA required in part (c) can be found from the base pairings given above. If the mRNA comprises AAG, then the complementary tRNA triplet is UUC, where U = uracil and C = cytosine. The most likely error is that candidates will give the answer TTC, having forgotten that the thymine of DNA is replaced by uracil in RNA.

10 (a) DNA (deoxyribonucleic acid) is a polynucleotide, i.e. a linear polymer of nucleotide monomers.
(i) What are the three chemical groupings within each nucleotide monomer?
(ii) What is the difference between the four types of nucleotide?
(iii) Briefly describe the role of each of the three chemical groupings within the nucleotide monomers in the function of the DNA molecule.   (3)
(b) Explain the term **semi-conservative** replication, as applied to DNA.   (1)
(c) What is the evidence for the semi-conservative nature of DNA replication?   (4)
(d) (i) Why is it important that DNA replication should produce two exact copies of the original DNA molecule?
(ii) What is the result of occasional mistakes in replication?
(iii) What are the implications of such mistakes for the organism, the species and evolution?   (7)
Mark allocation 15/70   Time allowed 35 minutes   *Northern Ireland 1983 Paper 1, No. 5*

Part (a) of this structured question tests the candidate's factual knowledge of DNA. The 'three groupings' referred to in (a) (i) are shown in Figure 3.6b. They are phosphoric acid (phosphate) H_3PO_4, the five carbon (pentose) sugar – deoxyribose and organic bases (adenine, thymine, cytosine and guanine). As the question refers to DNA, care must be taken to name the pentose as deoxyribose (not ribose) and to include thymine rather than uracil in the list of organic bases.

The differences required in (a) (ii) are the different organic bases, adenine, thymine, cytosine or guanine, that are present on the nucleotides.

In (a) (iii) the word 'briefly' is important. With just three marks for the whole of part (a) it would be foolish to be other than brief if the question were to be completed in the time allowed. The deoxyribose sugar and the phosphate groups make up a structural framework or backbone which support the organic bases that project sideways (see Fig. 3.7a). The bases themselves have the role of carrying genetic information for it is the sequence of these bases along the deoxyribose-phosphate framework which controls which proteins a cell makes and so determines the nature of each cell.

The semi-conservative replication referred to in part (b) is the ability of the double helix of DNA to unwind (unzip) through the breaking of the hydrogen bonds linking complementary bases. Free nucleotides then pair with their complementary organic bases on the two separate DNA strands, so forming two new double helices. Any description should include explanatory diagrams similar to those in Figure 3.8.

The evidence for semi-conservative replication (part c) comes largely from the experiments of Meselsohn and Stahl which are explained in the essential information.

In (d) (i) the importance of producing two exact DNA copies is that during growth and repair when cells divide mitotically, the daughter cells should be identical to the parent ones if they are to function in the same way as the cells they complement or replace. As the DNA ultimately determines the nature of a cell, it is essential that the daughter cells each receive an identical copy of the original parental DNA.

Any mistakes in replication (part (d)(ii)) will produce imperfect DNA copies. The mistakes are often the substitution of one organic base by another (say thymine instead of adenine). The sequence of bases along the DNA strand form a code whereby each triplet of bases codes for one amino acid (see essential information for details). Any change in the sequence of bases could result in the wrong amino acid being introduced into a polypeptide chain. This imperfect polypeptide chain may then be unable to perform some essential function in the body. The candidate could include a specific example, e.g. the substitution of valine (DNA code = CAA) for glutamic acid (DNA code = CTT) in the two β-polypeptide chains of haemoglobin result in sickle cell anaemia, which in the homozygous state is fatal. Any 'mistake' during replication that results in an alteration in the organic base sequence of DNA is termed gene mutation.

The implications of such mistakes ((d)(iii)) for the organism depend upon the precise nature of the error. If the polypeptide chain forms an enzyme, for example, the incorrect amino acid in the chain could affect hydrogen bonding in such a way that the molecular shape is altered. The lock and key hypothesis of enzyme action depends on the enzyme having a precise molecular shape. The alteration could render the enzyme useless and so ineffective in its particular metabolic pathway. The disruption of any pathway is likely to have detrimental, if not fatal, consequences. The candidate may prefer to discuss the sickle cell anaemia trait at this point rather than in (d)(ii) above. However, there is nothing to be gained and much to be lost in discussing it twice. Paradoxically, while some mutations prove detrimental to an individual organism others provide long term security for the species. As climatic and other environmetal conditions inevitably change, or as individuals disperse, they encounter new environments which make different demands on them. The development of entirely new characters by mutation increases the variety of genes in a gene pool and consequently makes the individuals of a species more various. It is through the continual natural selection of the individuals best suited to the environments that species adapt to changing conditions and so survive. Such adaptations are essential to the evolution of a species. These mistakes during the replication of DNA are therefore the raw material of evolution.

The highly structured nature of this question and the answer provided make an answer plan unnecessary.

11 An electron micrograph of active ribosomes is represented in the diagram below.

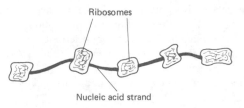

Ribosomes

Nucleic acid strand

Fig. 3.11 Arrangement of ribosomes

(1) The nucleic acid strand is
    A DNA   B messenger RNA   C transfer RNA   ·   D ribosomal RNA

(2) The advantage of the nucleic acid strand passing through five ribosomes at one time is that
    A ribosomes can be linked together
    B five identical protein molecules can be made from one set of instructions
    C five different types of protein molecule can be made at one time
    D more RNA molecules can be produced quickly

(3) Ribosomes such as those shown are likely to be found
    A in the nucleus   B in mitochondria   C in the vacuoles   D on endoplasmic reticulum
Mark allocation 4½/75   Time allowed 4½ minutes   *In the style of Scottish Higher I*

These three multiple choice questions test candidates' knowledge of cell ultrastructure and protein synthesis. In (1) option A can be discounted because, although there is some evidence for extranuclear DNA, in most cells it is almost entirely confined to the nucleus. As ribosomes occur in the cytoplasm the nucleic acid strand is not likely to be DNA. The nucleic acid shown is attached to the ribosomes; it cannot therefore be transfer RNA (option C) as this associates with amino acids, nor can it be ribosomal RNA (option D) which is part of the ribosomes. This leaves option B as the correct answer; messenger RNA wraps itself around ribosomes in the way shown in the diagram.

For (2) the role of ribosomes in protein synthesis needs to be known. The correct response is B because as the messenger RNA passes over each ribosome its code is 'read' and a new protein is produced. The messenger RNA codes for one protein so that all five molecules produced are the same.

Question (3) tests recall of factual information. Of the four options given ribosomes occur only 'on the endoplasmic reticulum' (D).

4 Cell Division and Genetics

4.1 MITOSIS AND MEIOSIS

Assumed previous knowledge The structure of DNA (see Section 3.3)

Basic cell ultrastructure (see Section 2.1)

Underlying principles

If a tissue is to be extended by growth, or damaged cells in it are to be replaced, it is important that the new cells are exact copies of the originals. In the same way a species that is successful colonizing a particular habitat may benefit from rapid multiplication to produce identical individuals which will also be suited to that environment. In both cases a rapid form of cell division that produces identical daughter cells is needed. This is mitosis, which maintains the constancy of a species.

The long-term survival of a species is, however, dependent on its ability to adapt to the constantly changing climatic, edaphic and biotic environment in which it lives, and on its ability to colonize new and different habitats. To achieve this the offspring need to be different from their parents so that each has the potential to survive in a different environment and thus continue the species. A method of producing new characteristics by the mixing of genetic material within and between organisms has therefore evolved. If the genes of two individuals are to be combined while maintaining the total chromosome number, the genetic material from each parent must be halved. This is meiosis, which creates variety within a species.

Points of perspective

Both practical and theoretical examinations often contain questions involving identification from photographs of the stages of mitosis and meiosis. A careful study of photographs of all stages of both types of cell division is therefore invaluable.

Experimental details of how to obtain microscope preparations of the stages of meiosis and mitosis are often required (see question analysis). It is clearly an advantage if the candidate has performed these experiments.

Essential information

Structure of chromosomes Chromosomes are rod-like structures, consisting of nucleic acids and protein, located within the nucleus. During cell division chromosomes change their length as a result of coiling and uncoiling, dividing to form paired, joined **chromatids.** Each chromosome has somewhere along its length a well-defined region where the chromatids are particularly closely associated and which seems to be the point at which force is exerted in the separation of dividing chromosomes. This structure is called the **centromere.** Along the length of the chromosome there may be distinct constrictions and bumps, the pattern of which is quite constant for a particular chromosome from cell to cell. These 'bumps' are called **chromomeres** and they are probably caused by the coiling within the chromatids. The number of chromosomes per nucleus is normally constant for all the individuals of a species, e.g. man has 46, rat 42, garden pea 14, tomato 24. The number of chromosomes characteristic of a species gives no indication of its level of organization.

Chromosomes are present in pairs and therefore it is often convenient to speak of the chromosome number of a particular species in terms of the number of pairs (i.e. 23 for man). The members of each pair are alike but the different pairs are distinguishable. Every body (**somatic**) cell contains the characteristic number of chromosomes but mature germ cells (**gametes**) contain only half the usual number, one member of each pair. The gametes are described as **haploid** in chromosome number and the somatic cells as **diploid.**

Mitosis In mitosis each chromosome duplicates itself and the duplicates are separated from each other at cell division, one going into the nucleus of one daughter cell and the second going into the other. The daughter cells are therefore identical with each other and with their parent cell in chromosome constitution.

When the cell is preparing to divide it is said to be in *interphase.* At this stage

1 the DNA replicates so that there is sufficient for two daughter cells.

2 the cell builds up its store of energy to provide sufficient for cell division.

3 the cell forms new cell organelles to supply the daughter cells.

Cell division itself is a continuous process but for ease of description four main stages are recognized: prophase, metaphase, anaphase, telophase. The following series of diagrams represents mitosis in an animal cell and shows only two different chromosomes.

(a) **Interphase**

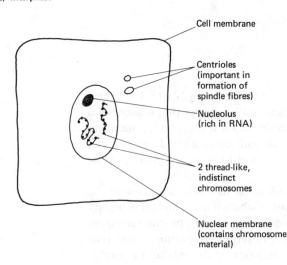

Cell membrane

Centrioles
(important in
formation of
spindle fibres)

Nucleolus
(rich in RNA)

2 thread-like,
indistinct
chromosomes

Nuclear membrane
(contains chromosome
material)

(b) **Early prophase**

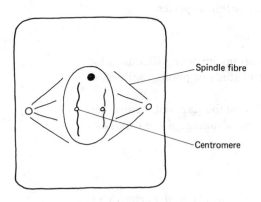

Spindle fibre

Centromere

Chromosomes more distinct as they
contract. Centromere visible.
Spindle starts to form from centrioles
lying at its poles. Nucleolus shrinks.

(c) **Late prophase**

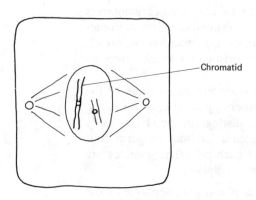

Chromatid

Chromatids visible as chromosomes shorten
and thicken. Nucleolus disappeared.
Nuclear membrane disappearing.

(d) **Early metaphase**

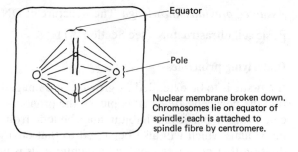

Equator

Pole

Nuclear membrane broken down.
Chromosomes lie on equator of
spindle; each is attached to
spindle fibre by centromere.

Late metaphase:
chromatids start to move apart.

(e) **Early anaphase**

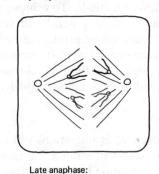

Chromatids move apart to
opposite poles, probably by
contraction of spindle fibres
(a process requiring energy).

Late anaphase:
chromatids reaching poles.

(f) **Early telophase**

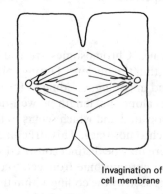

Chromatids assemble at poles
and cell membrane invaginates
(cell plate starts to form across
cell in plants).

Invagination of
cell membrane

(g) **Late telophase**

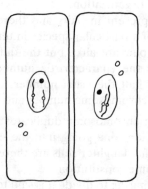

2 daughter cells formed.
Spindle fibres degenerate.
Nuclear membrane and
nucleolus reform.
Chromosomes regain
thread-like form.

Fig. 4.1 Stages of mitosis

Meiosis This results in the formation of haploid daughter cells since each receives only one of each type of chromosome instead of two. The same basic stages are recognized as in mitosis but they occur twice, i.e. first meiotic division followed by second meiotic division. The following series of diagrams is based on an animal cell containing two pairs of chromosomes.

(a) **Interphase**

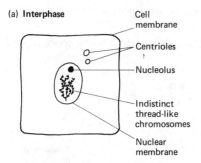

Cell membrane

Centrioles

Nucleolus

Indistinct thread-like chromosomes

Nuclear membrane

5 subdivisions of **prophase** are recognized leptotene, zygotene, pachytene, diplotene and diakinesis

(e) **Diplotene**

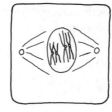

Chromatids continue to move apart as they shorten and thicken.

Diakinesis: Shortening and thickening continues; chiasmata move to ends; crossing over has occurred (see later); nuclear membrane breaks down.

(i) **Prophase II**

2 daughter cells.

Spindles start to form, usually at right angles to one formed in meiosis I.

In the following diagrams only one of the daughter cells is shown.

(b) **Leptotene**

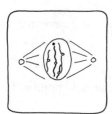

Chromosomes appear. Spindle starts to form.

(f) **Metaphase I**

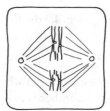

Homologous pairs of chromosomes align themselves on the equator of the spindle.

(j) **Metaphase II**

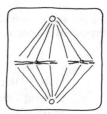

Chromosomes arrange themselves on the equator.

(c) **Zygotene**

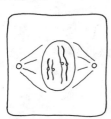

Nucleolus has disappeared. Homologous pairs of chromosomes (2 chromosomes determining the same features) associate, forming a bivalent: a process known as synapsis.

(g) **Anaphase I**

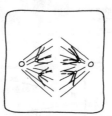

Homologous chromosomes move to opposite poles attached to spindle fibres by centromere (chromatids do *not* separate)

(k) **Anaphase II**

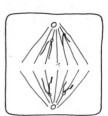

Chromatids pull apart and move to opposite poles.

(d) **Pachytene**

Chromatids become visible as they move apart from each other; they remain in contact at points called chiasmata.

(h) **Telophase I**

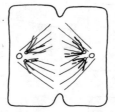

Chromosomes reach poles. Cell membrane invaginates.

There may be a short interphase or cells may move straight into the second meiotic division in which separation of the chromatids takes place.

(l) **Telophase II**

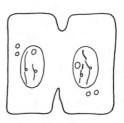

Cell membrane invaginates. Nuclear membrane and nucleolus reform. 2 daughter cells formed, each with half the number of chromosomes present in original parent cell.

Fig. 4.2 Stages of meiosis

Details of crossing over

Chiasmata may form between any two of the four chromatids and there may be up to eight chiasmata in a bivalent.

Chiasmata have two functions:

1 to hold homologous chromosomes together while they move into position on the spindle prior to segregation.
2 crossing over (or exchange of genetic material) occurs at the chiasmata leading to increased variation, the raw material of evolution.

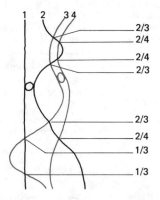

Fig. 4.3 (a) Some possible chiasmata forming in a bivalent. The four chromatids are numbered for reference

Fig. 4.3 (b) Crossover showing exchange of genetic material

Significance of meiosis

1 Halving the chromosome number ensures that when gametes with the haploid number fuse to form a zygote the normal diploid number is restored.
2 Meiosis leads to increased variation:
  (a) when the haploid cells fuse at fertilization there is recombination of parental genes.
  (b) during metaphase I homologous chromosomes are together at the equator of the spindle but they separate into daughter cells independently of each other.
  (c) chiasmata and crossing over can separate and rearrange genes located on the same chromosome.

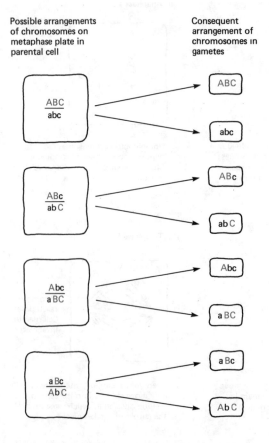

Fig. 4.4 Recombination of chromosomes in gametes

Table 4.1 Differences between mitosis and meiosis

Mitosis	*Meiosis*
One division of nucleus and one of chromosomes	Two divisions of nucleus and one of chromosomes
Chromosome number remains constant	Chromosome number halved
No association of homologous chromosomes	Homologous chromosomes associate in pairs
No chiasmata or crossing over	Chiasmata and crossing over occur
Two daughter cells formed	Four daughter cells formed (tetrad)
No variation (unless mutation occurs)	Variation due to exchange of genetic material
Chromosomes shorten and thicken	Chromosomes coil but remain longer than in mitosis
Chromosomes form a single line at the equator	Chromosomes form a double row at the equator
Chromatids move to opposite poles	Chromosomes move to opposite poles

Link topics

Section 2.1 The cell
Section 3.3 Genetic code
Section 4.2 Heredity and genetics
Section 4.3 Genetic variation and evolution
Section 5.1 Types of reproduction and life cycles

Suggested further reading

John, B. and Lewis, K. R., *Somatic Cell Division,* Oxford/Carolina Biology Reader No. 26, 2nd ed. (Packard 1981)
John, B. and Lewis, K. R., *The Meiotic Mechanism,* Oxford/Carolina Biology Reader No. 65, 2nd ed. (Packard 1983)
Kemp, R., *Cell Division and Heredity,* Studies in Biology No. 21, 2nd ed. (Arnold 1985)
Wheatley, D. N., *Cell Growth and Division,* Studies in Biology No. 148 (Arnold 1982)

QUESTION ANALYSIS

1 The diagrams below show two cells from the same animal in the process of division.

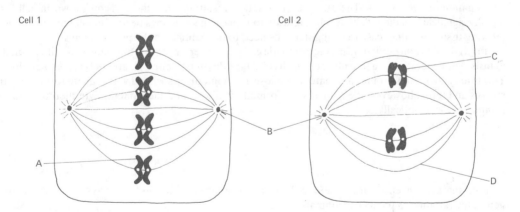

Fig. 4.5 Cells in the process of division

(a) Name the structures labelled A to D   (4)
(b) (i)   What type of division is shown in cell 1?   (1)
     (ii) What stage of division is shown in cell 1?   (1)
(c) (i)   Describe the differences shown by the two cells   (2)
     (ii) What is the significance of these differences?   (2)
(d) Which type of cell division occurs in a meristematic cell of the flowering plant?   (1)
(e) State the generation of a fern in which the type of division shown in cell 2 does *not* occur.   (1)
Mark allocations 12/200   Time allowed 9 minutes   *London January 1982, Paper 1, No. 1*

In addition to recall of factual information, this structured question tests the candidate's ability to make comparisons and understand the principles underlying mitosis and meiosis. Some application of knowledge of cell division to specific examples is also needed.

Many candidates fail to appreciate the detail and accuracy needed on a label in order to ensure maximum credit. Part (a) illustrates this point well. Label A would probably be labelled as 'chromosome' by many candidates. Careful examination of the drawing shows that the structure labelled is *single* stranded whereas whole chromosomes are double stranded. In addition a centromere is visible on *each* of the strands indicating that the chromosome has divided into separate chromatids which are in the process of segregating. The label A therefore refers to a **chromatid**. The argument that the structure A may be a chromosome of an homologous pair with the separate chromatids close together and therefore not distinguishable may be discounted when cell 2 is observed. It would be unreasonable of examiners to represent a structure such as a chromosome differently in the two diagrams. Two homologous pairs of chromosomes are clearly shown in cell 2 and as these differ in arrangement from the pairs in cell 1, the latter pairs must comprise two chromatids and not two chromosomes. The lesson here is to study carefully *both* diagrams and all labels before committing oneself to any answers. Further proof may be obtained from counting the numbers of structure A in cell 1. They number 8. As you are told the cells are 'from the same animal' it follows that if A is a chromatid then 8 chromatids should be visible in cell 2. This is the case although in cell 2 they are arranged in pairs in each of the four chromosomes shown. By the same reasoning, structure C is a chromosome but as it is paired with another of similar shape a fuller answer would be **'one of a pair of homologous chromosomes'**. Strictly speaking the diagram does not show conclusively that they are 'homologous' but similarity of shape, the fact that the cell is dividing and the chromosomes are paired, make this a reasonable assumption.

As label B points to a definite structure, rather than a general region, the answer **'centriole'** seems more appropriate than pole. D is a **spindle fibre** comprising a protein microtubule.

The type of cell division shown in cell 1 is mitosis. There are only two types of cell division: mitosis and meiosis. In the latter the homologous pairs of chromosomes come together (synapsis). As this pairing is only evident in cell 2 and not in cell 1 it is clear that cell 1 is undergoing **mitosis**.

Cell 1 has the chromosomes arranged in a single plane known as the metaphase plate or equator. This stage is therefore metaphase. As the single chromatids are clearly segregating 'late metaphase' is a more complete answer. In answering (c)(i), Table 4.1 in the essential information lists differences between mitosis and meiosis, but only those differences 'shown by the two cells' must be included, i.e. association of homologous chromosomes in cell 2, but not in cell 1, chromosomes forming a double row in cell 2, a single one in cell 1, the segregation of chromatids in cell 1 only and the shorter, fatter appearance of the chromosomes in cell 2.

The significance of these differences (c)(ii) is that association of homologous chromosomes allows the formation of chiasmata and hence crossing over to occur. This exchange of genetic material provides greater variety of offspring and so gives greater evolutionary potential in providing a wider range of offspring from which those better adapted to the prevailing conditions can be selected. This is important as the division occurring in cell 2 is commonly associated with gamete formation and so continuation of the species. The absence of variation resulting from the division shown in cell 1 is equally significant as the resulting cells are used to replace missing ones or to complement others in a growing tissue. In either case the daughter cells need to be identical to the parental ones.

In (d) the division occurring in a meristematic cell of a flowering plant is that shown in cell 1, namely **mitosis**. The generation of a fern plant in which cell 2 type division (meiosis) is absent is the gametophyte. This is likely to confuse some candidates as meiosis is commonly associated with gamete formation. In mosses and ferns, however, the gametes are formed by mitosis and the gametophyte therefore never produces cells meiotically.

2 Outline the main events during meiotic divisions in organisms. Discuss the ways in which meiosis generates variation in plants and animals.
Mark allocation 20/60   Time allowed 35 minutes
Welsh Joint Education Committee June 1980, Paper A2, No. 3

This is an essay question testing knowledge of meiosis and its importance in bringing about variation.

With about 35 minutes available for this question the main problem is to condense the process of meiosis sufficiently without loss of essential detail. The question says 'outline', a clear guide to the candidate that full details of each stage are not required. In addition 'main events' indicates that some stages may be omitted provided the most important ones are included.

Ideally all twelve drawings making up Figure 4.2 in the essential information should be included, but it is unlikely that the average candidate would manage this in the time without some loss of neatness, detail or clarity. Drawing (a), on interphase could be omitted (although a short sentence should be written) and the description started with leptotene (b), except that as it is now the first drawing, chromosomes, nuclear membrane, cell membrane, centrioles, spindle, centromere and disappearing nucleolus should all be labelled. Provided adequate explanations of events are given at each stage, it is unnecessary to label these parts on subsequent drawings, if they are drawn recognizably the same. All new structures should be clearly labelled as they arise. Zygotene (c) should be included as it shows homologous chromosomes associating, likewise pachytene (d) showing chromatids and crossing over.

Diplotene could be limited to the note on the diagram, omitting the actual drawing, and diakinesis should consist of the notes as given. Metaphase I, anaphase I, telophase I and prophase II should all be as shown in Figure 4.2 and the description finished with telophase II, showing the full tetrad of four cells rather than just two shown in drawing 1. Make clear that metaphase II and anaphase II are essentially the same as metaphase I and anaphase I.

In the second part the word 'discuss' is important and implies that more than a list of ways in which variation occurs during meiosis is required. Use the points made under point 2 in the 'significance of meiosis' section of the essential information as a guide but expand them to show precisely how the variation occurs. For example, it is clear that if two parental chromosomes mix, the offspring will assume some of the features of each, and thereby be different from both parents. However, if the offspring are to have the same number of chromosomes as each of their parents, they can only acquire half of each parent's chromosomes – hence variation by this means is only possible through meiosis occurring at some point prior to syngamy in an organism's life cycle. Use Figures 4.3a, b, and 4.4 to discuss how random distribution of chromosomes on the metaphase plate and crossing over also bring about variation during meiosis (points 2b and 2c under 'significance of meiosis' in the essential information).

Answer plan

Outline of meiosis – ideally draw all 12 diagrams and notes comprising Figure 4.2 in the essential information, but at least include: prophase I – leptotene (b), zygotene (c) and pachytene (d), metaphase I (f), anaphase I (g), telophase I (h), prophase II (i) and telophase II (l). Include the short written notes on any stage where the drawing is omitted.

Meiosis and variation – discuss how it allows:

(i)   sets of chromosomes to be mixed (fusion of two parental gametes).
(ii) individual chromosomes to be mixed (random distribution of chromosomes on the metaphase plate).
(iii) material between chromosomes to be exchanged (crossing over).

3 (a) Give an account of what you would do in order to observe the chromosomes of an onion bulb.

(b) The table below gives the amounts of DNA in a cell at various stages of cell division. The least amount of DNA present at any stage is taken as 1.0 and this is used as a basis for comparison of the other stages.

DNA content of cell	Examples of stages of cell division
1.0	meiosis, late telophase II
2.0	mitosis, early interphase
	mitosis, late telophase
	meiosis, metaphase II
4.0	mitosis, prophase
	meiosis, anaphase I

Explain the differences in DNA content between:

(i)   mitosis early interphase and mitosis prophase.
(ii) mitosis prophase and mitosis late telophase.
(iii) meiosis anaphase I and meiosis metaphase II.
(iv) meiosis metaphase II and meiosis late telophase II.

Mark allocation 9/100   Time allowed 15 minutes   *In the style of the Joint Matriculation Board*

This is a question testing ability to interpret data and apply knowledge of cell division as well as the ability to recall experimental work that has probably been conducted by the candidate in a laboratory. It is common to find questions on practical biology in theory papers where:

(a) the examining board has no practical examination.
(b) the experiments are too long to be conducted in a normal practical examination.
(c) the results of experiments are unpredictable or difficult to obtain.

Chromosomes are only visible during cell division and so, since the onion bulb is a vegetative structure, the first necessity is to initiate growth. The onion should be placed on a suitable water-filled container such as a gas jar, with the base just in the water. Within a few days young roots should develop. When they are about 1–2 cm long a few should be removed, placed in a mixture of 10 drops of aceto-orcein stain and 1 drop of molar hydrochloric acid in a watch glass. This should be warmed gently over a low flame for 5 min and then the 2 mm next to the tip of each root should be placed on a microscope slide with 2 drops of aceto-orcein. A coverslip should be placed over the specimens and gentle downward pressure applied (avoiding lateral movement of the coverslip). The slide should be warmed again for a few seconds before being viewed under a light microscope with an oil immersion lens if possible. The chromosomes should appear as dark red threads in some of the cells.

In the second part of the question the examiner is testing candidates' ability to apply their knowledge of mitosis and meiosis to the data in the table.

Candidates may find it easier to think of the DNA content as being directly proportional to the number of chromatids.

In (i) early interphase marks the end of a previous division in which the chromatids of each chromosome separate. By the following prophase they have replicated to give pairs again – hence the DNA content doubles from 2.0 to 4.0.

In (ii) the mitosis prophase DNA content of 4.0 becomes halved to 2.0 by late telophase because the chromatids have separated and the cell has divided into two, halving the total DNA per cell.

(iii) In meiosis anaphase I the DNA content is the original quantity of 4.0 because, although exchange of DNA may have taken place during a previous stage, no alteration in total content occurred. By metaphase II however the content is halved to 2.0 because the cell has divided into two at telophase I.

(iv) By late telophase II another division has further halved the DNA content at metaphase II to 1.0.

Answer plan

(a) Grow onion, remove root tips, stain (aceto-orcein + HCl), warm, place on slide with stain, warm, cover and view under high power of microscope.

(b) (i) 2.0–4.0 due to replication of chromatids during interphase.

(ii) 4.0–2.0 due to division of cell at telophase.

(iii) 4.0–2.0 due to division of cell at telophase I.

(iv) 2.0–1.0 due to second division of cell at telophase II.

4.2   HEREDITY AND GENETICS

Assumed previous knowledge Section 4.1 Mitosis and meiosis

Underlying principles

Organisms of the same species vary from each other within certain broad limits. Some characteristics in a population, e.g. height and weight, change gradually from one extreme to another with every conceivable intermediate represented. This is continuous variation and is a result of the environment acting on the organism's hereditary constitution. Other characteristics, such as eye colour and blood groups, fall into a few distinct groups with few, if any, representatives of intermediates. This is discontinuous variation and is the result of the organism's hereditary constitution alone.

In sexual reproduction a new individual results from the fusion of two gametes which between them contain all the necessary information for the development of a similar, but not identical individual. If each gamete contributes information on every characteristic of the organism it follows that at least two factors must control each characteristic in the offspring. These factors may or may not provide similar information on the character. If they do not, then either one or other factor manifests itself in the offspring or an intermediate state between the two extremes becomes apparent. If a factor does not show itself in any one generation it nevertheless retains the capacity to do so in a later one.

Points of perspective

Questions on this topic often involve the fruit fly *Drosophila*. The ability to recognize the different mutant types of *Drosophila* and to have carried out genetic crosses between them in the laboratory would be a distinct advantage.

As the results of genetic crosses involve analysis of numerical data, it is an advantage to have an understanding of basic statistics, including normal distributions, standard deviations, chi-squared tests and possibly binomial theory.

Some examples of, and the theory behind, selective breeding as a means of achieving better crops and animal types is useful and this could be extended into the moral implications of this and other genetic engineering in humans.

Essential information

Organisms are recognizably similar to their parents (heredity); however they are not identical (variation)

Genotype, which is the genetical make up of organisms, often determines limits, e.g. maximum height.

Phenotype is the actual appearance of an organism. It is caused by interaction between genotype and environment.

Mendelian inheritance The Austrian monk Mendel (1822–1884) observed clearly defined characters in the garden pea. Initially he studied only one pair of contrasting characters (monohybrid inheritance). He isolated plants that had 'pure bred' for several generations, artificially pollinated them and observed and counted the offspring (first filial or F_1 generation). He then crossed the F_1 plants with each other to give an F_2 generation. The characteristic apparent in the F_1 was termed **dominant** and the opposing character which disappeared in the F_1 and reappeared in the F_2 was termed **recessive**. Mendel knew nothing of meiosis, genes or chromosomes. He spoke of a pair·of factors (now called **alleles**) determining a character. For example, the gene for height is composed of the allele for tall and the allele for short. It is usual to denote the dominant allele with a capital letter (upper case) and the recessive with a lower case letter.

Let T = tall   t = short

	Pure bred tall		Pure bred short	
	TT	×	tt	
gametes:	T and T		t and t	

Punnett Square

TT

	gametes	T	T
t		Tt	Tt
t		Tt	Tt

F_1 generation

tt

All the offspring have the genotype Tt and the phenotype tall. Now cross two F_1 plants, i.e. Tt × Tt

Tt

	gametes	T	t
T		TT	Tt
t		Tt	tt

Tt

F_2 generation

TT = homozygous dominant, phenotype is tall
Tt = heterozygous, phenotype is tall
tt   = homozygous recessive, phenotype is short
        Therefore ratio of phenotypes is 3 tall : 1 short

This 3 : 1 ratio will only be apparent if samples are large enough. From these crosses Mendel formulated

The Law of Segregation
An organism's characteristics are determined by internal factors which occur in pairs. Only one of a pair can be represented in the gametes of the organism.

Test cross (back cross)
An organism with a dominant phenotype may be homozygous or heterozygous. In order to find the genotype of an organism it is crossed with a homozygous recessive individual of the same species. The following ratios are expected.

TT × tt

TT

	gametes	T	T
t		Tt	Tt
t		Tt	Tt

tt

All tall

Tt × tt

Tt

	gametes	T	t
t		Tt	tt
t		Tt	tt

tt

1 tall : 1 short

Dihybrid inheritance
Inheritance of two pairs of characteristics.
Let T = tall; t = short; C = coloured; c = white
Pure bred tall plants with coloured flowers (TTCC) crossed with pure bred short plants with white flowers (ttcc).

TTCC

	gametes	TC	TC
tc		TtCc	TtCc
tc		TtCc	TtCc

ttcc

All F_1 are heterozygous with the phenotype tall and coloured. When these are selfed to give the F_2 generation:

TtCc

gametes	TC	tc	Tc	tC
TC	TTCC	TtCc	TTCc	TtCC
tc	TtCc	ttcc	Ttcc	ttCc
Tc	TTCc	Ttcc	TTcc	TtCc
tC	TtCC	ttCc	TtCc	ttCC

TtCc

Therefore in the F_2 generation there are 9 tall, coloured

3 tall, white

3 short, coloured

1 short, white

From these results Mendel formulated

The Law of Independent Assortment

Each of a pair of contrasted characters may be combined with either of another pair.

In modern terms: each member of an allelic pair may combine randomly with either of another pair.

Genetics of sex

In humans the female sex chromosomes are XX; therefore, all female gametes contain an X chromosome (homogametic). The male sex chromosomes are XY; therefore, the gametes may contain either an X or Y chromosome (heterogametic).

In birds the female is the heterogametic sex.

In *Drosophila* the female is XX and the male XY but, unlike many species, the Y chromosome is not shorter than the X, but is a different shape.

Sex linkage

Two common examples of recessive alleles linked to the X chromsome are red/green colour blindness and haemophilia.

Let X^h represent the X chromosome carrying the allele for haemophilia.

Let X^H represent the X chromosome carrying the normal allele.

Female carrying haemophilia $= X^H X^h$

Normal male $= X^H Y$

$X^H X^h$

gametes	X^H	X^h
X^H	$X^H X^H$	$X^H X^h$
Y	$X^H Y$	$X^h Y$

$X^H Y$

This cross gives rise to 1 carrier female ($X^H X^h$)

1 normal male ($X^H Y$)

1 haemophiliac male ($X^h Y$)

1 normal female $X^H X^H$

Autosomal linkage

For only 23 pairs of chromosomes to determine all the different characters in a human, it is necessary that each chromosome possess numerous alleles. All alleles on the same chromosome are said to be linked; they move together from generation to generation. Linkage is the association of two or more alleles so that they tend to be passed from generation to generation as an inseparable unit and fail to show independent assortment.

Crossing over and recombination

In practice the situation is not so simple because chiasmata formation and crossing over between homologous pairs during the first prophase of meiosis (see Section 4.2) mean that linked alleles can be separated.

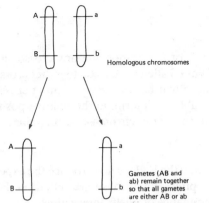

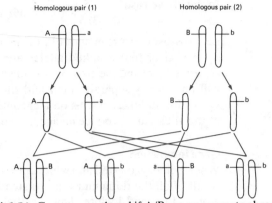

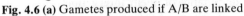

Fig. 4.6 (a) Gametes produced if A/B are linked

Fig. 4.6 (b) Gametes produced if A/B are on separate chromosomes

Fig. 4.7 Recombination of linked alleles due to crossover

If crossing over always occurred in the above case then all the gametes would be recombinants, i.e. Ab and aB. None would be AB and ab. This is most unlikely. If crossover occurred in 50% of cases then half the gametes would be Ab and aB (recombinants), and the remaining half would be the original AB and ab. With crossover in 25% of cases only 25% would be recombinants etc.

How can recombinants be detected?

The test cross (back cross) is used. Take an organism of genotype AaBb. If back crossed to the homozygous recessive aabb the following offspring could arise:

	Gametes of heterozygous parent			
	AB	Ab	aB	ab
Gametes of homozygous parent ab	AaBb	Aabb	aaBb	aabb
Number of offspring in typical sample	125	45	40	115
	both dominant features visible	one dominant feature (A) visible	one dominant feature (B) visible	no dominant feature visible

The offspring can be determined phenotypically.

If the alleles for characters A and B were on separate homologous pairs the offspring ratio should be 1 : 1 : 1 : 1, but it clearly is not here. There are three possible explanations:

(1) A/a is not segregating in a 1 : 1 ratio as expected. To test this add up all the offspring from a gamete containing a, and this should be approximately equal to those arising from a gamete with A.

a   gamete aB(40) + gamete ab(115)   = 155
A   gamete AB(125) + gamete Ab(45) = 170

As these figures are approximately equal, this is not the explanation.

(2) B/b are not segregating in a 1 : 1 ratio as expected. To test this follow the above procedure.

b   gamete Ab(45) + gamete ab(115)   = 160
B   gamete AB(125) + gamete aB(40) = 165

These figures are also approximately equal; therefore this is not the explanation.

(3) Alleles are linked. To test this add the offspring from the homozygous gametes (AB and ab), i.e. (125 + 115) = 240, and compare with the offspring from the heterozygous gametes (Ab and aB), i.e. (45 + 40) = 85. These are clearly not in the ratio of 1 : 1 therefore the alleles are linked and Ab and aB are the recombinants.

The crossover value $= \dfrac{\text{Number of recombinant offspring}}{\text{Total number of offspring}} \times 100$
(COV)

In the above case $\dfrac{85}{325} \times 100 = 26\%$ approx.

i.e. crossover occurs in about ¼ of the meiotic divisions.

It is clear that if linked alleles are at opposite ends of a chromosome any crossover will separate them and the crossover value will be high. If linked alleles are close together crossover will only rarely separate them and the crossover values will be low. By obtaining crossover values for all linked alleles on a chromosome it is possible to determine the relative positions of the alleles and produce an accurate map of a chromosome, i.e. **chromosome mapping.**

Gene interaction

When homozygous red and white antirrhinums are crossed the F_1 does not produce the expected result of all the dominant type. Instead all the F_1 are pink. When these are selfed the F_2 has 2 pink, 1 red and 1 white. This is called **incomplete dominance** or **allelic interaction.**

Certain breeds of poultry can be distinguished by the shape of the comb.

Rose comb   ×   Pea comb
RRpp   rrPP

The F_1 are all walnut comb (RrPp), i.e. incomplete dominance. In the F_2 four types of comb arise:

gametes	RP	Rp	rP	rp
RP	RRPP	RRPp	RrPP	RrPp
Rp	RRPp	RRpp	RrPp	Rrpp
rP	RrPP	RrPp	rrPP	rrPp
rp	RrPp	Rrpp	rrPp	rrpp

R–P– = walnut comb
R–pp = rose comb
rrP– = pea comb
rrpp = single comb

Multiple alleles

Sometimes more than two alleles control a particular characteristic. Only two alleles can occupy a locus on a pair of homologous chromosomes at any time. The ABO blood group system is controlled by three alleles:

    allele A = production of antigen A on erythrocyte.
    allele B = production of antigen B on erythrocyte.
    allele O = production of no antigens.

Therefore the possible combinations are AA; AO; AB; BB; BO; OO. A and B alleles show equal dominance but both are dominant to O.

Blood group	Possible genotype
A	AA or AO
B	BB or BO
AB	AB
O	OO

Transmission of the alleles is in the normal Mendelian fashion.

Link topics
Section 3.3 Genetic code
Section 4.1 Mitosis and meiosis
Section 4.3 Genetic variation and evolution

Suggested further reading

Clarke, C. A., *Human Genetics and Medicine* New Studies in Biology, 3rd ed. (Arnold 1987)
Harrison, D., *Problems in Genetics,* (Addison-Wesley 1970)
Srb, A. M. and Owen, R. D., *General Genetics,* 2nd ed. (Freeman 1965)

QUESTION ANALYSIS

4 (a) How is sex genetically determined in birds and humans?   (4)

(b)(i) A woman has a haemophiliac son and three normal sons. What is her genotype and that of her husband with respect to this gene? Explain your answer. (8)

(ii) Could she have a haemophiliac daughter? Explain your answer giving your reasons. (5)

(c) A population of human beings will contain many more colour-blind individuals than haemophiliacs although the genes are transmitted in the same way. Explain the difference in frequency. (3)

Mark allocation 20/100   Time allowed 30 minutes   *London June 1983, Paper 2, No. 2*

This structured essay question requires candidates to know and understand the principles underlying genetics, especially sex-linkage. A knowledge of genetic laws and conventions is also essential.

In part (a) some candidates may fail to appreciate that the mechanism of sex determination is different in birds and mammals. The first point however is common to both groups, namely that sex is determined not by alleles as other characters are, but by chromosomes. In both groups these sex chromosomes are of two types, the X chromosome and the Y chromosome. In mammals the female has two X chromosomes, and so produces gametes all of which carry an X chromosome (the homogametic sex). Male mammals are the heterogametic sex as they have both X and Y chromosomes and so produce two different types of gamete. In birds the situation is reversed with the male being XX (homogametic sex) and the female being XY (heterogametic sex). This information might reasonably be presented as a diagram but care is essential to indicate clearly the parents and the gametes they produce. Good candidates might further point out that it is effectively the presence or absence of the Y chromosome that determines the sex in each case.

Before attempting any genetics question it is essential to appreciate the type of problem involved and to be sure you are able to answer all parts of the question. This advice may appear obvious, and yet numerous examination papers each year include abandoned genetics questions, where a candidate has decided, after much effort and wasted time, that he cannot solve the problem and has had to start an alternative question. Candidates can score highly on genetics problems, but only if their genetic knowledge is good and the technique and presentation is sound. Equally no marks at all may be scored where the question is misunderstood or the genetics is inaccurate.

Three points about the allele for haemophilia must be appreciated in part (b):

(i) It is a recessive allele   (ii) It is sex-linked   (iii) It is carried on the X chromosome

All three points should be clearly stated at the beginning of the answer and not just assumed. Likewise it should be stated that as any son has both X and Y chromosomes and the Y chromosome could only have been contributed by the father, the X chromosome must always come from the mother. Haemophilia is thus transmitted by mothers to their sons. In the case described the mother has both haemophiliac and normal sons. She must therefore be able to contribute both normal and haemophiliac alleles, i.e. she must be heterozygous (a carrier). Explanations are crucially important. Each stage must be explained logically in prose or detailed annotations of diagrams. All too often answers comprise genetic symbols arranged in flow diagrams or Punnett squares with no clear explanation of why they were chosen, what they refer to, why particular crosses were made, where meiosis occurred and what the significance of the offspring is. A summary diagram of the information would be useful provided it is clearly annotated and a key is given for all the symbols used. A suitable diagram can be found under 'sex-linkage' in the essential information.

Part (b)(ii) has two possible answers depending on whether the father is haemophiliac or not. As the question provides no information about this it is probably best to consider both possibilities, assuming time permits. In any case examiners will usually give full credit for either answer if relevant evidence is given to support the one chosen. The answer should begin by stating that a haemophiliac daughter must be homozygous for the trait and the only possible genotype is therefore X^hX^h. One of each X^h must have been inherited from each parent. If the husband is haemophiliac (X^hY) he could transmit the X^h chromosome to his daughter. If this were to combine with an X^h chromosome from the mother the daughter could be haemophiliac. If the husband is normal (X^HY) he can only transmit the X^H chromosome to his daughter and she can either be normal (if the mother provides another X^H chromosome) or a carrier (if the mother provides an X^h chromosome). In no instance could she be haemophiliac.

Genetic diagrams could be used to summarize the information:

Let X^h represent the chromosome carrying the haemophilia allele.

Let X^H represent the chromosome carrying the normal allele.

If the father is normal his genotype is X^HY.

Part (b)(i) showed the mother to be X^HX^h.

parental genotypes: male X^HY female X^HX^h
                                    meiosis   meiosis
gametes: X^H and Y X^H and X^h

		female	
	gametes	X^H	X^h
male	X^H	X^HX^H	X^HX^h
	Y	X^HY	X^hY

None of the offspring have the genotype X^hX^h and therefore none are haemophiliac daughters.

If the father is haemophiliac his genotype is X^hY and the gametes he produces are X^h and Y. With the mother's genotype unchanged the cross is:

		female	
male	gametes	X^H	X^h
	X^h	X^HX^h	X^hX^h
	Y	X^HY	X^hY

One in four offspring (half the daughters) are haemophiliac (X^hX^h).

The main problem with part (c) is that most candidates will settle for one explanation. With three marks available it is wise to look for three separate points or at least expand fully the one or two points made. The different frequencies are the result of greater selection against the haemophilia allele than the colour blind allele. Due to the potentially lethal nature of the disease haemophiliacs have less chance of surviving to sexual maturity and so passing it on to their offspring. Even though modern medical advances make death from haemophilia less likely, some haemophiliacs or carriers of the disease choose not to have children in case their offspring are affected. As haemophiliac females rarely if ever survive to have children the disease is only passed on by carrier females. Colour blindness may however be passed on by carrier and colour blind females (although these are rare).

5 Coat colour in rabbits is determined by multiple alleles. Chinchilla coat (c^{ch}) is dominant to Himalayan coat (c^h). The allele for full coat colour (C) is dominant to both these, whereas the albino allele (c) is recessive to all the others.
The inheritance of coat colour follows normal Mendelian principles for autosomal genes.
(a) What is meant by 'multiple alleles'?
(b) State the number of alleles for coat colour that would be found in the cells of an individual rabbit.
(c) List all possible genotypes for the following rabbits:
   (i) Heterozygous for full coat
   (ii) Homozygous for Himalayan coat
   (iii) Heterozygous for chinchilla coat
(d) What offspring phenotypic ratios would be expected from the following crosses:
   (i) $c^hc^{ch} \times c^hc^{ch}$
   (ii) $Cc \times c^hc^{ch}$
(e) A breeder suspects his chinchilla buck is not homozygous at the C locus and so mates it with a pure-breeding Himalayan doe. He obtains a litter of 7 chinchilla rabbits only. How certain could he be that his buck is homozygous at the C locus?
Mark allocation 12/60   Time allowed 18 minutes   *In the style of the Cambridge Board*

This is a longer type of structured question testing understanding of the basic Mendelian principles of autosomal inheritance. In common with all genetic questions the candidate must use the symbols given in the question, even though these may differ from those he has previously used. Working should also be shown since a small error may produce a completely wrong answer, but if the examiner is able to see that all the work following the error is consistent and follows accepted genetic principles and techniques, a considerable amount of credit may still be given.

Before answering this question it would help to eliminate errors if the order of dominance is written in a prominent position, such that each allele is dominant to those below it, but recessive to those above it, e.g.

C full coat
c^{ch} chinchilla coat
c^h Himalayan coat
c albino

For (a) the candidate is required to give a definition of 'multiple alleles', i.e. a set of more than two possible alleles at a single locus on a chromosome, that determine a single character. Only two of each set can, however, be present in each somatic cell (this answers part b). In part (c) (i) the full coat allele (C) must be present since this is the phenotypic coat colour but as it is heterozygous the second allele must be different. Because (C) is dominant to all the others any of the remaining three can constitute the second allele giving the answers: Cc^{ch}; Cc^h; Cc.

Only one answer is possible in (c) (ii). The coat is Himalayan; hence the allele c^h is present; since it is homozygous, the second must also be c^h. In (c) (iii) the chinchilla allele c^{ch} can be combined with either of its two recessive alleles, c^h and c.

In (d) the working should be logically set out and the parents, gametes and offspring clearly stated.

When combining gametes remember to put the dominant allele first where it appears with one recessive to it.

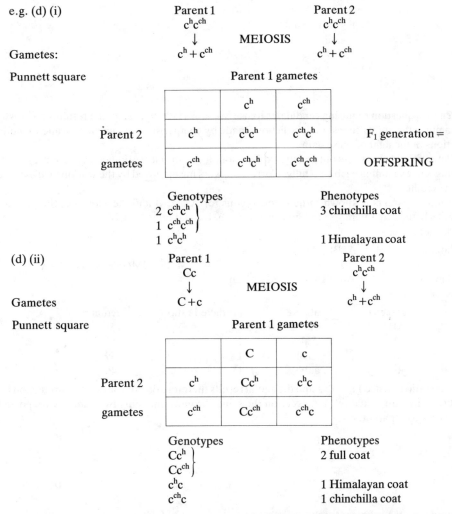

e.g. (d) (i)

Parent 1 — $c^h c^{ch}$ Parent 2 — $c^h c^{ch}$

↓   MEIOSIS   ↓

Gametes: $c^h + c^{ch}$ $c^h + c^{ch}$

Punnett square   Parent 1 gametes

	c^h	c^{ch}
Parent 2 c^h	$c^h c^h$	$c^{ch} c^h$
gametes c^{ch}	$c^{ch} c^h$	$c^{ch} c^{ch}$

F_1 generation = OFFSPRING

Genotypes   Phenotypes

2 $c^{ch} c^h$ ⎱
1 $c^{ch} c^{ch}$ ⎰   3 chinchilla coat

1 $c^h c^h$ 1 Himalayan coat

(d) (ii)

Parent 1 — Cc   Parent 2 — $c^h c^{ch}$

↓   MEIOSIS   ↓

Gametes   C + c $c^h + c^{ch}$

Punnett square   Parent 1 gametes

	C	c
Parent 2 c^h	$C c^h$	$c^h c$
gametes c^{ch}	$C c^{ch}$	$c^{ch} c$

Genotypes   Phenotypes

$C c^h$ ⎱
$C c^{ch}$ ⎰   2 full coat

$c^h c$ 1 Himalayan coat

$c^{ch} c$ 1 chinchilla coat

In part (e), the candidate should first work out the cross. If the chinchilla coat buck is heterozygous its genotype can be $c^{ch} c^h$ or $c^{ch} c$. The pure-breeding Himalayan doe must be $c^h c^h$.

Gametes of buck (if $c^{ch} c^h$)   Gametes of buck (if $c^{ch} c$)

	c^{ch}	c^h
Gametes c^h	$c^{ch} c^h$	$c^h c^h$
of doe c^h	$c^{ch} c^h$	$c^h c^h$

	c^{ch}	c
c^h	$c^{ch} c^h$	$c^h c$
c^h	$c^{ch} c^h$	$c^h c$

1 : 1 chinchilla : Himalayan   1 : 1 chinchilla : Himalayan

In both cases the probability of getting a chinchilla rabbit = ½. Hence with a litter of 7 chinchillas the breeder could expect 3 or 4 Himalayans if his buck was heterozygous. It is therefore more likely that it is actually homozygous which would produce all chinchilla rabbits. A more discerning candidate may like to put a figure on this probability. The probability of getting 1 chinchilla rabbit if his buck was

heterozygous = ½. The chances of getting 7 is therefore $(½)^7 = \dfrac{1}{128}$. Since the probability of being either

heterozygous or homozygous = 1 (certainty), the chances of being homozygous must be $1 - \dfrac{1}{128} =$

$\dfrac{127}{128} = 99.2\%$ certain.

6 Mice from a group known to be heterozygous for genes A and B were mated with a group known to be homozygous recessive for both genes. The resultant offspring are shown in the table below

Genotypes	Numbers
AaBb	151
aabb	146
Aabb	53
aaBb	59

(a) If the genes A and B were independently inherited, in what ratio would you have expected the four genotypes above to have occurred?

(b) Do the actual results obtained confirm your predictions?

(c) Could the actual results be explained if

  (i) the two genes were linked

  (ii) the two genes were not linked but selection operated against those offspring which were heterozygous for one of their two genes?

Explain your answers fully.

Mark allocation 9/100   Time allowed 15 minutes   *In the style of the Joint Matriculation Board*

This is a genetics question requiring candidates to use Mendel's laws to predict the results of a cross. It further examines the ability to test the predictions made by comparison with given data and to modify the predictions in the light of these data.

Candidates should always lay out genetic explanations logically, clearly indicating parents, gametes and offspring and explaining each step fully. Correct answers unsupported by the working will be given little, if any, credit.

One group of parents is known to be heterozygous for A and B and therefore has the genotype AaBb. Similarly the other parents are homozygous recessive and must be aabb.

The cross is therefore:

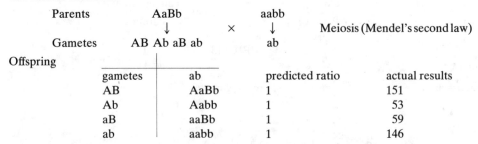

gametes	ab	predicted ratio	actual results
AB	AaBb	1	151
Ab	Aabb	1	53
aB	aaBb	1	59
ab	aabb	1	146

The predicted ratio is hence 1 : 1 : 1 : 1 and the actual results therefore do not confirm this prediction (b).

If the two genes are linked (c), i.e. occur on the same chromosome, only two gametes are possible for the heterozygous parent, e.g.

When crossed with the aabb parent the offspring are:

gametes	ab
AB	AaBb
ab	aabb

From the results these two genotypes (AaBb and aabb) represent the majority but not all offspring. Crossing over could account for the others, e.g.

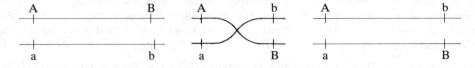

Therefore gametes Ab and aB represent recombinants (see crossing over and recombination in the essential information). The results could therefore be the result of A and B being linked.

If selection were operating against offspring that were heterozygous for one of their genes then those offspring with one pair of genes in the heterozygous state (i.e. Aabb and aaBb) should be present in smaller numbers than expected. This is the case and so could also explain the result. In practice such selection normally operates where two enzymes (produced by two genes) are needed for a vital metabolic process. If these two enzymes are only produced when their respective dominant genes are present, then at least one A and one B gene are required in the genotype. Aabb and aaBb lack one dominant gene, in each case, and so the metabolic pathway is affected and the organisms could be at a disadvantage. In such an example the genotype aabb might be expected to be similarly disadvantaged and selected against. The results given do not however confirm this, which casts doubt on (ii) as a likely explanation. However in the absence of detailed information on the mechanism and nature of genes A and B, both explanations must remain as possibilities.

4.3 GENETIC VARIATION AND EVOLUTION

Assumed previous knowledge Cell division and chromosome replication.
Mendelian genetics.
Chromosomes as carriers of genetical information.
Evolution as a process of gradual change by natural selection resulting from selective pressures in the environment.

Points of perspective

A knowledge of the historical background to evolution and in particular the work of Charles Darwin. The evidence for evolution.

Underlying principles

If a species is to survive it must adapt to the changing environmental conditions which occur over a period of time. As organisms disperse themselves they will encounter new conditions, to which they too must adapt. The adaptations involve structural, physiological and behavioural changes to individuals. While interbreeding among individuals of a species continues these changes become combined in the offspring increasing variety while remaining a single species. If however groups become genetically isolated because interbreeding fails to take place, each group may adapt to its own conditions and become sufficiently different from its neighbouring groups that interbreeding becomes impossible. Two species will then have been formed.

The mechanism by which the adaptations arise was thought by Lamarck to be the passing on to the offspring of features acquired during the lifetime of the parents. Darwin argued that chance mutations gave rise to a population of varied individuals in which only those suited to the prevailing conditions survived and passed on their genetic features to the next generation.

Essential information

Mutations These are changes in genes or chromosomes responsible for much of the genetic variation which is the prerequisite for the process of evolution. They may occur as alterations in the structure of a gene or as alterations in the structure or number of chromosomes. Most variations in gene structure are the results of errors made during the complex replication process which occurs during every cell division. Mutation rates for a particular gene occur at a constant rate; but all genes have a low frequency rate. Frequencies vary from 1 in 10^3 to 1 in 10^7 replications per locus per generation. Local environmental conditions, such as ionizing radiations or carcinogenic chemicals, increase the rates. In diploid types, the mutant gene may be dominant, recessive or intermediate in its effect. The most common mutants are recessive or partly recessive and they are deleterious (they reduce the viability of the organism and are deleted from the gene pool).

There are several ways in which chromosomes become altered, therefore giving rise to mutation. Basically, they involve a change in either structure or number.

Structural change This takes place when the number of chromosomes is unaltered but the linear arrangement within one or more of them is changed. Four types of structural mutation are important during evolution.

Deficiency
This results in loss of genes from one or more chromosomes.

A   B   C   D   E   F   G   H   I   J   Normal chromosome

A   B   C   D   G   H   I   J   Deficiency chromosome for the genes E and F

This gives a deficiency heterozygote and the deletion usually has deleterious consequences.

Duplication
This results in addition of genes, where a number of genes are carried twice in one of a pair of chromosomes.

A   B   C   D   E   F   G   H   I   J   Normal chromosome

A   B   C   D   E F   E F   G   H   I   J   Duplication chromosome for the genes E and F

This gives a duplication heterozygote. Duplication of a gene is not usually serious but animals may show abnormalities.

Inversions

These can take place when chromosomes break in two places and the segment rejoins in an inverted position.

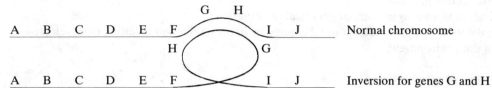

Normal chromosome

Inversion for genes G and H

This gives an inversion heterozygote and causes a decrease in the frequency of crossing-over during meiosis.

Translocation

This is an actual change of pieces of chromosomes between two non-homologous pairs.

A B C D E F	Normal	A B C J K L	Translocated
G H I J K L	non-homologous chromosomes	G H I D E F	chromosomes

The translocation heterozygote is sterile because only those gametes with a full number of genes can give viable zygotes after fertilization.

Numerical change This results in a change in the normal number of chromosomes for a particular species. The most important is **polyploidy** in which the number of full sets of chromosomes is increased. Instead of somatic cells being diploid, they become tetraploid or even octoploid. A tetraploid individual will produce diploid gametes. Triploids will produce gametes with irregular numbers of chromosomes and so will not be self-perpetuating. Polyploids are much larger than diploids of the same species and tetraploids have been very important during the evolution of plants by artificial selection. Sometimes individual chromosomes are deleted or duplicated, giving one fewer $(2n-1)$ or one more $(2n+1)$ than the normal complement. In Down's syndrome (mongolism) there are $(2n+1)$ chromosomes, i.e. 47 in humans.

The results of selection pressures on mutations

Industrial melanism Industrial melanism in moths is an example of genetic selection induced by pollution. It was recognized in soot-polluted Manchester in 1850, when the black, melanic form of the peppered moth, *Biston betularia,* was first seen. By 1900, it had almost replaced the typical, mottled form. The increase from 1% to 99% took 50 generations (the moth has an annual life cycle). During the second half of the nineteenth century, the rest of Britain began to receive the melanic (black) form as a result of dispersal from Manchester and natural selection in other cities. The most substantial selective pressure for melanic forms is visual predation by moth-eating birds, which remove the moths from trees and other vertical surfaces when they rest there during daylight. Where surfaces are blackened by soot the melanic form is better camouflaged and predation is largely restricted to the mottled form. In areas not polluted by soot the reverse is true and the mottled form predominates. There are more than 150 species of moths known to exhibit industrial melanism.

Metal tolerance in plants At least 21 species of plants have been recognized as having metal tolerant strains. Species of *Festuca* and *Agrostis* have become genetically adapted to high concentrations of zinc and lead and are tolerant on sites with a high concentration of the metals but may not survive where the concentration is low. Where several metals are present, such as nickel and copper, plants may be tolerant to both but tolerance to one metal is genetically independent of another.

Insecticides The widespread use of insecticides has provided a selection pressure which has resulted in adaptive changes within a very short time. It has taken only a few years, and in some cases only months, to produce strains of insects which have become completely resistant to many insecticides. At least 225 resistant species of insects have been recognized. Many of these show resistance to DDT, dieldrin, aldrin, lindane, malathion and fenthion. The housefly *Musca domestica* has developed strains in some parts of the world which are resistant to every insecticide which can be 'safely' used, with the exception of pyrethrum compounds. In some types of resistance enzyme induction has been responsible for the insects' ability to break down the insecticides, i.e. mutant strains 'switch on' genes which help control the synthesis of oxido-reductase enzymes to render the insecticides harmless. Many species of insects which are vectors of disease have become resistant in this way; among them the yellow fever mosquito *Aedes aegypti* and the malaria-carrying mosquitoes *Anopheles gambiae* and *A.culicifacies*.

Antibiotics The resistance of bacteria to antibiotics has arisen by the selection of resistant mutants. As with types of insecticide resistance, some resistance to antibiotics has evolved by 'switching on' genes to control the production of antibiotic-splitting enzymes. All naturally occurring penicillin-resistant strains of *Staphylococcus,* as well as penicillin-resistant strains of many other species, owe their resistance to their ability to produce penicillin B lactamase. Resistance to antibiotics has already destroyed the usefulness of several drugs. The first signs of staphylococcal resistance appeared soon after penicillin became extensively used in hospitals.

Table 4.2 Incidence of penicillin-resistant infection at a general hospital

Date	Total patients	Patients with penicillin-resistant strains
Apr – Nov 1946	99	15
Feb – June 1947	100	38
Feb – June 1948	100	59

By 1950, the majority of staphylococcal infections in all British general hospitals were penicillin-resistant. Resistance has now appeared in certain staphylococci to all major antibiotics, often as a triple resistance to penicillin, tetracycline, and streptomycin.

Adaptive immunity in mammals

Micro-organisms and insects reproduce at such an alarming rate that it is relatively easy to visualize how they can adapt faster than man can combat them. Prolifically breeding mammals such as the brown rat and the rabbit have also evolved resistant strains capable of withstanding warfarin and myxomatosis virus, respectively.

Warfarin was developed as a rat poison. It contains di-coumarol which interferes with normal clotting mechanisms in the blood. In 1960, strains were recognized in Shropshire which were immune to warfarin. The incidence of resistant rats in parts of Wales had risen to 50% by 1970 because of the intensive selective pressure of the widespread use of warfarin. The resistance is due to a single mutant gene and individuals homozygous for this gene are weaker than normal warfarin-sensitive rats.

Myxomatosis-resistance in rabbits became widespread in the mid 1950s. Prior to this, the virus was responsible for up to 90% mortality of rabbits in certain areas. The genetic response to this selective pressure was the production of mutants which spent a greater proportion of time above the ground, like hares. Normal rabbits live in crowded warrens underground where the vector of the virus, the rabbit flea, can spread rapidly throughout a population. Those that spent most time above ground were favoured, whereas previously they were selected against because of predation. Even though the rabbits were not physiologically resistant, their altered behaviour offered them protection. In the Australian population, however, a genetically resistant strain emerged after the initial epidemics. Most British rabbits are now genetically resistant to myxomatosis.

Sickle-cell anaemia

This is caused by a mutation which resulted in the incorporation of an incorrect amino acid at one point in the protein chains of the haemoglobin molecule. The mutated gene is known as Haemoglobin S or HbS and is recessive. Thus only those persons having the sickling gene from both parents (i.e. homozygous for the gene) suffer acutely from the disease. The disease is an often fatal form of anaemia which is relatively common in West Africa. Whenever the blood cells of a victim encounter a low level of oxygen, as in the venous blood of tissues, they are liable to collapse to a sickle shape and may form blockages and other complications in blood vessels. Sufferers have only a 20% chance of surviving to maturity as compared with normal people. Heterozygous individuals suffer sickling also when the oxygen tension falls below a critical level. Given this information, one would not expect the mutation to be favoured. Indeed, such harmful mutations are frequently eliminated. This happens rapidly where it is a dominant gene, more slowly where it is a recessive, as selection acts most often on a mutation when it appears in the phenotype. New HbS genes arise spontaneously, so they will always be present in a population, but a percentage of them should be eliminated at each generation.

However, the observations do not bear out this hypothesis. There are large areas of Africa and Asia where the gene occurs, usually in a single dose, in 15–20% of the population. There are even communities within these regions with frequencies of 40%. A heterozygote can be recognized because the blood cells show some sickling when the person is artificially exposed to low oxygen pressure in the laboratory. Clearly, the gene is not being removed from the population. We can only conclude that in some way HbS confers an advantage over the normal Hb gene, enough to redress the losses of genes at each premature death from anaemia. The answer lies in resistance to malaria. Children with HbS have been shown to have a 25% better chance of surviving malarial attacks than those with normal genes. Selection is not acting to remove the genes causing sickle cell anaemia. They are removed when homozygous but this selection is outweighed by their selective advantage when heterozygous, due to better resistance on the part of heterozygotes to malaria.

EVOLUTION AND CHARLES DARWIN (1809–1882)

Darwin became the naturalist on HMS Beagle which sailed in 1832 to South America and Australasia. During the voyage he was influenced by his knowledge of fossils and noted their similarity to present day forms. He also noticed the differences in South American animals and plants which could be related to differences in their environment. In the Galapagos Islands he observed that while the finches there were unique they had a general resemblance to finches which existed on the mainland of South America. He considered that originally a few finches had strayed from the mainland to these islands, and, as they bred, they produced new types with differences that allowed them to fare better in the new conditions. After the voyage he spent over twenty years developing his views on evolution.

It is problematical whether he would have published his findings had not his theory been anticipated by Alfred Wallace. Wallace sent his theory of how evolution may have come about to Darwin, who found it was in essence, the same as his own. Darwin was advised to read a joint paper on the subject to the Linnaean Society. This he did in 1858 and the following year published his book *On the Origin of Species by Means of Natural Selection and the Preservation of Favoured Races in the Struggle for Life.*

The Darwinian Theory

Darwin's theory is concerned with three observable facts plus two deductions.

Over-reproduction Malthus attempted to show that all organisms tend to increase in a geometric ratio. Evidence for the enormous reproductive capacity includes:

1 The very slow-breeding elephant: if it were to bring forth six young in a life time, and if their descendants continued to breed at this rate, in 750 years there would be 19 million elephants.
2 Many fish are capable of laying millions of eggs, e.g female cod lays about 2–3 million eggs per year.
3 Flowering plants generally produce many hundreds of seeds.

Relative constancy of the numbers of species Darwin's second fact was that despite this tendency towards a geometric increase in numbers, the numbers of a given species tended to remain constant. Hence, many plants and animals must be destroyed at some stage of their lives.

Struggle for existence From these two facts, Darwin arrived at the first deduction. It is obvious that if the parents produce vast numbers of offspring or over many years are capable of doing so, and yet the number of a given species remains constant over many years, then there must be some sort of struggle between the organisms.

Variation among the offspring The third observable fact was that among the offspring of any two parents (animals and plants), there are usually slight and perhaps hardly noticeable differences. Generally, no two offspring were identical.

Natural selection/survival of the fittest This is the final deduction arising from the third fact above. If the variation confers upon the individual some greater ability to withstand the hazards and competition of the struggle for existence, then that organism will stand a better chance of survival, and so live to breed. He anticipated inheritance by noting that **like produces like,** i.e. that the advantageous variation will be inherited, resulting in more offspring with this variation.

It is important to qualify this: the inheritance of one small variation will not by itself produce a new species. However, the production of variations in a particular direction over many generations and their inheritance will gradually lead to the evolution of a new species.

Darwin thought that it was by the production of many small continuous variations and their **cumulative inheritance** that new species arose. Modern studies of the mechanism of natural selection via genetic variation have shown that it can be considered as strong evidence for evolution. Other evidence is as follows.

A summary of the evidence for evolution

Classification If the species present on the earth were descended from a few simpler forms, then we could expect to be able to classify them into phyla, classes, orders, families, genera and species just as is now done. On the other hand, it is difficult to imagine how this would have been possible were the organisms not related by descent, but each specially created according to individual plans.

Homology When the anatomy of one vertebrate is compared with that of another, in general, resemblances are more obvious than differences. Structures in individuals of two different species are said to be homologous when they show the same type of structure, the same relation to other organs of the body, and a close similarity in their embryology and development. An example is the pentadactyl limb, which is found throughout the Mammalia.

Vestigial organs Many organisms have obvious structures which seem to serve no useful purpose but which continue to be developed from one generation to the next. For example, in man, the caecum, appendix and coccyx. Vestigial hind limbs occur in some snakes in the form of two small bones on each side, one representing the ilium and the other the femur. Slugs have vestigial shells embedded in their dorsal surfaces. The ancestors of these types possessed these as well developed and functional structures.

Comparative physiology Studies of groups of animals have shown that the relationships which their structural characters indicate are the same as those which are indicated by the degree of similarity between the chemical and physical processes which go on in their bodies. Certain diseases affect only closely related animals or plants. For example, poliomyelitis can be transmitted to primates. Among plants, wheat rust attacks many members of the Graminae. When bacteria invade organisms they are immobilized by chemicals formed in the blood called precipitins. These are produced when any foreign protein, such as that from tissues of any other animal, is injected. Therefore, if blood or serum from a dog is injected into a rabbit, the blood of the rabbit will develop precipitins and when the blood of this rabbit is added to the blood serum of a dog, precipitation of proteins will occur. The precipitins developed in the blood depend on the particular kind of foreign protein which is injected. Thus, in the above example, the precipitins developed in the rabbit's blood in response to the presence of the dog's serum are most effective in causing precipitation in dog's serum, and are either entirely ineffective or much less so when added to horse's serum. Precipitin tests thus provide a means of studying relationships between animals from a physiological point of view.

Embryology The theory of recapitulation states that organisms pass through developmental stages corresponding to adult stages of organisms lower in the scale of evolution. Each individual in its development goes through a series of changes corresponding to the series of stages which classification, comparative anatomy, and comparative physiology indicate as stages which were passed through in the evolution of the species. The appearance of gill clefts in mammalian embryos can be explained because the gills were developed in their ancestors, where they were present not only in the embryo but also in the adult. The closer the relationships of groups of animals and plants through classification, the more do their embryos resemble each other, especially in the earlier stages.

Artificial selection Our most useful plants have been cultivated and our most useful animals domesticated by man since prehistoric times. There are vast differences between present-day varieties and their ancestors. The varieties have evolved as a result of man selecting examples with the most desirable qualities and maintaining the qualities from one generation to the next through selective breeding. The principle is the same as natural selection except that it is very much quicker and the features selected may not be of survival value in natural populations. It is really a practical way of demonstrating selection in the evolution of species.

Geographical distribution A very large amount of information is available which shows that plants and animals are by no means evenly distributed throughout the earth. Not only have certain zones, determined by latitude, elevation, temperature, etc. their own characteristic flora and fauna, but places with the same climatic conditions in different regions of the world do not always possess the same animal forms. Elephants for example, live in India and Africa but not in South America. New Zealand and Britain have similar climatic conditions but the native fauna and flora are entirely different. The present-day distribution of flora and fauna is not completely explained by the hypothesis of 'suitability to environment'. It can be explained if two conditions are accepted:

1 that existing animals and plants are the descendants of extinct populations which were of a more generalized type, and that these dispersed from their places of origin, becoming adapted to meet new selective pressures of the environment.

2 that with changes in the topography of the land, certain groups became isolated due to the formation of barriers, e.g. sea, mountains, etc., and evolved along different paths but were prevented from interbreeding with other populations, eventually becoming so different that they could not do so even if brought together again.

Paleontology Geologists are able to place layers of sedimentary rocks in order of their formation and they can estimate the age of each layer from the thickness of the rock above and from carbon-dating techniques. In this way, a geological time scale can be made (see Figure 4.8). Most sedimentary rocks contain fossils. Vertebrate fossils form clear evidence of evolution. Fish, which first appeared about 360 million years ago could have given rise to amphibians whose earliest representatives appeared about 280 million years ago. These could have been the ancestors to the reptiles which made their appearance around 250 million years ago. The earliest mammals appeared as fossils in rocks about 150 million years ago, with the earliest known bird fossils present in rocks, 140 million years old. Fossil evidence is supported by specimens which appear to form a transition between two groups. For example, *Seymouria* is a fossil which shows a combination of reptilian and amphibian features. Fossils of all the main invertebrate groups are found in rocks younger than 500 million years. The most convincing fossil evidence is found in cases where, in successive layers from one locality, a series of fossils exhibits gradual change. Good examples of this are seen in the evolution of the oyster and in a species of sea-urchin, where complete records of most intermediate forms are seen as fossils in successive strata.

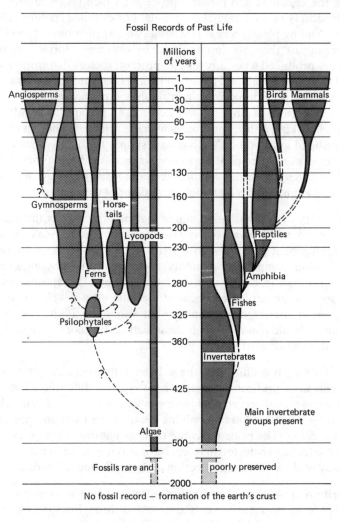

Fig. 4.8 A geological time scale

Lamarck's·theory

Lamarck published his theory of evolution in 1809 in his *Philosophic Zoologique*. He believed that the organic world is the result of the action of vital forces different from any of the physical forces operating in the organic world. His theory is developed from a series of postulates.

1 Life, by its own efforts, tends continually to increase the volume of every body which it possesses and to increase the size of its parts up to a limit which life itself determines.

   We know today that growth is the result of internal and external forces of the interaction of the organism with its environment.

2 The formation of a new organ in the body, is the result of a new need which has arisen and continues to be felt by the animal and of the new activity which this need initiates and prolongs.

   This is directly contrary to embryological evidence. For example, the vertebrate eye starts to develop in the vertebrate embryo long before there is any need for an eye, and the mammalian eye develops in the darkness of the uterus.

3 The extent of development of organs and their power of action is proportional to their use.

   It is true that use tends to increase the development of an organ and lack of use leads to degeneration.

4 All changes in or acquisition to the bodily organization of an individual during its life are transmitted to its descendants by the process of reproduction.

   There is no evidence of inheritance of characters acquired during the life of an individual.

The theory of Lamarck fails to give an adequate explanation of evolution because the postulates on which it is based cannot be accepted in the light of present knowledge.

Suggested further reading

Ayala, F. J., *Genetic Variation and Evolution*, Oxford/Carolina Biology Reader No. 126 (Packard 1983)
Bradshaw, A. D. and McNeilly, T., *Evolution and Pollution*, Studies in Biology No. 130 (Arnold 1981)
Dowdeswell, W. H., *Evolution: A Modern Synthesis* (Heinemann 1984)
Parkin, D. T., *An Introduction to Evolutionary Genetics* (Arnold 1979)

QUESTION ANALYSIS

7 All the pupils from Form I to V in a school which had some immigrants were asked to roll their tongues. The results are shown below.

Rollers 378   Non-rollers 72   Total 450

The ability to roll one's tongue is determined by a dominant gene R and the lack of this ability is due to a recessive gene r.

(a) Assuming that the Hardy–Weinberg principle applies, calculate the proportions of the school population which have the genotypes RR, Rr, rr.

(b) Suggest two reasons why the results would have to be viewed with caution before coming to a conclusion.

Mark allocation 7/100   Time allowed 10 minutes   In the style of the Welsh Joint Education Committee

(a) By using the equation $p^2 + 2pq + q^2 = 1$, it is possible to find the probable frequency of each genotype in a population even though they cannot be distinguished as, for example, in heterozygotes and homozygotes due to the dominance of one gene over the other.

   The ability to roll the tongue is not shared by everyone. In this investigation 84% are tongue rollers and 16% non-rollers. Given that the gene for rolling (R) is dominant over that for non-rolling (r), of the 84% rollers, some must be homozygous (RR) and others heterozygous (Rr). We can find out how many of these genotypes are in the population by the Hardy–Weinberg equation.

Let p = frequency of R   16% or 0.16 were non-rollers (rr)
q = frequency of r   therefore $q^2 = 0.16$ or $q = \sqrt{0.16} = 0.4$
By Hardy–Weinberg   since $p + q = 1$
$p^2 + 2pq + q^2 = 1$ $p = 1 - q = 1 - 0.4 = 0.6$
and $p + q = 1$

As we know the values of $p + q$ we can then work out the frequencies of each genotype

$$RR = p^2 = 0.6^2 = 0.36 = 36\%$$
$$Rr = 2pq = 2 \times 0.6 \times 0.4 = 0.48 = 48\%$$
$$rr = q^2 = 0.16 = 16\%$$

(b) The results would have to be viewed with caution because the Hardy–Weinberg equation only applies to large populations in which random mating occurs in the absence of mutation, migration, selection and genetic drift. The explanation of these four factors is as follows:

Mutation

If gene R recurrently mutates to r, the gene frequencies p and q must change. Assuming r to be as fit to survive as R, R would eventually disappear from the population. Mutation rates of most genes are very low, however, so the chance of this happening is remote. Reverse mutations r to R also occur but not necessarily at the same rate, so that equilibrium is established.

Migration

You are told, in the question, that there are some immigrants in the sample population. Gene frequencies can be altered if members emigrate or if individuals with different genotypes immigrate into the population.

Natural selection

Darwin's theory of natural selection depends on the fact that some combinations of alleles help survival more than others. The individuals who possess different genotypes do not transmit their genes to the next generation with equal frequency as the Hardy–Weinberg law requires. Those individuals which survive and become parents form a non-random sample of the population. Theoretically it might be possible for the tongue rolling gene to be linked with genes which are of a greater survival value.

Genetic drift

The number of individuals carrying a particular allele will vary from one generation to another so that gene frequencies fluctuate about a mean. This occurs because mating will not be random and the gametes which unite are a mere sample from the parental gene pool. Sampling error therefore produces a random shift of gene frequencies or genetic drift. In a small population this effect is significant.

8 The following graph shows the emergence of bacterial strains which are resistant to an antibiotic used. The figures relate to one hospital.

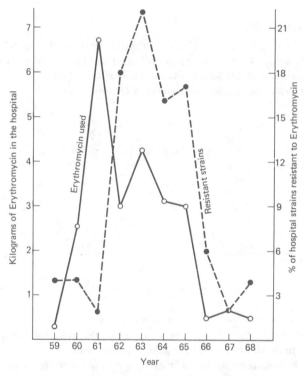

Fig. 4.9 The development of erythromycin-resistant
strains of bacteria at a London hospital

(a) Briefly explain how each of the following could contribute to the rapid emergence of resistant strains.

(i) DNA in a bacterium is haploid. (2)

(ii) Bacteria usually reproduce asexually by fission. (2)

(iii) Fission may take place as often as every 30 minutes. (2)

(b) The graph may be used to demonstrate evolution in action. What selection pressure occurs and what is the result of such pressure? (2)

Mark allocation 8/125   Time allowed 10 minutes   *In the style of the Scottish Higher II*

This structured question requires candidates to interpret a graph and answer questions related to the information provided by the graph. Candidates should begin by studying the graph from which they should appreciate that the number of resistant strains reflects the amount of antibiotic used after a time lag of two years. Question (a) requires explanations regarding the rapidity of the development of the antibiotic resistant strains. In (i) the DNA being haploid means that all alleles are effectively dominant.

In diploid cells recessive mutations which occur do not express themselves immediately, if at all, because they are initially masked by the normal dominant allele. If cells are haploid, recessive mutations express themselves immediately. The asexual fission referred to in (ii) means that the mutation (resistance to erythromycin) is passed on to every cell arising from the mutant parent cells. There is no chance of genetic variation arising from the mixing of genotypes and so the mutation is not masked. The rapidity of cell divisions mentioned means that millions of copies of the mutant genes for resistance can be made in a single day. The selection pressure referred to in (b) is the antibiotic erythromycin itself. It will destroy all susceptible (i.e. normal) strains of bacteria but not the mutants with genes for resistance. These mutants survive to reproduce their own kind and have the potential for further mutation into other resistant types. The greater the quantity of antibiotic used, the greater this selection pressure and the more rapidly resistant strains develop.

9 What do you understand by a species? Outline the ways in which new species arise.
Mark allocation 20/100   Time allowed 35 minutes   *London June 1981, Paper II, No. 8*

This is very much an open-ended essay type question and will involve a critical appreciation of the species concept. As with most evolutionary topics, a wide background of reading will be essential to gain marks for this type of question. It is unlikely that examiners will have an objective marking scheme for such a topic.

The smallest group that a biologist distinguishes is called a species. It is not possible to find a completely satisfactory definition of a species. While the many members of a given species resemble one another more closely than they do the members of any other species, the most widely used and most satisfactory definition of a species is not based upon the criterion of appearance at all. It depends on the closeness of their relationship with one another. A species can be described as a population of related organisms, individuals of which are able to mate (if reproduction is sexual) and produce viable offspring which can interbreed. The major drawback in using this concept of a species is apparent when one considers those organisms which now exist only as fossils. It is obviously impossible to tell whether Fossil X can interbreed with Fossil Y. So we can never be sure whether Fossil X is a different species from Fossil Y.

The second part of the question relies on the much broader concept of evolution of a species. Geographical isolation of populations and the effects of natural selection on their gene pools will be the major theme of the answer. The way in which variation, natural selection, and adaptation lead to evolution of a new species is summarized in Figure 4.10.

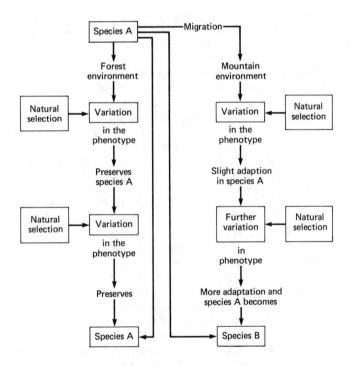

Fig. 4.10 Evolution of a new species

10 (a) Genetic variations may arise in living organisms by
   (i) changes in the structure of chromosomes.
   (ii) changes in the number of chromosomes.
   (iii) gene mutations.
Give a short account of how these mechanisms operate.

(b) Give **one** example of a chromosome number abnormality and **one** example of a gene mutation, each of importance to man.

(c) Using **one** example show how variation either by chromosomal changes or by gene mutation can become established in a population.

(d) Describe how you might determine how the variation in length of stem found in a population of plants was due to continuous or discontinuous variation.

Mark allocation 20/100   Time allowed 40 minutes   *In the style of the Joint Matriculation Board*

The answer will depend on how well you can recall factual information and apply it to the question:

Part (a) requires a straightforward treatment of the various forms of mutations which can be recognized in chromosomes and genes (see essential information).

In order to be equipped to answer this type of almost 'general knowledge' question (b), it will be essential to have read widely about the topic. The subject of evolution lends itself to questions which involve discussions and essay-type questions, and a prerequisite to answering them is to be as widely read as possible. Perhaps the most well-known example of an abnormality in the chromosomes number of the human is Down's Syndrome where individuals have 47 chromosomes instead of the normal 46. Note that the question merely asks you to give or state one example and so do not waste valuable time giving more than a straightforward statement. An example of gene mutation in man could be haemophilia.

In order to give a complete answer to (c), it would be wise to revise the information given in the essential information 'The results of selection pressures on mutations'. Several examples are summarized which illustrate how gene mutations have become established in a population, e.g. metal tolerance in plants or resistance to insecticides.

Part (d) is slightly different from the previous parts. Although it involves recall, it relies on knowledge of practical techniques and the correct interpretation of data obtained from an investigation. Those candidates who have had experience in carrying out this or similar practical work will be at an obvious advantage. The answer will depend on an understanding of the concept of continuous and discontinuous variation.

The variation of a character is said to be continuous when any of the infinite number of forms which are intermediate between the extremes may be found. Discontinuous variation is more readily recognized as a 'freak' mutant which readily stands out among the normal individuals in a population. A variation in stem length in a population can be measured statistically. Continuous variation can be described graphically, by its mean, modal and median values plus its standard deviation. The typical Gausian, bell-shaped curve will indicate perfect continuous variation if numbers of individuals are plotted against classes of lengths.

11 Read the passage and answer the questions which are based on it.

Gene Banks (from M. E. Clarke *Contemporary Biology,* Saunders 1979)

Because our crop plants have been so highly bred for specific characters, they now lack the genetic diversity that exists in wild populations. They are often readily susceptible to new diseases that may infest them. Wheat, for example, is particularly susceptible to a fungus, the wheat smut. Naturally, plant geneticists breed for resistant plants, but, micro-organisms have a high mutation rate, and new infective forms soon evolve. It is now necessary to develop new rust-resistant strains of wheat every few years to maintain yields. In 1970, a fungal infection of corn leaves, due to a newly mutated pathogen, spread across the United States, destroying 15 per cent of the total crop. Such highly destructive attacks by pathogens and other pests are bound to recur, since agro-ecosystems cannot be isolated from the rest of the environment. Where will the genes come from to generate new, resistant plants?

Wild plant populations, with their great genetic diversity, are seldom severely damaged by pathogens. There are always enough resistant individuals that survive and replenish the population. Many varieties of crop plants, some of them hundreds if not thousands of years old, are still being grown by peasant farmers throughout the world. But today, there is a rapidly spreading tendency to replace these with modern, highly selected, genetically uniform varieties. Furthermore, as human activities encroach ever further into remaining areas where wild relatives of our crop plants still thrive, these natural reservoirs of genetic variety will also disappear.

These facts have led many geneticists to push for the establishment of **gene banks.** Genetic resource centers are being created to collect and store seeds from various parts of the world. The question is whether the effort will be sufficiently widespread and complete to preserve the natural genetic variability, not only of our current crop plants but also of other plants that could be used for human food, before they disappear for ever. Many seeds now in storage could never be replaced because the plants already have disappeared from their original homelands.

(Reproduced from 'Higher Paper II Interpretation Question' Arnold, B., Mills, P. R., Aberdeen College of Education Newsletter 37 May 1981 by permission of the editor)

(i) Give two examples, other than disease resistance, of specific characteristics for which certain of our crop plants have been so highly bred.   (2)

(ii) What is meant by genetic diversity?   (1)

(iii) What effect is selective breeding likely to have on the genetic variation found in a particular plant variety?   (1)

(iv) Why can a pathogen do more damage to a crop than to a wild population?   (2)

(v) Explain carefully how a fungal infection of corn leaves can lead to reduced crop yield.   (3)

(vi) State two ways in which an agro-ecosystem usually differs from a natural ecosystem.   (2)

(vii) The failure to isolate agro-ecosystems from the rest of the environment would cause 'highly destructive attacks to occur'. Explain why this is so. (2)

(viii) Why has the breeding of resistant crops not provided the answer to attack by pathogens? (2)

(ix) Briefly explain why gene banks are necessary when we already have crops which give adequate yields. (2)

Mark allocation 17/125   Time allowed 20 minutes   In the style of the Scottish Higher II

These questions are designed to test understanding of the passage, but questions (v) and (vi) make demands which go beyond the passage itself. With an ever increasing volume of scientific literature being produced each year it is vital that individuals are able to comprehend the information the writer is trying to convey. For this reason comprehension questions are finding their way on to biological examination papers. Two readings of the passage should be made even before the questions are read. The first reading should be made without pause and with a view to obtaining the basic ideas the writer is trying to express. The second reading should be made more slowly and with a view to obtaining a deeper understanding of these ideas and absorbing some of the evidence in favour of them. The questions should then be attempted. After each question the passage should be examined for the answer or the information on which the candidate can compile the answer from his own knowledge. An outline of acceptable answers to the questions is given below:

(i) High yield. Ease of harvesting or all ripen together. Any other acceptable example, e.g. short growing season, climatic tolerance, early ripening (any two).

(ii) Measure of the variation of genotypes within a species.

(iii) Reduces it or makes it more uniform.

(iv) Crops are usually monocultures. Crop plants have low genetic variability. High density planting of crops (any two).

(v) May destroy photosynthetic tissue. May use products of photosynthesis reducing the quantity available to the corn for growth. May reduce supply of photosynthetic substrates, e.g. may prevent carbon dioxide diffusion by blocking stomata or reduce water availability by blocking small xylem cells.

(vi) Producers are usually monocultures. Intensive management of physical and chemical factors in the environment.

(vii) Pathogens in wild populations spread to nearby crops.

(viii) Pathogens have a high mutation rate. New varieties of pathogen can exploit genetic weakness in crop plants.

(ix) Conditions might change. New genetic material will then be useful.

5 Reproduction and Development

5.1 TYPES OF REPRODUCTION AND LIFE CYCLES

Assumed previous knowledge Basic structure of antheridia and archegonia in mosses and ferns. Structure of sporangia in ferns; capsule in mosses. Method of spore dehiscence and dispersal in fungi, mosses and ferns.

Details of one hemimetabolous (incomplete metamorphosis) insect and one holometabolous (complete metamorphosis) insect. Details of frog life cycle with the various tadpole stages.

For information on all these see 'Suggested further reading'.

Underlying principles

No living organism can continue indefinitely even if it escapes predators, disease and other potentially fatal circumstances; therefore reproduction is essential for the survival of the species. Asexual reproduction, which is rapid and produces many offspring, is the best method for increasing the numbers of a species, but it suffers the disadvantage of producing identical offspring which have no scope for adapting to new situations. Most life cycles therefore include meiosis, which produces some variety of offspring (see Section 4.1). In addition, many organisms involve two parents in a sexual process to further increase variety by the mixing of two genotypes.

Life cycles are adapted to the organism's mode of life; resistant phases are used to overcome adverse conditions. Larval stages are used in parasites to invade intermediate hosts as a means of returning to the primary host, and in free-living organisms as a means of exploiting food supplies and aiding dispersal. Further adaptations of life cycles occurred in plants as they became terrestrial, with a shift from a dominant (water dependent) gametophyte to a dominant sporophyte able to withstand desiccation.

Points of perspective

Knowledge of the functional significance of larvae in life cycles would be an advantage. Changes in life cycles associated with the evolutionary change from aquatic to terrestrial environment should be considered.

Essential information

Asexual reproduction This involves only one organism. The individuals produced are genetically identical, i.e. belong to a clone. Asexual reproduction ensures rapid multiplication. There are five broad types of asexual reproduction: fission; budding; fragmentation; sporulation; vegetative propagation.

Fission

This is typical of bacteria and protists. The cell divides into two or more equal parts. In **binary fission** (splitting in two) growth of the population is exponential, i.e. one cell divides into two, two into four, four into eight, etc. This is often a very rapid form of reproduction, e.g. many bacteria divide every 20 minutes. In many parasitic protista **multiple fission** occurs giving an even greater rate of reproduction, e.g. in the malarial parasite *Plasmodium,* where the nucleus divides into thousands of parts. This splitting is called schizogamy and each resultant cell is a schizont.

Budding

The parent produces an outgrowth, or bud, which detaches to become a separate individual, e.g. in *Saccharomyces* (yeast), *Hydra, Obelia.*

Fragmentation

This only occurs in simple organisms where the tissue is relatively undifferentiated, e.g. sponges and filamentous green algae, where, if small parts break off an organism, they will form whole new organisms.

Sporulation

This is the formation of small unicellular bodies which detach from the parent and, given suitable conditions, grow into new organisms. Spores are formed by bacteria, protozoa and many lower plants. Spore-bearing structures and spores vary but the spores are generally small, light, easily dispersed, have resistant walls and are produced in vast numbers.

Vegetative propagation

Part of a plant becomes detached and grows into a new plant, e.g. the leaves of African violet develop into new plants. Some organs of vegetative propagation are also known as perennating organs since they enable the plants to survive adverse conditions, e.g. the development of a stem into a corm (crocus) or stem tuber (potato); the swelling of a root in carrot or dahlia; the swollen buds of bulbs such as onion.

Sexual reproduction This involves the fusion of specialized haploid cells called **gametes** which generally arise from two individuals. Usually the gametes differ in structure, size and behaviour (heterogametes), one being small, highly motile and produced in large numbers, the other being larger, non-motile and produced in small numbers. Some lower organisms, e.g. certain algae and fungi, produce identical gametes (isogametes). Fusion of the gametes is called **syngamy** (for details of the mammal see Section 5.2 and for details of the flowering plant see Section 5.3).

Table 5.1 Differences between asexual and sexual reproduction

Asexual	Sexual
No mixing of genetic material; therefore less variation in offspring, therefore less evolutionary potential	Genetic mixing; therefore increased variation and great evolutionary potential
No gametes	Gametes
Usually more offspring	Fewer offspring
One parent	Usually two parents
Rapid; therefore takes advantage of favourable conditions	Longer process
May be resistant phases, e.g. spores, perennating organs	Rarely special resistant phase; may overcome difficult periods by delayed implantation (e.g. bats) or prolonged gestation (e.g. beavers)
The structures produced in asexual reproduction are more often used as a means of dispersal in animals than in plants	The structures produced are more often used as a means of dispersal in plants than in animals

Parthenogenesis This is the development of a new organism from an unfertilized egg. Sometimes this egg has been produced by meiosis and is therefore haploid; it develops into the new individual e.g. drones in honeybee colonies. Aphids produced by parthenogenesis are diploid because the eggs were produced by mitosis not meiosis.

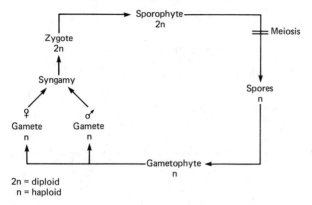

2n = diploid
n = haploid

Many plants show a regular alternation of generations between a haploid gametophyte and a diploid sporophyte.

The relative significance of each stage in the life cycle varies with species.

(a) Generalized plant life cycle

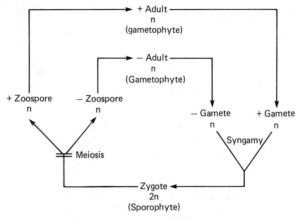

The zygote is the only diploid cell and meiosis occurs during its germination. The haploid phase multiplies extensively by mitosis giving a large number of unicellular individuals.

(b) Algae e.g. *Chlamydomonas*

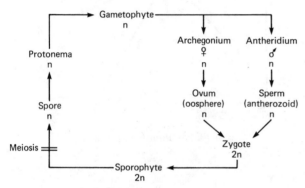

Gametophyte and sporophyte are almost equally conspicuous: the sporophyte grows out of the leafy gametophyte.

(c) Mosses e.g. *Funaria*

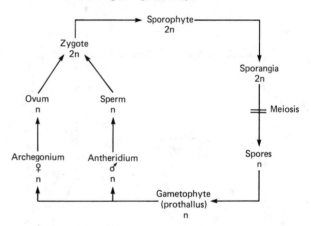

(d) Homosporous ferns, e.g. *Dryopteris*

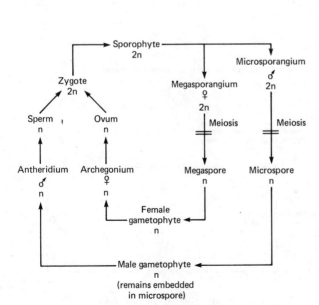

(e) Heterosporous ferns, e.g. *Selaginella* (Club Moss)

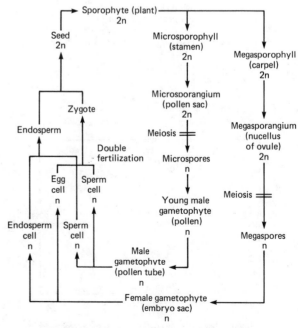

The angiosperms show maximum development of the sporophyte with a very reduced gametophyte stage.

(f) Angiosperms

Fig. 5.1 Life cycles in plants

All cells of multicellular animals are diploid
except the gametes which are haploid.

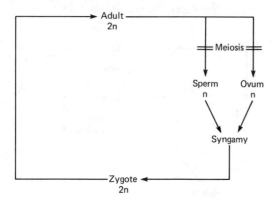

There is no true alternation of generations in animals.

(a) Generalized animal life cycle

Sometimes, although asexual reproduction is most common,
sexual reproduction occurs to produce a resistant phase
capable of withstanding adverse conditions, e.g. *Hydra*.

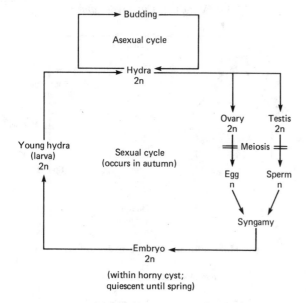

(within horny cyst;
quiescent until spring)

(c) Coelenterates, e.g. *Hydra*

(a) Incomplete metamorphosis, e.g. dragonfly.

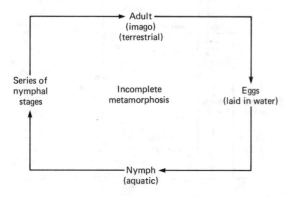

The nymph differs from the adult mainly in the absence of wings.

(e) Insects

Some animals do not show sexual reproduction, e.g. *Amoeba*.

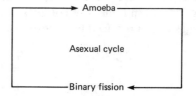

(b) Protozoans, e.g. *Amoeba*

Parasitic organisms have very specialized life cycles,
e.g. *Fasciola hepatica* (liver fluke).

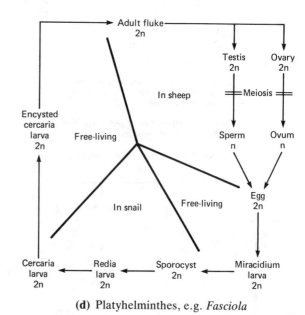

(d) Platyhelminthes, e.g. *Fasciola*

(b) Complete metamorphosis, e.g. cabbage white butterfly.

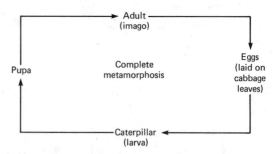

The larva is markedly different from the adult.

(e) Insects

Metamorphosis is also shown in the life cycles
of some chordates, e.g. frog.

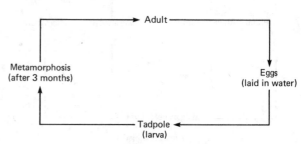

(f) Chordates, e.g. frog

Fig. 5.2A Life cycles in animals

Link topics

Section 4.1 Mitosis and meiosis
Section 5.2 Reproduction in mammals
Section 5.3 Reproduction in flowering plants

Suggested further reading

Roberts, M. B. V., *Biology, A Functional Approach,* 3rd ed. (Nelson 1982) (Chapters 24 and 25)

QUESTION ANALYSIS

1 (a) Explain what is meant by heterothallism.

(b) Explain why water is required for successful reproduction in mosses and ferns.

(c) Compare the modes of nutrition of a moss sporophyte and a fern sporophyte.

(d) Name the cells of a fern which undergo meiosis and name the structure inside which these cells occur.

(e) State 2 ways in which the spores of a gymnosperm differ from the spores of a fern.

(f) Explain why it is inaccurate to describe the flower of an angiosperm as its sexual reproductive apparatus.

Mark allocation 10/100   Time allowed 12 minutes

Associated Examining Board November 1979, Paper I, No. 2

This is a longer style structured question on many aspects of reproduction and life cycles. It involves some recall of factual information but mostly tests a deeper understanding of the principles underlying this topic by requiring explanations of certain points.

The word heterothallism in (a) is derived from two Greek words: *heteros* meaning 'other' and *thallos* meaning 'young shoot'. It refers to the thalli, cells or mycelia of plants which appear identical but are sufficiently different to prevent sexual reproduction occurring between similar strains. The strains are usually designated + and − and sexual reproduction can only occur between these two types. A well-known example is the mycelia of the pin mould *Mucor*.

To answer (b) it must be remembered that mosses and ferns occupy an intermediate position in the evolution of plants from aquatic to terrestrial modes of life. While they can survive without being totally immersed in water, they still retain a gametophyte stage in their life cycle that reveals their aquatic ancestry. Most aquatic organisms release one or both of their gametes into the surrounding water and the sperm swim to find and fertilize the egg. In the same way the gametophyte stage of mosses and ferns still requires a film of water to transfer the sperm (antherozooids) to the egg and so effect fertilization. In the absence of water the sperm are not normally released or, if they are, they dry out and die before fertilizing the egg.

Clear comparisons must be made in (c). Although the moss sporophyte may photosynthesize a little of its nutritional requirements, it is almost totally dependent on the gametophyte for its nutrition, which it obtains in organic form (amino acids, glucose) with some inorganic materials such as minerals and water. The fern sporophyte photosynthesizes the nutrients required from simple inorganic substrates such as carbon dioxide, water and minerals.

In (d) the fern cells that undergo meiosis are the spore mother cells found within the sporangium.

One problem that could arise in (e) is that the pteridophyta are often referred to as 'ferns' although the true ferns belong to only one subdivision, the Filicales. A candidate would have difficulty in finding differences between the spores of a gymnosperm and a pteridophyte, but if the spores of the Filicales are compared the task is easier. Possible differences are that the fern spores only possess a single nucleus whereas those of gymnosperms bear two (tube nucleus and generative nucleus); fern spores do not have the characteristic wings of gymnosperm spores and all filicales are homosporous whereas all gymnosperms are heterosporous.

The answer to (f) requires the candidate to fully understand the nature of a flower and appreciate that the plant that bears it is the sporophyte generation. Sexual reproduction is always concerned with the production of haploid gametes, and the structures that produce them are hence borne on the gametophyte generation. Since angiosperm plants are sporophytes they cannot produce gametes directly, and the flower cannot therefore be the 'sexual reproductive apparatus'. The flower is in fact the asexual apparatus that produces spores, usually of two types: microspores (pollen) and megaspores (one of which becomes the embryo sac). These spores contain nuclei which function as gametes. Since the flower only indirectly produces the gametes via spores, it is not technically the sexual apparatus, although it is easy to see how such confusion can arise.

2 Figure 5.2(b) illustrates the life cycle of a bryophyte.
   (a) Name the structures labelled A–E (5)
   (b) State clearly at which stage in the life cycle meiosis occurs. (2)
   (c) (i) What is meant by 'alternation of generations' in relation to this life cycle? (2)
       (ii) State TWO ways in which the alternation of generations shown by this life cycle differs from that found in flowering plants. (2)

Mark allocation 11/200   Time allowance 8 minutes   *London Board 1985, Paper 1, No. 5*

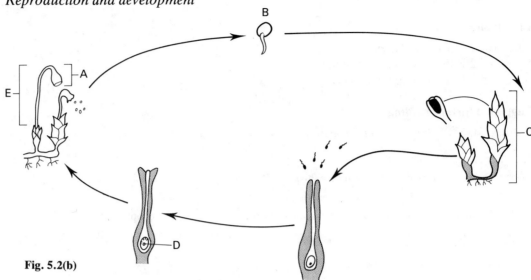

Fig. 5.2(b)

In answering (a) the candidate should pay careful attention to the bracket labels. A is clearly the capsule and B a germinating spore; the word 'germinating' should not be omitted. C is the leafy gametophyte and D is the ovum in the venter of the archegonium. E is an example of where the bracket label is important. Some candidates may be tempted to label it as the seta, but a single line label would then be appropriate. The bracket label indicates that the total structure, ie seta and capsule, is being considered. The answer is therefore the sporophyte generation.

The answer to (b) can be deduced from Fig. 5.1(c) in the essential information. In (c)(i) the candidate should make reference to there being two distinct types of organism, a haploid gametophyte and a diploid sporophyte, which each give rise to the other, i.e. they alternate. The differences referred to in (c)(ii) include the fact that flowering plants have a dormant sporophyte and no archegonia or antheridia, whereas bryophytes have a dominant gametophyte which possesses archegonia and antheridia.

3 With reference to a named fern and a named moss, explain the significance of the term alternation of generations.
Mark allocation 20/60   Time allowed 35 minutes
Welsh Joint Education Committee June 1979, Paper A2/B, No. 10

In this essay question, the key word is 'significance' indicating that, while accounts of the life cycles of your two chosen examples are needed, they should show the importance and purpose of each stage. The examples chosen will depend on those that have been studied as part of the syllabus and learned for the examination; in the present case *Dryopteris* will be used as the fern example and *Funaria* as the moss.

Throughout the answer remember that plants slowly evolved from an aquatic to a terrestrial mode of life and that the moss and fern life cycles represent intermediate stages in this transition. As part of the transition, plant life cycles changed from dominant water-dependent gametophyte generations to dominant sporophyte generations with much less dependence on water. This type of discussion would form a useful introduction to the answer. Make clear the meaning of the term 'alternation of generations'; see 'Life cycles in plants' under essential information and include the generalized life cycle in Figure 5.1a. Although the question mentions fern first, it is perhaps more logical to begin with the moss. Draw the life cycle diagram (Fig. 5.1c) and stress the dependence of the sporophyte generation on the gametophyte, the latter therefore being considered dominant. Correlate this to the mosses' dependence on water and their consequent restriction to moist habitats. For the fern draw Figure 5.1d and stress the reduction of the gametophyte to the relatively small prothallus and the dominance of the large leafy sporophyte. Correlate this to the fern being less dependent on water than the moss and its consequent wider terrestrial distribution. If time allows, the life cycle diagrams would benefit from being illustrated. As well as discussion on the significance of the differences between the moss and fern life cycles the examiner is also looking for the significance of the many similarities. For instance, the candidate should show that alternation between haploid and diploid stages allows meiosis to take place; the resulting increase in variety and evolutionary importance should be stressed (Sections 4.1 and 4.3). The significance of having the haploid and diploid stages as separate individuals in the fern is that two habitats, moist for the gametophyte and drier for the sporophyte, may be exploited. The cycle may be adapted to overcome seasonal climatic changes with the gametophyte thriving in the wetter periods and the sporophyte withstanding the drier ones.

In both mosses and ferns the significance of the sporophyte generation is in the production of spores by meiosis. These are dispersed under dry conditions, over long distances, increasing the chances of encountering new habitats. This capitalizes on the variety produced as a consequence of meiosis and extends their geographical range. In both examples the gametophytes are significant in that the gametes they produce may fuse with those of another plant of the same species thereby mixing two genotypes and adding to the variety produced by meiosis in the sporophyte generation.

Answer plan

Transition from aquatic to terrestrial life in plants.
Meaning of alternation of generations and its importance in this transition.
Basic alternation of generations life cycle (Fig. 5.1a).
Alternation of generation in moss (Fig. 5.1c) – stress dominance of gametophyte/restriction to moist areas.
Alternation of generation in fern (Fig. 5.1d) – stress dominance of sporophyte/can survive dry conditions.
Significance of the stages of each cycle – survival under varying climatic conditions; variety through meiosis (spores) and syngamy (gametes); dispersal by spores capitalizing on evolutionary potential.

4 'Non-sexual reproduction produces offspring identical with the parent, whereas sexual reproduction produces variation'. Discuss this statement.
Mark allocation 20/100 Time allowed 35 minutes   *London June 1980, Paper II, No. 8*

This is an essay question testing candidates' ability to discuss the degree of variability produced in sexual and asexual reproduction. Some candidates may think the question to be easier than it is because they consider it to be accurate in all cases. Only those candidates with a good knowledge of reproductive biology and genetics are likely to spot the circumstances where it is incorrect. The examiner will give most credit to the discerning candidates who, while broadly agreeing with the statements, appreciate those instances where it is untrue and are able to support both arguments with full explanations and particular examples. A logical approach is to deal with non-sexual and sexual reproduction separately and for each in turn show where the statement is true and then discuss the examples where it is not. Taking non-sexual reproduction, a definition should be given that stresses the absence of gametes, one parent only being involved and offspring usually being produced by mitosis. These factors mean that offspring are frequently identical. Examples should be given: e.g. binary fission in *Amoeba;* budding in *Hydra;* spore production in *Mucor.* The statement is therefore basically correct. In mosses and ferns, however, spores are produced non-sexually by meiosis, and (as shown in Section 4.1) this leads to variety. Mutations occur randomly and, although uncommon, offspring produced non-sexually may result directly from a cell in which a mutation has occurred and these will not be identical to the parent. In addition, even offspring genetically identical may adapt to different environmental conditions and hence vary in their appearance (phenotype). The statement is therefore not always true.

For sexual reproduction, the definition should stress the involvement of two parents in many cases and the production of gametes, usually by meiosis. These factors produce variety by the mixing of two genotypes from different parents and, in meiosis, crossing over and random segregation of pairs of homologous chromosomes. The way these factors produce variation may be discussed more fully if time allows, but candidates must not neglect the occasions when sexual reproduction does not produce variety; e.g. in mosses, ferns and some algae, gametes are produced mitotically and the resulting offspring will be genetically identical if these gametes self fertilize (mutations excepted). In hermaphrodite organisms self fertilization reduces the variety of offspring, but if the gametes are produced meiotically some variety remains. It is debatable whether identical twins can be considered as an exception to the statement. While superficially they appear to be the identical offspring of sexual reproduction, they are not directly derived from the fusion of gametes, but only indirectly through the mitotic division of a zygote and the subsequent separate development of the two cells.

    Conclude the answer along the lines that the statement is generally true, but not invariably so.

Answer plan

Non-sexual:

Offspring identical because:

  1 one parent – no mixing of genotypes.
  2 mitosis (usually) involved.

Offspring may vary because:

  1 mosses and ferns produce spores non-sexually by meiosis.
  2 mutations may arise.
  3 environment makes them phenotypically different.

Sexual:

Offspring vary because:

  1 two parents usually – genotypes mixed.
  2 meiosis usually involved giving variation due to crossing over and random segregation of homologous pairs of chromosomes in metaphase I.

Offspring may be identical because:

    in mosses, ferns and some algae – gametes produced mitotically and self fertilization occurs.

5.2   Reproduction in mammals (and vertebrate breeding cycles)

Assumed previous knowledge Meiosis
Basic O-level details of courtship, copulation and development of the embryo, birth and parental care (see 'Suggested further reading').

Underlying principles

Reproduction is concerned with the perpetuation of a species. Any mechanisms that increase the chance of fertilization and survival of the offspring will be of evolutionary advantage. Mammals owe much of their evolutionary success to the development of such mechanisms, which include:

1 Development of secondary sex characteristics to allow sexually mature individuals to recognize and mate with each other.
2 Seasonal breeding cycles that restrict copulation to times that will ensure birth at seasons most favourable to the survival of offspring.
3 Female receptiveness to the male only when ovulation is taking place, or even ovulation being stimulated by the act of copulation.
4 Internal fertilization bringing sperm and egg close together within the relative safety and stability of the female genital tract.
5 Internal development of the embryo in stable and protected conditions.
6 The placenta acting largely as a barrier to harmful substances and an exchange mechanism for beneficial ones.
7 Suckling as a means of providing the newly born with a fairly secure source of food ideally suited to its early development.
8 Parental care allowing development of the young in controlled and protected conditions with the maximum use of learned behaviour, which has the advantage of being adaptable to meet varying circumstances.

Points of perspective

A knowledge of the biological basis for the various methods of birth control would be an advantage.

It is essential to know about the reproductive processes of mammals other than man, if only to illustrate those points not applicable to humans, e.g. delayed implantation, yearly oestrous cycles and copulation inducing ovulation.

The importance of an understanding of breeding cycles in mammals to the development of animal husbandry should be appreciated, in particular in respect of man's domesticated animals.

A background knowledge of the medical aspects of reproduction would be useful, including the biological significance of ante- and post-natal care.

Essential information

Sexual reproduction involves the fusion of two haploid gametes to form a zygote. The male gamete (spermatozoon) is small, highly motile and produced in large numbers whereas the female gamete (egg cell) is larger, non-motile and fewer are produced. Egg cells, or ova, are produced in the ovary by a process known as oogenesis; spermatozoa are produced in the testes by spermatogenesis (see Figures 5.3 and 5.4).

Fertilization When the head of the spermatozoon comes into contact with the vitelline membrane, the acrosome opens releasing a chemical which softens the membrane and the inner membrane of the acrosome inverts to form a filament which pierces the egg membranes. This process is called the acrosome reaction. Following this the cortical granules apply themselves to the inner surface of the vitelline membrane, thickening it and preventing the entry of further sperm. The head and middle piece move through the cytoplasm and the nuclei of the sperm and egg cell fuse, thus restoring the diploid number.

Development of the zygote The zygote moves down the oviduct to the uterus, a journey which takes about one week in humans. By the time it has reached the uterus it has divided mitotically to form a hollow ball of cells – a blastocyst. The blastocyst becomes implanted in the lining of the uterus and the outer layer of cells forms trophoblastic villi which project into the uterine wall. As the embryo develops it becomes surrounded by a series of extra-embryonic membranes: the amnion, which encloses a fluid-filled cavity which eventually fills the entire uterus, and the allantois and chorion, which unite to form the allanto-chorion which develops into the placenta. The only connection between the embryo and the wall of the uterus is the stalk of the allantois which becomes the umbilical cord. This contains the umbilical artery and vein conveying foetal blood to and from the placenta.

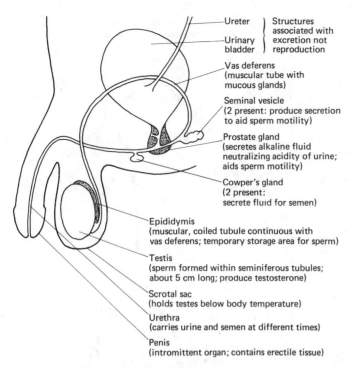

Fig. 5.3 (a) Male reproductive system, lateral view (human)

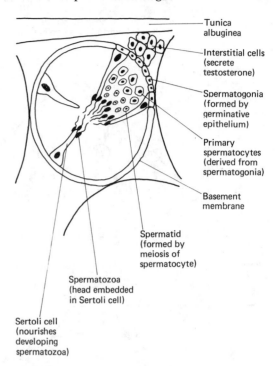

Fig. 5.3 (b) TS seminiferous tubule (Each testis contains about 1000)

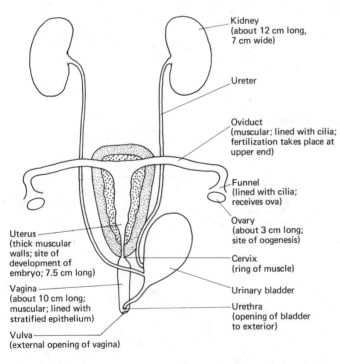

Fig. 5.3 (c) Female urinogenital system, ventral view (human)

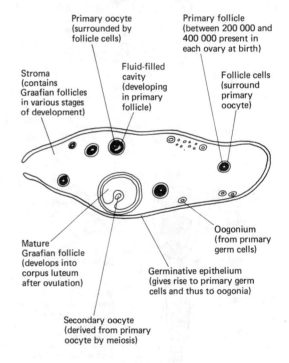

Fig. 5.3 (d) LS ovary

The placenta

Exchange of materials takes place between the maternal blood spaces in the uterine wall and the foetal capillaries in the chorionic villi.

The functions of the placenta

1 Oxygen, water, soluble food and salts pass from maternal to foetal blood.
2 Carbon dioxide and nitrogenous waste pass from foetal blood to maternal blood for removal by the mother.
3 During processes **1** and **2** it prevents mixing of foetal and maternal bloods since the foetus may inherit the father's group which, if incompatible with the mother's, would damage the foetus if mixing occurred.

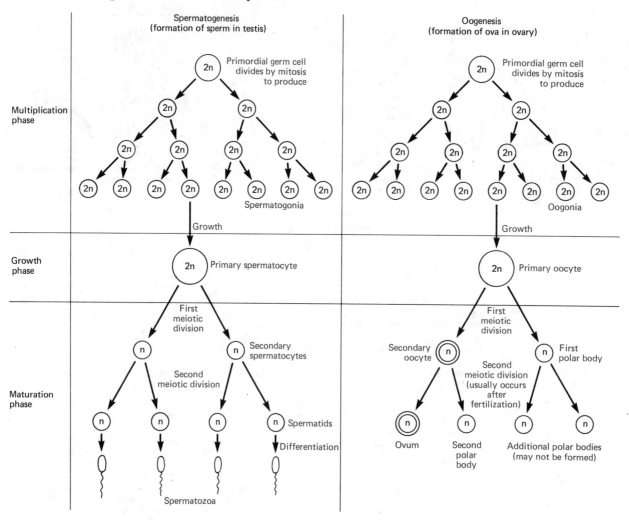

Fig. 5.4 (a) Gametogenesis

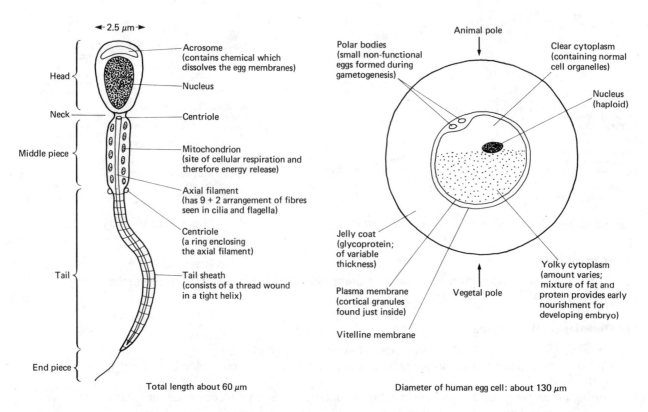

Fig. 5.4 (b) Human spermatozoon,
based on electron micrograph

Fig. 5.4 (c) Generalized animal egg cell

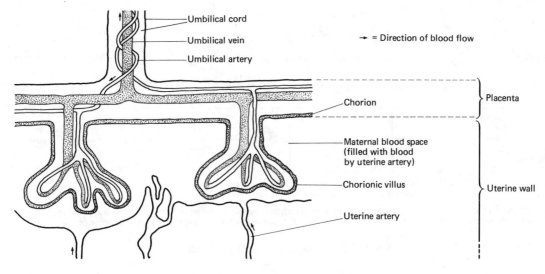

Fig. 5.5 Blood flow in the placenta

4 During pregnancy the placenta progressively takes over the role of producing the hormones that prevent ovulation and menstruation.

5 It allows maternal and foetal blood pressures to differ from one another.

6 It prevents the passage of some pathogens from the mother to the foetus, although some viruses and bacteria and many antibodies can cross between the two.

7 While allowing some hormones across, it prevents the passage of those maternal hormones that could adversely affect foetal development.

Female sexual cycle (oestrous or menstrual cycle) In humans this is a twenty-eight day cycle controlled by hormones secreted by the pituitary gland and the ovary. The main hormones involved are:

(a) Follicle stimulating hormone (FSH) which
(i) causes Graafian follicles to develop in the ovary.
(ii) stimulates the tissues of the ovary to produce oestrogen.

(b) Oestrogen which
(i) repairs the uterine wall following menstruation.
(ii) builds up in concentration during the first two weeks of the menstrual cycle until it stimulates the pituitary gland to produce luteinizing hormone.

(c) Luteinizing hormone (LH) which
(i) brings about ovulation.
(ii) causes a Graafian follicle to develop into a corpus luteum which produces progesterone.

(d) Progesterone which
(i) inhibits FSH production and therefore stops further follicles developing (a fact made use of in the development of the contraceptive pill).
(ii) causes development of uterine wall prior to implantation.

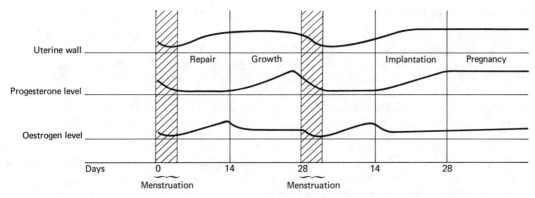

Fig. 5.6 (a) Human oestrous cycle

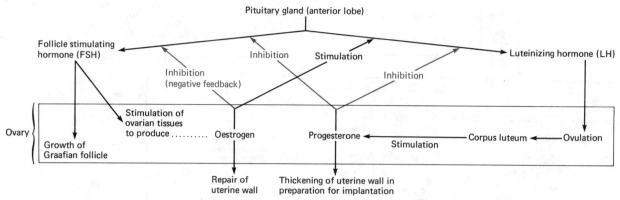

During pregnancy: progesterone, produced initially by the corpus luteum and later by the placenta, is present in high concentrations; therefore, LH production and hence ovulation are inhibited; as FSH production is also inhibited, ripening of follicles and production of oestrogen cease.

Fig. 5.6 (b) Hormonal control of the oestrous cycle

Male sexual cycle There is no cycle in male humans but the gonads are regulated by hormones identical with those of the female. FSH, which is generally called interstitial cell stimulating hormone (ICSH) in the male, promotes spermatogenesis. LH, also called ICSH, causes the interstitial cells between the seminiferous tubules to secrete androgens, hormones which stimulate the development of male secondary sex characteristics.

Breeding cycles in vertebrates If the cycle is seasonal, environmental changes trigger the endocrine system at the most advantageous time of the year for breeding. Whereas in the stickleback *Gasterosteus aculeatus* daylength is the most important environmental trigger and temperature plays a subsidiary role, in the minnow *Couesius plumbeus* the cycle is most affected by temperature.

The breeding of birds is nearly always seasonal. In temperate latitudes increase in day-length in the spring acts through the pituitary gland to cause ripening of the gonads. Changes in behaviour associated with the development of the gonads vary with species. Other factors such as degree of activity and food eaten may be important, especially near the equator where there is little seasonal variation.

If the cycle is more or less continuous, as in most endotherms and some tropical ectotherms, the controls are usually within the organism and rely on a feedback mechanism. Usually the female will only receive the male during a strict period of oestrus (or heat) but in humans receptivity may occur throughout the cycle. In some animals, e.g. rabbit and ferret, coitus induces ovulation and fertilization usually takes place within a few hours. In certain bats copulation occurs in autumn and fertilization is delayed until spring. Implantation usually occurs about one week after fertilization, as in humans and rabbits, but in some animals, e.g. pig, cat and dog, implantation is delayed for about two weeks and in badgers mating takes place in the summer but implantation is delayed until late in the winter.

Link topics

Section 4.1 Mitosis and meiosis
Section 5.4 Growth and development
Section 5.5 Control of growth
Section 10.1 Hormones and homeostasis

Suggested further reading

Cohen, J. and Massey, B., *Biology of Animal Reproduction* Studies in Biology No. 163 (Arnold 1984)
Dale, B., *Fertilization in Animals* Studies in Biology No. 157 (Arnold 1983)
Freeman, W. H. and Bracegirdle, B., *An Atlas of Embryology*, 3rd ed. (Heinemann 1978)
Hart, G., *Human Sexual Behaviour*, Oxford/Carolina Biology Reader No. 94 (Packard 1978)
Mackean, D. G., *Introduction to Biology*, 5th ed. (Murray 1980)
Rhodes, P., *Birth Control*, Oxford/Carolina Biology Reader No. 4, 2nd ed. (Packard 1976)
Slater, P. J. B., *Sex Hormones and Behaviour*, Studies in Biology No. 103 (Arnold 1978)

QUESTION ANALYSIS

5 The placenta serves as a link between foetus and mother. At the same time it acts as a barrier between them. By reference to the functions of the placenta, explain what these statements mean.
Mark allocation 8/100   Time allowed 12 minutes

Southern Universities Joint Board June 1979, Paper I, No. 9

This is a shorter style structured essay testing knowledge of placental functions. The question gives two statements each of which needs explaining in relation to the functions of the placenta. For planning purposes a list of these functions should be made (see 'Functions of the placenta' in the essential information), but not used as such in the essay. The list can be marked showing whether the function is a means of linking the foetus and mother, acting as a barrier between them or performing neither function. The key words are 'link' and 'barrier'.

The actual answer should explain each of the statements in turn illustrating their meaning by reference to the relevant function of the placenta. For instance, in discussing the placenta as a link between foetus and mother, functions 1 and 2 in the essential information can be used, while functions 3, 5, 6 and 7 illustrate its role as a barrier between them. Do not list or describe the functions alone, or make drawings of the placenta. The examiner requires functions to be related to the relevant statement and hence function 4 can be omitted since it is not relevant to either. Remember to give the answer the examiner requires, not the one you think he should have. His mark scheme will allow him to give more credit to the candidate who selects the six relevant functions from the list and relates each to the appropriate statement than by the one who lists all seven regardless of the statements given.

Answer plan

Placenta as a link: illustrate by reference to (numbers refer to detailed list in the essential information)
1 oxygen, water, food and salts from mother to foetus
2 carbon dioxide from foetus to mother
6 (part) antibodies from mother to foetus

Placenta as a barrier: illustrate by reference to

3 prevents bloods mixing
5 prevents high maternal blood pressure affecting foetus directly
6 (part) filters out some pathogens
7 filters out some hormones

6 Make labelled drawings only to illustrate the essential organs of reproduction of either a named flowering plant or a named mammal. In the example you choose briefly describe the events which lead up to (a) the production and release of gametes, (b) fertilization (c) development, nutrition and protection of the embryo.
Mark allocation 28/40   Time allowed 35 minutes   *Oxford Local June 1980, Paper II No. 10*

This is an essay question examining knowledge of reproduction and development of either a mammal or a flowering plant.

The key words in the first part are 'labelled drawings only'. The labels should be as comprehensive as possible, see Figures 5.3a, 5.3c, 5.7a for relevant examples of the type of labelling required. For the mammal Figure 5.3d of the ovary and Figure 5.3b of the testis could be included provided time permits; likewise Figure 5.8a of the anther and Figure 5.8c of the ovule could be included for the flowering plant. The word 'only' indicates that separate written accounts and explanations of the structures will be ignored, so keep to short, detailed annotations. The word 'essential' means that only structures directly concerned with reproduction should be included. In Figures 5.3a and 5.3c 'kidney', 'ureter', 'urinary bladder' and 'urethra' (female only) should be excluded or, if included for reference, a note made that these are not reproductive structures. If the flowering plant is used, the labels for Figure 5.7a should be used in their entirety since all play an important role in reproduction. This diagram is of a generalized plant and should not be drawn as such but used only as a guide to labelling. Draw a specific example that you yourself have studied and learned. Remember to do the flowering plant or the mammal, not both and do not forget to name them. If a mammal other than human is chosen some modification will be needed to Figures 5.3a and 5.3c.

In the second part the 'events which lead up to' the processes stated are required, rather than the processes themselves. According to the example chosen these will be:

Flowering plant In (a) the events leading up to the production and release of gametes are stated under the headings 'production of pollen' and 'structure and development of the ovule' in Section 5.3. An outline only of these processes is required.

In (b) the events to be described include those above (but do not restate them here) and pollination, the details of which will depend on the example chosen. A general account is given under 'pollination', Section 5.3. In addition an outline of the growth of the pollen tube (under 'fertilization' in the same section) is required.

In (c) the events will again depend on the example and again an outline is given under 'fertilization' and 'seeds and fruits' in Section 5.3.

Mammal: In (a) the events include gametogenesis (see Fig. 5.4a) and courtship behaviour and mating which will vary according to the example chosen.

For (b) in addition to the events above, the swimming of the sperm through the uterus and the movement of the ovum down the oviduct, should be detailed. The account of sperm penetration is given under 'fertilization' in the essential information.

For (c) the information under 'development of the zygote' should be summarized, and presented with the importance of maternal behaviour in protecting the foetus. This latter point is frequently omitted by candidates who mention only the physical protection afforded by the fluid-filled amniotic cavity and the mother's abdominal wall. The protective functions of the placenta must not be forgotten.

Whichever example is chosen, the account should only be an outline since the time allowance does not permit a detailed account. In view of this, and the fact that detailed diagrams are given in the first part, it is doubtful that the inclusion of further diagrams here is either possible or desirable. For the same reasons no account of meiosis would be expected in (a).

Answer plan

Follow one or other scheme depending on the organism chosen. Other details will vary according to the specific example selected.

Mammal	**Flowering plant**
Labelled drawings = Figures 5.3a and 5.3c (5.3d and 5.3b only if time allows)	Labelled drawings = Figures 5.7a (guide to labels only; 5.8a and 5.8c only if time allows)
(a) Figure 5.4a + courtship and mating	(a) From Section 5.3: 'production of pollen' and 'structure and development of ovule' (summary)
(b) Section 5.2: 'fertilization' (summary) + sperm and ova movement	(b) Section 5.3: 'pollination' (summarize and adapt to chosen example), 'fertilization' (pollen tube growth)
(c) Section 5.2: 'development of the zygote' (summary) + protection by maternal behaviour and placenta	(c) Section 5.3: 'fertilization' (last part) and 'seeds and fruits' (summary)

7 The graph below shows the thickness of the uterus wall throughout the menstrual cycle in an adult human female.

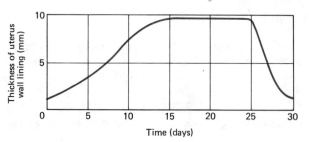

Fig. 5.6 (c) Variations in thickness of uterine wall

(a) From the graph state the day
  (i) ovulation is most likely to happen
  (ii) assuming sperm are present, when fertilization is most likely to occur
  (iii) the corpus luteum begins to break down
  (iv) menstruation begins

(b) (i) Describe how the menstrual cycle is controlled by hormones, stating for each hormone where it is produced and exactly what it does.
    (ii) Give two medical uses of the hormones you have mentioned.

Mark allocation 20/100   Time allowed 40 minutes   *In the style of the Joint Matriculation Board*

This is a highly structured essay-type question testing factual recall, the ability to interpret graphical information and the medical application of female sex hormones.

Part (a) requires sufficient knowledge of uterine changes during the menstrual cycle to be able to correlate changes in the wall thickness to other events that occur in the cycle. In (i), for instance, the candidate should reason that the uterine wall thickens in order to receive and nourish the newly fertilized ovum. It follows that ovulation should occur at a time that ensures that the ovum reaches the uterus when the wall is at maximum thickness. The journey from the ovary to the uterus may take up to a week and therefore release of the ovum 5 or 6 days before the wall reaches maximum thickness will ensure it reaches the uterus when it is fully thickened. Later dates of release will also ensure this, but clearly early implantation is preferable in case menstruation occurs early. On the graph the probable days are 8–15 with 9 or 10 as the most likely. Note the words 'from the graph' and do not use remembered days since these may not correspond with the ones in this question. For instance day 14 is usually cited as the date for ovulation, because most cycles use the start of the menstrual flow as day 1. This graph uses the end of the menstrual flow as day 1 and so is displaced by about 5 days compared with the usual graphs. Candidates using memory rather than their ability to interpret the graph given will therefore fare badly.

For (ii) the candidate needs to know that an ovum is usually fertilized 2 or 3 days after release, making the most likely time for fertilization between days 11 and 13 on the graph.

The corpus luteum breaks down shortly before menstruation. It is then easy to see that, since menstruation involves the breakdown of the uterus wall and it begins on day 25, the corpus luteum must begin to break down on about day 22 or 23.

The hormones required in (b) (ii) are FSH, LH, progesterone and oestrogen. The information required to answer this question is presented in Figure 5.6b. Where each hormone is produced is clear, but in giving 'exactly what it does' candidates must be sure to include the effect of each in stimulating or inhibiting the production of other hormones, as well as the effects on the actual reproductive organs. A key word in this section is 'controlled'; details of the actual cycle are not required except where they are referred to in respect of hormonal control.

In (b) (ii) one medical use is the oral contraceptive where carefully balanced quantities of oestrogen and progesterone are taken for 21 days of the menstrual cycle in order to inhibit ovulation. A second use is in fertility pills for certain females who have failed to conceive. The gonadotrophins (FSH and LH) are used in such pills.

Answer plan

(a) (i) day 9 or 10 (range 8–15)
   (ii) days 11 to 13
  (iii) days 22 or 23
  (iv) day 25

(b) (i) FSH, LH, oestrogen and progesterone. Use Figure 5.6b for details (include inhibition or stimulation of production of each other)
   (ii) Oestrogen and progesterone in oral contraceptives.

LH and FSH in fertility pills.

8 (a) In a mammalian foetus there is an opening called the foramen ovale, between the left and right atria.
   (i) What is the function of the foramen ovale?
   (ii) It is normal for the foramen ovale to close at birth. If this did not happen what symptoms might be experienced by the baby? (4)

(b) (i) Give an account of the role of the placenta as an endocrine organ in mammals.
   (ii) State three features of such a placenta that suit it to its function of exchanging materials.
  (iii) Why does foetal haemoglobin pick up oxygen at the placenta at partial pressures which cause the maternal haemoglobin to release its oxygen? (10)

(c) (i) Explain the differences in the amounts of calcium (Ca) required by three different women as shown in the table below

	Ca requirements (mg day^{-1})
Adult female	700
Pregnant female	1650
Nursing mother	2000

   (ii) Why is the enzyme rennin more common in young mammals than older ones?
  (iii) The young of carnivorous mammals are frequently helpless at birth and require a long period of time before they acquire independence from their parents. In contrast the young of herbivorous mammals are well developed at birth and soon attain independence from their parents. Explain these differences. (6)

Mark allocation 20/100   Time allowed 40 minutes   In the style of the Joint Matriculation Board

This is a structured question covering mammalian foetal development and aftercare. As well as examining recall of factual knowledge, the question tests a deeper understanding of this knowledge by requiring the candidate to explain its importance to the foetus. It also requires the application of other biological topics (e.g. nutrition, behaviour, blood and circulation) to events during early mammalian development.

In (a) (i) the function of the foramen ovale is to allow blood to flow from the right atrium of the heart to the left atrium, thereby by-passing the lungs which do not function as a gaseous exchange surface before birth. Forcing large volumes of blood over the pulmonary capillary network would therefore be both pointless and wasteful of energy. In part (ii) it should be apparent to the candidate that the failure of the foramen ovale to close at birth will result in deoxygenated blood entering the systemic system causing inability to perform strenuous tasks, breathlessness and a bluish tinge to the skin due to large quantities of deoxygenated blood.

In (b) (i) the endocrine role of the placenta varies slightly from mammal to mammal; since the question states 'mammals' all examples should be covered rather than restricting the answer to one specific example. The main placental hormones are progesterone, which prevents ovulation, maintains the lining of the uterus and so prevents menstruation; oestrogen (especially in mouse and rabbit) which stimulates ovulation and builds up the uterine lining; human chorionic gonadotrophin (HCG) which functions in a similar way to luteinizing hormone (LH).

In part (ii) the candidate should choose the following placental features:

1 It has a large surface area over which maternal and foetal blood come into close contact without mixing.

2 Maternal and foetal bloods are constantly being replaced in the placenta to maintain a diffusion gradient.

3 The small distance between maternal and foetal bloods speeds diffusion.

In part (iii) a knowledge of haemoglobin dissociation curves is an advantage. Basically the foetal haemoglobin has a higher affinity for oxygen than maternal haemoglobin at a given partial pressure of oxygen (i.e. the foetal haemoglobin dissociation curve is shifted to the left).

In (c) (i) the candidate should appreciate the importance of calcium in the diet, in particular its role as a major component of bone. A pregnant female requires more than twice as much as a normal female in order to provide the necessary calcium for the development of the foetal skeleton. The nursing mother requires even more since the newly born baby, including the skeleton, grows very rapidly. The mother provides this essential calcium in the milk she produces when suckling the infant and hence must compensate for the loss by a greater intake of calcium in her own diet. In part (ii) the candidate needs to recall that rennin has the function of curdling the milk so that it remains in the stomach long enough to allow complete digestion. Since milk comprises a much more substantial proportion of a young mammal's diet than that of an older one, it is produced in proportionately larger quantites.

In part (iii) the candidate needs to understand how the behaviour of carnivorous and herbivorous mammals is correlated to their diet. The herbivore gains protection from predators largely by its speed of escape and the security of roaming in herds. A young herbivore would be vulnerable to attack from a predator unless it could run with the herd soon after birth, especially as herbivores are often found in shelterless open areas. By contrast the carnivorous mammal has few, if any, predators and is raised in the relative security of a nest or den in a sheltered place. In addition the carnivore needs practice in capturing its food efficiently; a long period in contact with its parents, learning by example and through play, gives it the necessary opportunity to acquire these skills.

5.3   REPRODUCTION IN FLOWERING PLANTS

Assumed previous knowledge Mitosis and meiosis (see Section 4.1)
The life cycle of angiosperms (see Section 5.1, Fig. 5.1f)

Underlying principles

The successful colonization of land was largely achieved by the reduction of the water-dependent gametophyte generation and the dominance of the sporophyte generation. One of the greatest problems to be overcome was the transference of the gametes in the absence of water. The angiosperms achieved this by reducing the gametophyte generation and then protecting it within the sporophyte. Methods of transferring the young male gametophyte (within the pollen grain) from one plant to another developed, e.g. wind and insects, while it is protected within a resistant wall. A mechanism for allowing the male gamete to reach the protected female gamete developed, i.e. the pollen tube and the protection of the resulting embryo by the ovary wall. The innovations of angiosperm reproduction to terrestrial existence can be summarized:

1 Development of an ovary, stigma and style, the latter providing a pathway guiding the male nuclei of the pollen tube to the embryo sac.
2 Development of stamens leading to various highly successful systems of pollination.
3 Reduction of the female gametophyte to a 7-celled embryo sac and enclosure within the ovary for protection.
4 Reduction of the male gametophyte to a pollen tube with the complete elimination of motile sperm.
5 Presence of a double fertilization resulting in the embryo and the endosperm. The latter is contained within the seed and forms the source of nourishment for the young embryo.
6 Development of the fruit, usually from the ovary wall.

Points of perspective

It would be useful to have knowledge of a wide range of pollination mechanisms with precise details of the adaptations of one flower example to each type.

The importance of the parallel evolution of insects and flowering plants should be appreciated.

The storage materials in seeds and the experimental tests for them as well as their economic importance as human food sources, e.g. cereals, peas, bean, soya, should be known.

A good candidate should know some details of the length of dormancy in seeds and the factors that induce and break it.

Some outline knowledge of the different types of fruit, e.g. dry, succulent and false fruits, would be helpful.

The candidate with a good detailed knowledge of a number of methods of dispersal, especially some of the less usual ones, e.g. squirting cucumber, will obviously have an advantage.

The candidate should have performed, or know how to perform, laboratory experiments to determine (i) factors affecting germination, (ii) the respiratory rate (RQ) of germinating seeds and (iii) the enzymatic activity of seeds, e.g. amylase action.

Essential information

A flower is a vegetative shoot whose parts are adapted for reproduction. The female part of the flower is referred to as the gynoecium and consists of the carpels. The male part is the androecium and consists of the stamens. The other parts of the flower are mainly concerned with protection and the attraction of insects.

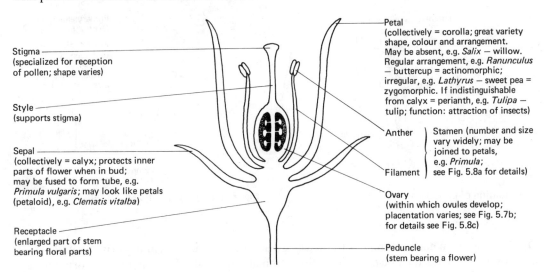

Fig. 5.7 (a) LS generalized flower

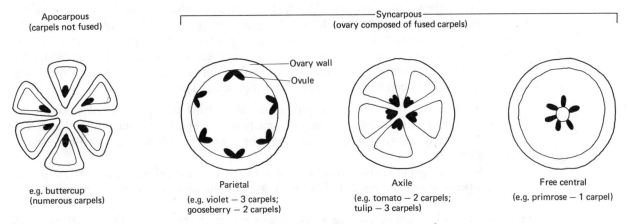

Fig. 5.7 (b) Common forms of placentation

Production of pollen The centre of the pollen sacs contains pollen mother cells each of which divides meiotically to give a tetrad (four) of pollen grains. These four cells later separate. Each young grain has one nucleus but this divides into two. One of these, surrounded by denser cytoplasm, forms the generative cell and this later gives rise to the male gametes; the other forms the tube nucleus. The pollen grain has a double wall, a delicate inner one of cellulose (the intine) and an outer one of variable thickness (the exine). Pollen grains vary in shape and in size from 3–300 μm. They are liberated by the longitudinal rupture of the anther lobe together with the breakdown of the wall between each pair of sacs.

Structure and development of an ovule One cell of the nucellus becomes enlarged and is known as the embryo sac mother cell. This divides meiotically to give four cells, three of which are crushed as the remaining one enlarges to form the embryo sac. The single nucleus within the embryo sac divides mitotically and the two nuclei move to opposite ends of the sac. Each of the two nuclei divides mitotically twice so that there are four haploid nuclei at each end of the sac.

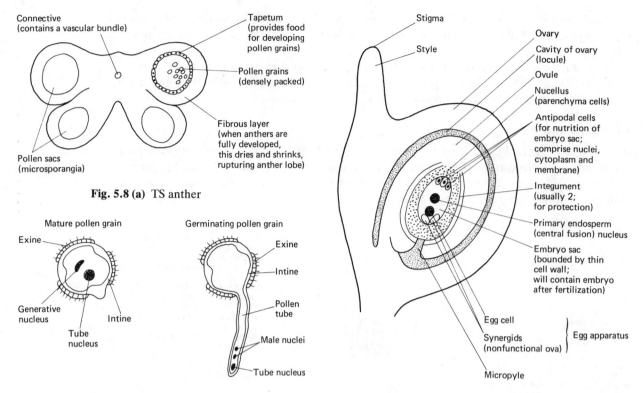

Connective
(contains a vascular bundle)

Tapetum
(provides food
for developing
pollen grains)

Pollen grains
(densely packed)

Fibrous layer
(when anthers are
fully developed,
this dries and shrinks,
rupturing anther lobe)

Pollen sacs
(microsporangia)

Fig. 5.8 (a) TS anther

Mature pollen grain

Exine

Generative
nucleus

Tube
nucleus

Intine

Germinating pollen grain

Exine

Intine

Pollen
tube

Male nuclei

Tube nucleus

Fig. 5.8 (b) Germination of pollen grains

Stigma

Style

Ovary

Cavity of ovary
(locule)

Ovule

Nucellus
(parenchyma cells)

Antipodal cells
(for nutrition of
embryo sac;
comprise nuclei,
cytoplasm and
membrane)

Integument
(usually 2;
for protection)

Primary endosperm
(central fusion) nucleus

Embryo sac
(bounded by thin
cell wall;
will contain embryo
after fertilization)

Egg cell

Synergids
(nonfunctional ova)

Egg apparatus

Micropyle

Fig. 5.8 (c) LS generalized carpel

One nucleus from each end moves to the centre and these fuse to form the primary endosperm nucleus (central fusion nucleus). The three remaining at the micropyle end form the egg apparatus and the three at the other end, the antipodal cells.

Pollination This is the transference of pollen from the anther to the stigma.

Self pollination: the transference of pollen from the anther to the stigma of the same flower, or a different flower on the same plant. Many plants are designed to self pollinate should cross pollination fail, e.g. in the Compositae as the flower ages the stigmas curl to touch the anthers. A few species of plants produce flowers which become self pollinated while still in the bud stage (= cleistogamy) but this usually only occurs in a small proportion of the flowers, e.g. *Viola* (violet), where open flowers may also be self pollinated by insects.

Cross pollination: the transference of pollen from the anther to the stigma of a flower on a different plant of the same species. Although some plants regularly show self pollination there is a general tendency towards cross pollination which reduces inbreeding and increases the variability of a population.

The two main agents of cross pollination are wind and animals (mainly insects). The main characteristics of wind- and insect-pollinated flowers can be summarized as in Table 5.2.

Fertilization As the female gamete in angiosperms is protected within the carpel, the male gamete can only reach it via the pollen tube, which provides a channel of entry and protection for the male nuclei. The pollen grain germinates within a few minutes of landing on the stigma and the pollen tube pushes between the loosely packed cells of the style. The entire contents of the pollen grain move into the tube, the tube (vegetative) nucleus moving first. Callose plugs block the older, empty parts of the tube as it grows. The pollen tube may secrete pectases to soften the middle lamellae of the cells of the style and the growth towards the micropyle is thought to be chemotropically controlled. The normal mode of entry to the ovule is through the micropyle. The pollen tube contains three nuclei, the tube nucleus and two male gametes derived from the generative nucleus, which keep near the tip as growth proceeds. When the pollen tube penetrates the embryo sac one male nucleus fuses with the egg cell (oosphere) and one with the primary endosperm nucleus. Since this second fusion involves three nuclei (the primary endosperm nucleus was derived from two polar nuclei), it is called triple fusion. Double fertilization is said to occur because two male nuclei fuse with the female nuclei. The fertilized oosphere gives rise to the embryo and the fertilized primary endosperm nucleus to the endosperm.

Table 5.2 Differences between wind- and insect-pollinated flowers

Wind	*Insects*
Enormous amounts of pollen produced because there is great wastage	Less pollen produced because mechanically more precise
Pollen small, light and smooth	Pollen larger and heavier often with projections
Often found in plants which occur in groups	Often found in plants which are more or less solitary
Often found in unisexual flowers (usually with an excess of male flowers)	Mostly in bisexual (hermaphrodite) flowers
Flowers dull, scentless and nectarless	Bright, scented flowers with nectar, therefore attractive to insects
Stigmas long and protrude above petals	Stigmas often deep in corolla
Stigmas often feathery or adhesive	Stigmas often small
Stamens long and protrude above petals	Stamens may be within corolla tube
Examples: grasses (e.g. *Festuca*); hazel (*Corylus*); willow (*Salix*); plantain (*Plantago*)	*Examples:* snapdragon (*Antirrhinum*); foxglove (*Digitalis*); buttercup (*Ranunculus*); clover (*Trifolium*)

In non-endospermic seeds the endosperm is used up to form the cotyledons before the seed is ripe but in endospermic seeds nuclear and cellular divisions of the fertilized primary endosperm nucleus give rise to an extensive endosperm.

Seeds and fruits The development of the fertilized ovule produces a seed and the ovary as a whole develops into the fruit.

The mature seed is covered by a hard leathery testa, which is the product of one or both integuments and may have a scar, or hilum, on it which marks the point of attachment to the ovary wall. The embryo within the testa consists of one cotyledon (monocotyledon) or two (dicotyledon) which may or may not store food. The cotyledons are attached to a central axis differentiated into a plumule (shoot) at one end and a radicle (root) at the other.

Examples of types of seeds:
    dicotyledon, non-endospermic: broad bean, pea, sunflower
    dicotyledon, endospermic: castor oil
    monocotyledon, non-endospermic: water plantain
    monocotyledon, endospermic: maize, onion

Dispersal of seeds and fruits Sexual reproduction ensures variability and a potential for the colonization of new habitats. If this potential is to be realized dispersal is essential. Dispersal also reduces the chances of backcrossing to the parents and the resultant problems of inbreeding. It also prevents overcrowding and competition and decreases vulnerability to epidemic attacks.

There are four main mechanisms of dispersal: mechanical, wind, animal and water.

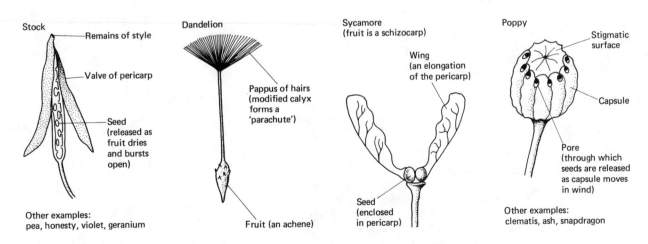

Fig. 5.9 (a) Mechanically-dispersed fruits and seeds

Fig. 5.9 (b) Wind-dispersed fruits and seeds

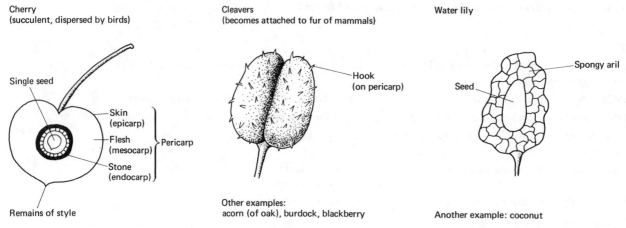

Cherry
(succulent, dispersed by birds)

Single seed

Skin
(epicarp)

Flesh
(mesocarp) } Pericarp

Stone
(endocarp)

Remains of style

Cleavers
(becomes attached to fur of mammals)

Hook
(on pericarp)

Other examples:
acorn (of oak), burdock, blackberry

Water lily

Spongy aril

Seed

Another example: coconut

Fig. 5.9 (c) Animal-dispersed fruits and seeds

Fig. 5.9 (d) Water-dispersed fruits and seeds

Germination A seed begins to develop after a period of dormancy and is dependent on the food reserves in the cotyledon until the first true leaves develop. Germination requires an adequate supply of oxygen and water and a suitable temperature. As the first stage of germination is the absorption of water and since testas are frequently impermeable to water, germination may be delayed until the testa is broken or decayed. The water renews the physiological activities of the endosperm and embryonic tissues. The enzymes in the seeds are activated and food changed from an insoluble to a soluble form. There are two main types of germination:

1 hypogeal in which the cotyledons remain below the ground, e.g. broad bean;
2 epigeal in which the cotyledons emerge above the ground, e.g. runner bean.

Link topics

Section 5.1 Types of reproduction and life cycles
Section 5.5 Control of growth

Suggested further reading

Black, M., *Control Processes in Germination and Dormancy*, Oxford/Carolina Biology Reader No. 20 (Packard 1972)
Rudall, P., *Anatomy of Flowering Plants* (Arnold 1987)

QUESTION ANALYSIS

9 Each of the lettered diagrams below are parts of a flowering plant, though not drawn to the same scale. The diploid number for this plant is 16.

A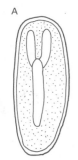

B

C

D

E

Fig. 5.10 Parts of a flowering plant

Give the letter which corresponds to the structure in which

1 all the nuclei have 8 chromosomes
2 some nuclei have 8 chromosomes and the remainder have 16 chromosomes
3 some nuclei contain 24 chromosomes
4 a cell is about to divide meiotically
5 the male gametophyte is contained.

Mark allocation 5/40   Time allowed 7½ minutes   *In the style of the Joint Matriculation Board*

This is a series of multiple choice questions testing knowledge of plant reproduction and the ability to apply that knowledge to a series of diagrams. The ability to interpret diagrams of stages in the reproductive cycle of a plant is also involved.

In this type of multiple choice question where a number of responses refer to the same set of choices it is important to remember that not all the choices need be used and that any choice may be used more than once. Begin by studying the drawings carefully and identifying them.

A – developing seed containing endosperm and embryo
B – pollen grain
C – early stage of ovule development
D – the embryo only, comprising an octant, suspensor and basal cell
E – mature ovule.

In **1** the structure in which all nuclei have 8 chromosomes is **B**, the pollen grain; since 16 is the diploid number, the structure required must contain only haploid nuclei. Although other structures have some haploid nuclei only **B** has them exclusively. In **2** only **E** has nuclei that are both diploid and haploid. The three nuclei at each pole are haploid whereas the central one (primary endosperm nucleus) is the result of the fusion of two of the polar nuclei and is hence diploid (16 chromosomes). In **3** the answer is **A** because the diploid primary endosperm nucleus of **E** fuses with one of the haploid nuclei derived from **B** to give a triploid cell that divides to form endosperm (the shaded area in **A**). Each endosperm cell therefore has a nucleus with 24 chromosomes. The cell about to divide meiotically is the megaspore mother cell shown in **C**. Any divisions about to occur in the cells of the other structures will be mitotic. The male gametophyte is contained in **B** (the pollen grain).

10 A pollen tube of a plant is

    A – the male gamete
    B – the embryo
    C – the male gametophyte
    D – a germinating spore.

Mark allocation 1/40   Time allowed 1½ minutes   *In the style of the Cambridge Board*

This is a multiple choice question testing understanding of the true nature of a pollen tube. In flowering plants the gametophyte stage is so reduced that stages of spore production, germination to a gametophyte and gamete production are so condensed that they are hard to recognize. Candidates are frequently confused by this situation and so find such questions difficult.

As with all multiple choice questions of this type the best procedure is to read the question or statement followed by each of the alternatives in turn noting whether they are true or false. Do this with all alternatives, even after you think you have the correct answer. Remember there is usually only one answer so never give two alternatives. Eliminate the obviously inaccurate ones and if you are left with more than one 'correct' answer choose the most likely or, if equally correct, choose one by guessing. Do not leave the question unanswered as there are not usually penalties for inaccurate responses, but always check the rubric carefully to be certain this is the case.

In this question **A** is reasonably close to the answer since the male gametes are the two male (sperm) nuclei which are contained within the pollen tube. However the pollen tube as a whole is more than just the male gametes – so mark **A** with a cross.

B is clearly incorrect since an embryo is a young organism in early stages of development. In a plant the embryo arises after fertilization and since this cannot occur until the pollen tube has released its contents, the pollen tube cannot be an embryo. Mark **B** with a cross.

C is 'the male gametophyte'. Since the tube bears the haploid male (sperm) nuclei which are the male gametes formed within the tube from the haploid generative nucleus, it follows that the pollen tube must be the male gametophyte (i.e. the haploid gamete-producing stage). Mark **C** with a tick.

D could easily mislead candidates since the pollen grain was originally a microspore and the pollen tube is derived from the pollen grain as it germinates. However it is essential to remember that the gametophyte is the part derived from the germination of a spore. Since the question makes no mention of the pollen grain, but only the tube derived from it, it must be considered a 'gametophyte' and not a 'germinating spore'. Mark **D** with a cross.

The correct answer is **C** – the male gametophyte.

11 (a) Compare wind pollinated flowers with insect pollinated flowers.   (6)
    (b) Describe briefly three mechanisms in flowers which favour inbreeding and three which favour outbreeding.   (8)
    (c) Discuss the genetical consequences of inbreeding and outbreeding.   (6)

Mark allocation 20/100   Time allowed 40 minutes   *In the style of the Joint Matriculation Board*

This is a question testing recall in (a) and (b) with some interpretation and deeper understanding required to answer (c).

The key word in (a) is 'compare' and the candidate should make clear, simple comparisons of one feature at a time. Under no circumstances should a description of wind pollinated flowers be followed by a description of insect pollinated ones. The comparisons to be made are summarized in table form in the essential information. Some explanations of each point would be needed with examples to illustrate them where possible. A table could be used but where the comparisons are to be fairly detailed (as they should be here) it is best to use it only as a last resort when time is short.

Part (b) should deal with inbreeding and outbreeding separately and should not at this stage make any reference to the significance or consequences of them since this is required in (c). Candidates should read and fully understand the whole question before answering. Too often they answer one part in such detail that there is little left to say in later sections. Remember that material must be included under the relevant section to be eligible for marks. The key word is 'brief'. The examiners are clearly warning candidates against getting carried away with long detailed accounts of pollination mechanisms in various flowers, so keep the answers to essential detail adequate to make the point. Small sketch drawings may help illustrate some of the mechanisms. Three mechanisms favouring inbreeding are:

1 Movement of stamens or stigma when the flower gets older so that the stigmatic surface touches the anthers or some part of the flower coated in pollen. The movement is often the result of uneven drying in the filaments or style that causes curling, e.g. Compositae (daisy family).

2 Cleistogamy. A number of flowers (never all) release pollen on to ripe stigmatic surfaces before the flower bud opens, thus favouring self fertilization and excluding all possibility of cross pollination, e.g. *Viola* (violet).

3 Floral structure. Some flowers are so arranged that when the corolla tube separates from the plant the anthers rub against the stigma transferring pollen, e.g. *Digitalis* (foxglove).

Three mechanisms favouring outbreeding should be selected from:

1 Dichogamy. Either the stamens ripen some time before the stigma and ovule (protandry) or vice versa (protogyny), e.g. *Geranium*.

2 Dioecious flowers. The plant has separate male and female flowers thus reducing the chances of self pollination, although not eliminating the possibility, e.g. *Zea* (maize). In some species the whole plant has flowers of one sex totally eliminating the possibility of self pollination, e.g. *Salix* (willow).

3 Floral shape. In some highly specialized flowers there are structures that operate to expose the stigma to an insect on entry, but cover it as the insect leaves, e.g. *Iris* (flag). In other types the positioning of the anthers and styles differs in plants of the same species in such a way that pollen is unlikely to be transferred between flowers of the same type, e.g. thrum-eyed and pin-eyed flowers of *Primula* (primrose).

4 Incompatibility. In some plants pollen only achieves fertilization if it lands on stigmatic tissue of a different genetical constitution, e.g. *Pyrus* (pear).

In (c) the genetic consequences of inbreeding refer to the problems associated with the development of a population of organisms with identical genetical constitutions. These include increased vulnerability to disease and its more rapid spread in the absence of resistant varieties; reduced potential for selective adaptation to changing environmental conditions and hence evolutionary change; a possible build-up of harmful recessive genes and their expression due to a greater chance of homozygous recessive offspring.

The 'consequences of outbreeding' are largely the reverse of the consequences of inbreeding, i.e. less vulnerability to disease, greater evolutionary potential and reduced chance of harmful recessive genes expressing themselves. These are all a consequence of the greater variety produced by the mixing of two genotypes.

Answer plan

(a) Expand table comparing wind and insect pollinated flowers (essential information).

(b) Mechanisms favouring:

Inbreeding: **1** Movement of stamens or stigma, e.g. Compositae.

            2 Cleistogamy, e.g. *Viola*.

            3 Floral structure, e.g. *Digitalis*.

Outbreeding: **1** Dichogamy, e.g. *Geranium*.

            2 Dioecious plants, e.g. *Salix*, and flowers, e.g. *Zea*.

            3 Floral shape, e.g. *Primula*.

            4 Incompatibility, e.g. *Pyrus*.

(c) Inbreeding reduces variation making the species more vulnerable to disease, with more mutant recessive genes and less evolutionary potential and adaptability. Outbreeding reverses these trends.

12 Make labelled drawings only to illustrate the essential organs of reproduction of either a named flowering plant or a named mammal. In the example you choose briefly describe the events which lead up to (a) the production and release of gametes, (b) fertilization, (c) development, nutrition and protection of the embryo.

Mark allocation 28/140   Time allowed 35 minutes   *Oxford Local June 1980, Paper II, No. 10*

The full answer plan to this question is given in Section 5.2 Reproduction in mammals.

5.4   GROWTH AND DEVELOPMENT

Assumed previous knowledge Mitosis (see Section 4.1)

Plant and animal tissues (see Section 2.2)

Underlying principles

There is a limit to the size of any individual cell, partly due to the distance over which a nucleus can exert its controlling influence. Therefore single celled organisms will absorb and assimilate materials until they reach a certain size, when they divide to give two separate individuals. The multicellular condition allows organisms to attain greater size. This brings its own problems but these are largely outweighed by such advantages as:

1 increased scope for differentiation, specialization of function and hence greater efficiency.
2 the ability to separate more easily processes requiring different conditions, e.g. different pH values.
3 increased ability to store materials.
4 the ability to replace damaged cells from those remaining.
5 greater competitive advantage, e.g. large plants have better access to light than small ones.
6 some security from predators that find a large organism more difficult to capture and ingest.

All multicellular organisms originate as a single cell, the zygote, and undergo three phases of development:

1 Growth: an irreversible increase in mass.
2 Differentiation: the development of differing cell structures and functions.
3 Morphogenesis: the development of the overall form of organs and hence the organism.

In addition to the importance of growth and development in the early stages of an organism's life, these events continue to manifest themselves later in such processes as regeneration, repair and gametogenesis.

Points of perspective

Knowledge of a wide range of methods for measuring growth would be helpful. It would be a considerable advantage to have carried out basic growth measurement experiments on a yeast colony, a stem or a root and a small mammal. In addition it would be useful to have carried out experiments to determine the effects of external factors such as temperature on the rate of growth.

It is necessary on some syllabuses to study the embryology of chosen examples by microscopic examination. Even where the syllabus does not require it, the candidate's understanding of the process would be improved by such a study.

Essential information

Growth Growth is an irreversible increase in size during development. It usually occurs in three distinct phases: cell division, cell assimilation and cell expansion.

Measurement of growth Using a single parameter for the measurement of growth may not take into account growth in all directions, e.g. measuring an increase in length does not take into account changes in girth. Changes in volume are difficult to measure in irregular organisms and changes in fresh weight may be complicated by temporary changes, e.g. drinking water. Dry weight gives a less misleading picture but, since the measurement of this involves killing the organism, numerous similar organisms are required. A further complication is that an overall measurement of growth does not take into account the fact that different organs may have their own peculiar growth rates (allometric growth).

Growth curves Fig. 5.11 represents a variety of growth curves in plants and animals.

Growth rate A graph of growth rate is drawn by plotting growth increments (i.e. increase in growth over successive periods of time) against time.

The rate of growth in plants and animals is affected by:

1 **Genotype:** dominant and recessive alleles determine such things as protein synthesis, metabolism and size.
2 **Hormones:** auxins and gibberellins in plants and thyroxine and somatotrophin in animals.
3 **Nutrition:** plants require carbon dioxide, light and water; animals need proteins, fats, carbohydrates, vitamins, mineral ions and water.
4 **Environment:** e.g. light for vitamin D synthesis in some animals and for photosynthesis in plants; effect of day-length on meristematic activity; temperature has different effects on ectothermic and endothermic animals; thermoperiodicity in plants influences flowering, etc.

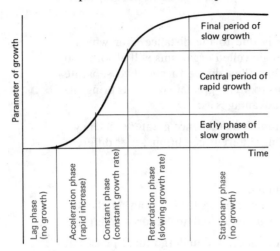

This sigmoid (S-shaped) curve can be applied to a population,
an individual or an organ of an individual,
although the pattern may be modified.

Fig. 5.11 (a) General growth curve

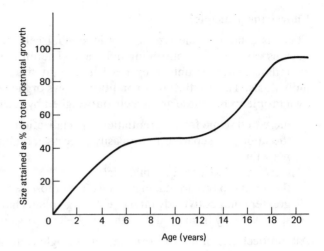

This resembles two flattened sigmoid curves on top of each other.
The first represents the early growth phase in a child,
the second the later growth phase in adolescence.

Fig. 5.11 (b) Human growth curve

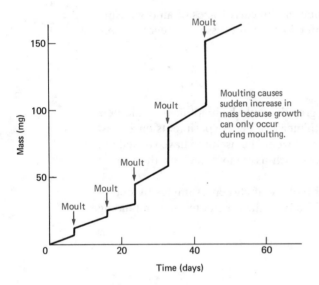

Moulting causes sudden increase in mass because growth can only occur during moulting.

This is known as intermittent growth.

Fig. 5.11 (c) Growth curve of an arthropod,
e.g. *Notonecta glauca*

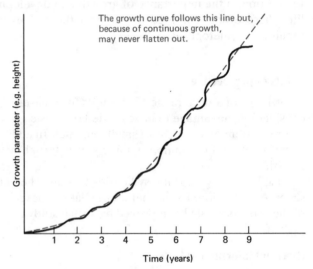

The growth curve follows this line but, because of continuous growth, may never flatten out.

This is known as intermittent growth.

Fig. 5.11 (d) Growth curve of a perennial plant
in temperate regions

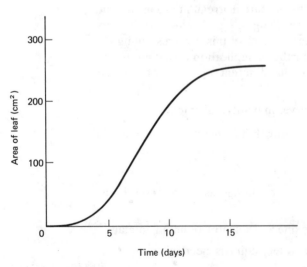

In plants growth of an organ often reflects
the growth of the whole organism.

Fig. 5.11 (e) Growth in area of a cucumber leaf

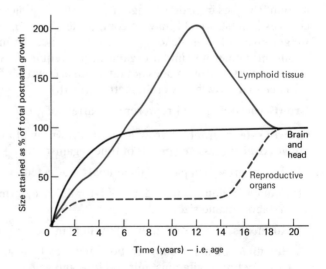

In animals growth of organs is often allometric.

Fig. 5.11 (f) Growth curves of some human organs

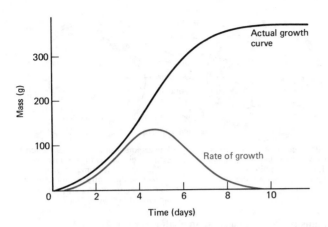

The growth rate curve of most organisms is bell-shaped.

Fig. 5.12 (a) Relationship between actual growth and rate of growth

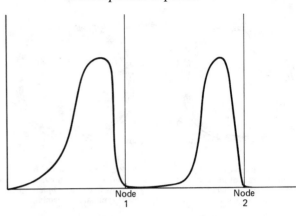

In *Tradescantia* maximum growth occurs at the base of each internode; there is no growth at the nodes.

Fig. 5.12 (b) Growth rate of *Tradescantia*

Growth may be limited by such negative factors as disease, pollution and parasites. These are the most important influences on growth but some organisms may be affected by stress, atmospheric pressure and gravity.

Development in plants In the early stages of development, cell division occurs throughout the embryo, but as it develops into an independent plant the addition of new cells is restricted to certain parts – **meristems**. The presence of meristems, whose activity permits growth throughout the life of the organism, distinguishes plants from animals.

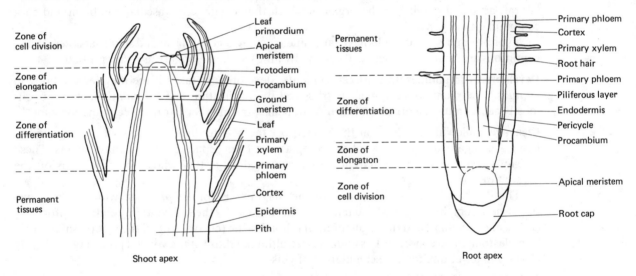

Mitosis of the cells of the apical meristem forms cells which are pushed farther away from the tip as more form. As they move back they elongate and differentiate.

Fig. 5.13 (a) Primary (apical) meristem in a shoot

Fig. 5.13 (b) Primary (apical) meristem in a root

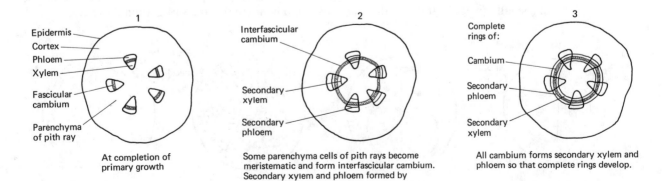

At completion of primary growth

Some parenchyma cells of pith rays become meristematic and form interfascicular cambium. Secondary xylem and phloem formed by fascicular cambium.

All cambium forms secondary xylem and phloem so that complete rings develop.

Fig. 5.13 (c) Secondary (lateral) meristems producing increase in girth in a shoot

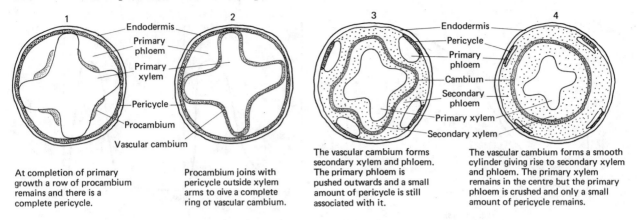

Fig. 5.13 (d) Secondary (lateral) meristems producing increase in girth in a root

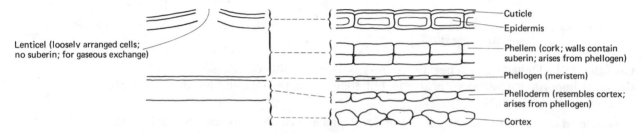

Fig. 5.13 (e) Lateral meristems in cork formations

Meristems may be apical, i.e. located at the tips of main and lateral roots and shoots, or lateral, arranged parallel to the organ in which they occur, e.g. vascular cambium and cork cambium (phellogen).

Typically meristematic cells have large nuclei, compact cytoplasm and small vacuoles. When conditions are favourable these cells divide by mitosis, then elongate and differentiate.

Development in animals In contrast to plants, development takes place all over the body in animals. Embryonic development is triggered by fertilization and the multiplication of cells mostly ceases after the organism reaches adult size; the number of organs remains constant.

Cleavage Following fertilization the nucleus of the zygote divides mitotically, each division being accompanied by cleavage of the cytoplasm to form separate cells, or blastomeres. These divisions give rise to an embryonic structure called a blastula whose form depends on the amount of yolk in the egg.

The description given below and in the diagrams is for *Amphioxus*, which is later distinguished from the frog, fowl and mammal. In *Amphioxus* the egg contains relatively little yolk (i.e. is microlecithal) and the segmentation is holoblastic (cleavage of all the cytoplasm occurs). The blastomeres are more or less equal in size although those at the vegetal pole may be slightly larger and fewer due to the accumulation of yolk.

Frog The egg is larger and contains more yolk. Cleavage is holoblastic but more unequal. The cells at the animal pole are smaller and more numerous than the yolky ones at the vegetal pole. The blastocoel is confined to the upper region of the blastula.

Fowl The very large amount of yolk (macrolecithal) permits only partial cleavage (i.e. it is meroblastic), which is confined to a small region at the animal pole and results in the formation of a small cap of blastoderm, the blastodisc, and beneath this the sub-germinal cavity.

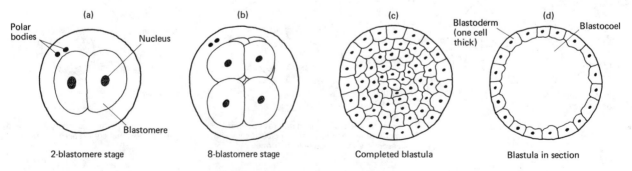

Fig. 5.14 Cleavage in *Amphioxus*

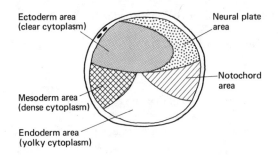

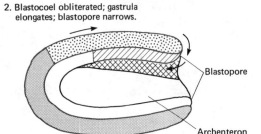

Fig. 5.15 (a) Presumptive areas of blastula in *Amphioxus*

Fig. 5.15 (c) Elongation of the gastrula

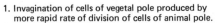

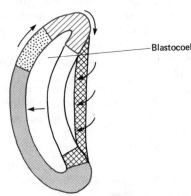

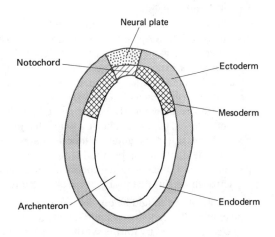

Fig. 5.15 (b) Movement of presumptive areas during gastrulation in *Amphioxus*

Fig. 5.15 (d) TS gastrula

Mammal The egg is minute and contains very little yolk (microlecithal). Cleavage is holoblastic but results in a solid ball, or morula, and not in a hollow blastula.

Gastrulation Gastrulation follows cleavage and results in the formation of a hollow gastrula comprising a wall of two distinct cell layers and a cavity, the archenteron. The blastoderm of chordates consists of definite regions (presumptive areas) destined to give rise to the ectoderm, endoderm, mesoderm, neural plate and notochord. During gastrulation the cells of these presumptive areas are moved or migrate to their correct positions in the gastrula. Gastrulation is modified by the amount of yolk present in the egg.

Frog The thickness of the blastoderm at the vegetal pole prevents the formation of a gastrula by simple invagination. The rapidly dividing cells of the animal pole migrate over the surface of the yolky cells at the vegetal pole, a process known as overgrowth or epiboly. At the same time, along a small crescentic area on the dorsal side, cells start to migrate beneath the outer layers. This groove is the beginning of the archenteron.

Fowl A primitive streak and groove appear on the blastodisc and in this region migration of cells occurs during gastrulation. The sub-germinal cavity becomes the archenteron.

Mammal The morula divides to form a hollow ball of cells whose wall is one cell thick and whose cavity encloses a mass of cells later known as the germinal disc. The process of gastrulation is similar to that of the fowl.

Link topics

Section 4.1 Mitosis and meiosis
Section 5.5 Control of growth

Suggested further reading

Black, M. and Edelman, J., *Plant Growth,* Part 1 (Heinemann 1970)
Clowes, F. A. L., *Morphogenesis of the Shoot Apex,* Oxford/Carolina Biology Reader No. 23 (Packard 1972)
Freeman, W. H., and Bracegirdle, B., *An Atlas of Embryology,* 3rd ed. (Heinemann 1978)
Hunt, R., *Plant Growth Analysis*, Studies in Biology No. 96 (Arnold 1978)

Northcote, D. H., *Differentiation in Higher Plants*, Oxford/Carolina Biology Reader No. 44, 2nd ed. (Packard 1980)
Sutcliffe, J., *Plants and Temperature*, Studies in Biology No. 86 (Arnold 1977)

QUESTION ANALYSIS

13 (a) Some factors affect growth in both mammals and flowering plants. Give an account of these factors.   (10)

(b) Describe with full practical detail how you would measure the rate of growth in (i) a yeast culture, (ii) the main root of a seedling.   (10)

Mark allocation 20/100   Time allowed 35 minutes   London Board June 1981, Paper II, No. 4

This is an essay question testing the ability to select relevant information from a list of learned facts and to recall experiments performed in the laboratory, adapt them if necessary or devise them using basic learned principles.

The key word in (a) is 'both'. Candidates should restrict themselves to those factors affecting mammals and flowering plants. It is important to cover all relevant factors and not restrict the answer to one group, e.g. climatic factors. The candidate should remember that growth is affected as much by internal as external factors and begin by looking at each in turn. Internal factors include the organism's genotype, i.e. dominant and recessive alleles that determine size, metabolism and the production of protein within individual cells. Another internal factor is hormones, e.g. thyroxine and somatotrophin in mammals; auxins and gibberellins in plants.

The external factors can be neatly divided into nutritional (fats, carbohydrates, proteins, vitamins, minerals and water in animals; minerals, water, light and carbon dioxide in plants) and environmental (light and temperature in plants and animals). A discerning candidate may also include such factors as parasites, disease and external chemicals (e.g. pollutants).

Throughout this first part there are three important considerations:

1 All statements should be qualified, i.e. it is not sufficient to say 'a source of nitrogen affects growth in plants'. The candidate should specify its importance and say 'a source of nitrogen, e.g. nitrate from the soil, is necessary for the synthesis of protein in plants, and so affects growth'.

2 Plants and animals should both be mentioned in respect to each chosen factor, e.g. 'light is necessary in plants for photosynthesis in order to build up carbohydrates from which the fats and proteins needed in growth are synthesized and in some mammals, e.g. humans, it is needed for the synthesis of vitamin D in the skin, necessary for normal growth of bones'.

3 Choose a few examples from the five categories genotype, hormones, nutrition, environment and disease/chemicals thereby ensuring a good range. One example from each will bring better marks than five examples from one category.

In (b) the key phrases are 'full practical details' and 'measure the rate of growth'. Remember that 'rate' means the increase in growth per unit time and it is therefore essential to state the units in which the growth is measured.

For yeast the best method is to count the increase in number of yeast cells on a daily basis using a graduated slide or counting chamber (haemocytometer). Points to be described in detail include the vessel and nutrients used to grow the yeast, how sterile conditions were maintained, the constant temperature used and how it was maintained, the size of sample taken (shake the flask first) and the calculations involved in producing the final units, e.g. number of yeast cells/cm^3 of culture solution, repetition of counts to give statistically accurate results and the presentation of the final results, e.g. graph of daily increase in yeast cells (log no./cm^3) against time (days).

If time allows, the candidate could mention problems such as whether to count budding yeast cells as one or many and how to make allowance for cells lying on the lines demarcating the counting area.

For the seedling the area to be measured should be clearly defined, i.e. length of whole of main root. Again it is important to include the conditions for growth (temperature used, how humidity is maintained, darkness), the time over which the experiment was run (1 week), intervals at which measurements were taken (daily), repetition for statistical accuracy and the units used and how presented, e.g. graph of daily increase in length (mm) against time (days).

Answer plan

(a) **Factors** (give details throughout)
Internal
Genotype: dominant/recessive alleles; protein synthesis in cells
Hormones: thyroxine/somatotrophin; auxins/gibberellins
External
Nutrition: carbon source, nitrogen source, minerals, vitamins, water
Environmental: light, temperature
Negative factors: parasites, disease, chemicals.

(b) **Yeast**

> Culture vessel and medium (source of carbon, nitrogen, phosphorus and sulphur)
> Sterile conditions (cotton wool plug)
> Volume used/temperature used (16°C)
> Counting chamber (haemocytometer) details
> Interval (daily for one week)
> Graph: daily increase in cells (log no./cm^3) against time (days).
>
> **Seedling**
> Region defined
> Conditions (dark, temperature, humidity)
> Interval (daily for one week)
> Graph: daily increase in length (mm) against time (days).

14 (a) (i) For a named chordate describe, using labelled diagrams, the main sequence of events in development from fertilization to the establishment of the primary germ layers of the embryo. (11)
(ii) Discuss the effect of the amount of yolk on the processes of cleavage and gastrulation. (5)
(b) Which of the germ layers in the chordate embryo gives rise to the following structures: (i) skin epidermis, (ii) epithelium of the intestine, (iii) muscles, (iv) brain? (4)
Mark allocation 20/80   Time allowed 35 minutes   *In the style of the Cambridge Board*

This is a structured essay testing recall of factual information. In (a) (i) remember to name the chordate; *Amphioxus* is the example used in the essential information since it is generally less complex than other chordates but the candidate should choose the one with which he is most familiar. Candidates often separate practice and theory, choosing not to include practical drawings in theoretical examinations. This is an example where relevant practical drawings could easily be included. The answer should basically consist of Figures 5.14a–d and 5.15a–d with some written account explaining the gradual changes to link each of the drawings.

In (a) (ii) the key word is 'discuss' which implies that more than a simple statement such as 'the presence of yolk inhibits cleavage' is required. The candidate should therefore mention its significance in later development. Where yolk is present in small quantities or absent altogether (e.g. *Amphioxus*) cleavage involves the whole egg and the resulting blastula has cells of about equal size. Where larger quantities of yolk are present (e.g. frog) the whole egg divides, but unevenly with fewer divisions at the yolky (vegetal) pole than at the non-yolky (animal) pole. This results in an egg with a few large cells (vegetal pole) and more, smaller cells (animal pole). In the presence of still larger quantities of yolk (e.g. chick) cleavage is confined to a small yolk-free region.

Another consequence of the difference in yolk is that where little is present the subsequent cell movements are simpler resulting in a circular blastopore. Where more yolk is present the blastopore is crescent-shaped and where a great deal is present the blastula takes on the shape of a disc (blastodisc) making invagination as such difficult. In these types cells move through the upper layer forming a second layer (primitive streak) which is homologous to the blastocoel.

The archenteron of a non-yolky egg fills the centre of the blastula. Confined to the animal pole of the blastula where there is a moderate amount of yolk, it forms a tiny cavity under the blastodisc where there is much yolk.

The answers to (b) are straightforward recall of learned facts. The answers are given in the plan below.

Answer plan

(a) (i) Name of chordate.
Diagrams 5.14a–d and 5.15a–d with linking explanations.
(ii) Yolk inhibits cleavage with the following consequences:

	Little/no yolk	*Moderate yolk*	*Much yolk*
Example	*Amphioxus*	Frog	Chick
Cell size	All equal	Large and small	About equal, on huge ball of yolk
Cell movements	Simple	More complex	Very complex
Blastopore	Circular	Crescentic	None; primitive streak instead
Archenteron in blastula	Fills centre	Towards animal pole	Tiny cavity under blastodisc

(b) (i) skin epidermis – ectoderm
(ii) epithelium of intestine – endoderm
(iii) muscles – mesoderm
(iv) brain – ectoderm.

15 The following results, expressed graphically, relate to a yeast population grown in a suitable medium. One of the vertical axes relates to the growth curve and the other to the growth rate curve.

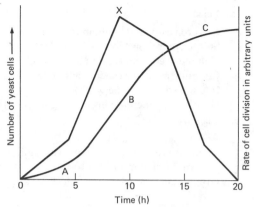

Fig. 5.16 Rate of cell division in yeast

(a) Explain the difference between a 'growth curve' and a 'growth rate curve'.

(b) Explain the shape of the growth curve in regions A, B and C.

(c) Suggest two factors which may cause a reduction in the growth rate after point X.

Mark allocation 7/100   Time allowed 8 minutes   Assoc. Examining Board Nov. 1980, Paper I, No. 13

This is a structured question testing the ability to interpret graphs and the candidate's understanding of the difference between 'growth' and 'rate of growth'.

Question (a) often proves difficult because candidates are not clear exactly what is meant by 'growth rate'. A brief explanation is given in the essential information for this section and Figure 5.12a illustrates the relationship. Basically the 'growth curve' measures the actual size of the yeast colony at fixed times whereas the 'growth rate curve' measures the increase in yeast cells during successive time intervals. For example, if the actual yeast cell numbers at 5-hourly intervals during the experiment were 10, 30, 70, 90, 100, they could be plotted to produce a curve similar to curve ABC in the question. If the increase in yeast cells during each 5-hourly interval is calculated, the following data are obtained

Hours	0	5	10	15	20	25
Number of cells	10	30	70	90	100	100
Increase in cells at 5-hourly intervals		20	40	20	10	0

If this last set of figures is plotted a 'growth rate curve', similar to curve X in the question, is obtained.

Growth curve = amount of growth time^{-1}

Growth rate curve = amount of growth time^{-2}

In answering (b) the candidate should be conversant with Fig. 5.11a which specifies the main phases of growth. In the question A refers to the acceleration phase where there is a rapid increase in cell numbers – probably a geometrical increase (i.e. 2, 4, 8, 16, 32, etc.). B refers to the constant phase where growth rate is constant (i.e. 32, 64, 96, 128, etc.). C refers to the retardation phase where growth is slowing (i.e. 128, 158, 168, 172, 173, etc.).

In (c) the factors reducing the growth rate could be a reduction in available food (e.g. sucrose) or some essential mineral (e.g. nitrogen or phosphate) because they have been used up in the earlier stages of growth. It could also be due to a build-up of waste products (e.g. alcohol) inhibiting further cell division. In the absence of other information it is better not to suggest a fall in temperature or similar external change but assume that these factors remain constant during the experiment.

16 Discuss the relative importance of hereditary and environmental factors in the growth and development of organisms.

Mark allocation 20/100   Time allowed 40 minutes   London June 1983, Paper 2, No. 8

This essay question tests the ability to apply knowledge of genetics, evolution and ecology to growth and development. It is a wide ranging topic requiring broad biological knowledge.

A good starting point is to list hereditary and environmental factors and then discuss the effect of growth and development. The most important hereditary factor is the gene and the factors it determines in the organism. The other hereditary factors lead on from this, for instance transcription of the genes' information, protein (enzyme) synthesis and the role of these enzymes in metabolism.

Discussion of the effects of these factors on growth and development should include frequency of any gene in the gene pool and whether it is dominant or recessive. This will determine the likelihood of the gene expressing itself. A mention of mutant genes and their possible effect on growth and development would be worthwhile. The expression of a gene, i.e. whether it is 'switched on' or not, should be included and the role of repressor and releaser substances in this expression. Finally the candidate should consider the long term evolutionary effects of heredity on growth and development, e.g. selection pressures altering gene frequency. Such a discussion could include the effects of changing environmental conditions on natural selection. This would give a reasonable lead into the second part of the question.

Environmental factors are so numerous that to list them all could be time-consuming and not particularly helpful. It would be better to consider those processes which affect growth and development in plants and animals and consider how the environment affects each process.

In the plant for instance a good starting point would be the germination of seeds. This process is affected by such environmental factors as water and oxygen availability and temperature. Remember to be **specific**, however, giving examples if possible of actual seeds and the precise conditions needed for germination. After germination it would be necessary to list the environmental factors influencing growth, photosynthesis, water and mineral uptake and flowering.

In animals growth and development would be affected by the availability of food, space and mates. The environmental factors affecting each of these processes and how each affects growth and development should be discussed in turn.

Answer plan

Heredity factor	**Effect on growth**
Expression of gene	Presence of substrate in a medium will 'switch off' gene controlling production of that substrate e.g. Tryptophane in *E. coli*. Repressors and releasers in general.
Protein (enzyme) synthesis	Concentration of substrate/product controls rate of protein synthesis.
Selection of genes	Conditions determine gene frequency through natural selection. Changing environment favours one gene or another. Stable environment allows expression of a variety of genes.

Environmental factor

Plants	**Effect on growth and development**
Temperature	Suitable temperature needed for photosynthesis, germination, growth, flower formation, uptake of water and minerals.
Light	Needed for photosynthesis, phototropic responses, germination in some species. Length of light period affects time of flowering.
Oxygen	Needed for germination and respiration.
Water	Necessary for photosynthesis, germination, uptake and transport of sugars and minerals.
Minerals	Magnesium in chlorophyll for photosynthesis; nitrogen in proteins.

Animals	
Food	Depends ultimately on the producer plant and hence the factors affecting its growth. Also factors affecting the food source directly, e.g. temperature, water supply, disease, parasites etc.
Space	Needed to provide food, territory etc. Space available depends on climatic conditions, vegetation, topography, altitude etc.
Mates	Light/temperature may affect some sexual cycles hence receptivity of mate. Food supply, temperature, water, disease etc affect population size and therefore availability of mate.

17 Describe the formation of vascular tissues (xylem and phloem) in a herbaceous dicotyledonous stem. Your answer should pay particular attention to:

(a) the location and structure of meristematic cells,

(b) the changes which take place in meristematic cells leading to the formation of vascular tissues,

(c) the structure of each type of vascular tissue,   (d) the position of the vascular tissues.

Mark allocation 17/125   Time allowed 20 minutes   *Scottish Higher II 1980, No. 12B*

This is a structured essay question testing knowledge of the origin and structure of vascular tissues in dicotyledonous stems. It makes straightforward demands under three headings. Candidates should appreciate that the subheadings in the question are unlikely to carry equal weighting. With the emphasis in the first sentence upon 'formation', parts (a) and (b) are likely to carry more marks than are (c) and (d). The descriptions may be in prose although the use of well-annotated drawings is to be recommended as these are often a clearer and sometimes more concise way of expressing information.

The location and structure of meristematic cells is dealt with in the essential information and careful reference should be made to 'development in plants' and the stem drawings in Figure 5.13.

The main changes referred to in (b) for xylem are elongation of the meristematic cell, production of a vacuole and death of the cell resulting in loss of the cytoplasm and the end walls. The cells then form continuous columns, the side walls becoming impregnated with lignin. In phloem the early stages are similar but the end walls form sieve plates, the cytoplasm is retained and companion cells develop. For details of the structure of each type of vascular tissue required in (c) refer to Section 2.2 especially Figure 2.9. The position of the vascular tissues called for in (d) is given in Figure 8.7b in Section 8.3.

Answer plan

(a) Meristematic cells form actively dividing tissue which gives rise to the new cells required for growth. The apical meristem is located at the stem tip and other dividing cells form a broken cylinder called the cambium which is situated inside the stem (Figure 5.13).

The cells themselves are small and regularly shaped, having dense cytoplasm and thin walls.

(b) To form xylem, newly formed cells elongate and vacuolate. They lose their cytoplasm, their end walls break down and lignin is deposited in the side walls.

To form phloem, sieve plates are formed at end walls and cytoplasmic contents are retained. Companion cells are formed alongside the sieve tubes.

(c) Vessels, tracheids, sieve tubes and companion cells should be described in some detail, including use of diagrams (see Figure 2.9, Section 2.2).

(d) The arrangement of the vascular tissues into bundles should be indicated briefly.

5.5 CONTROL OF GROWTH

Assumed previous knowledge Growth and development (Section 5.4).

Underlying principles

The size of an organism and its parts is broadly determined genetically, but changing environmental circumstances and the seasonal use of certain organs (e.g. reproductive ones) require variation in growth within these genetic limits. In addition plants have no contractile tissue to allow them to move, and yet their survival may depend on movement to or from stimuli such as light and water. Such movements can only occur as part of the growth process. It is clear that this variation in growth must be carefully controlled if it is to be of value and the absence of such control may be detrimental (e.g. cancers). The control is largely exerted by hormones in animals because their effects are more general than the specific responses of the other major controlling system, the nerves. In addition growth is a relatively slow process often involving permanent change. This is more suited to hormonal control than the rapid and temporary

Scientist	Date	Experiment	Result	Inference
Darwin	1880	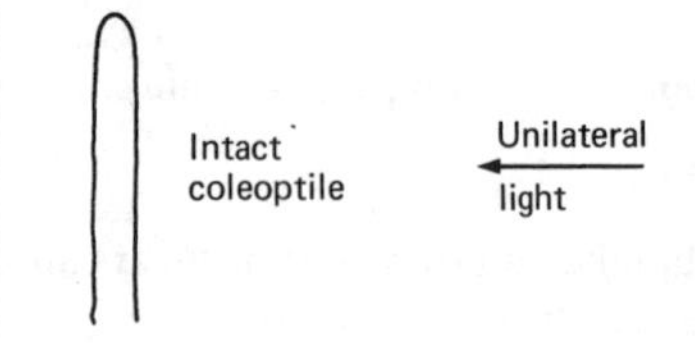 Intact coleoptile — Unilateral light	 Bends towards light	Coleoptile is positively phototropic. Bending occurs behind tip
		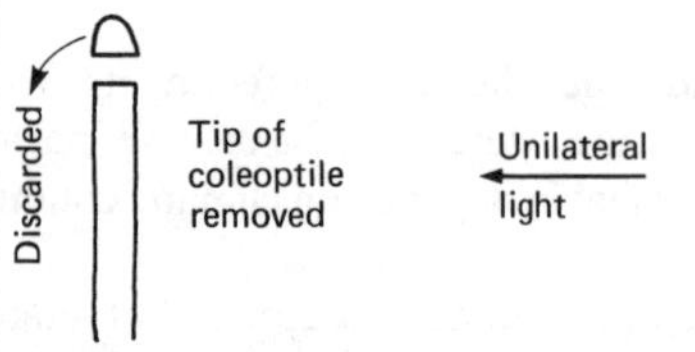 Tip of coleoptile removed — Unilateral light (Discarded)	 No response	Tip is needed to cause response; it either perceives the stimulus and/or produces the 'message'
		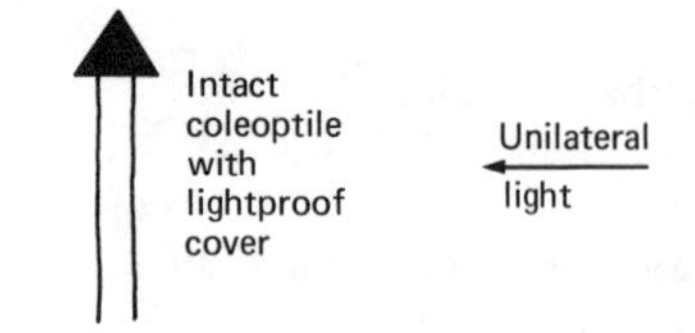 Intact coleoptile with lightproof cover — Unilateral light	 No response	The tip perceives the light stimulus
		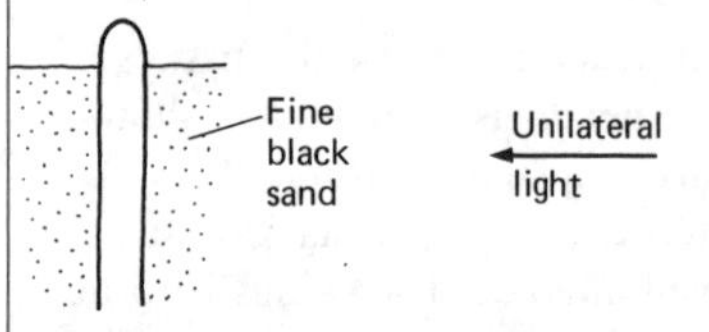 Fine black sand — Unilateral light	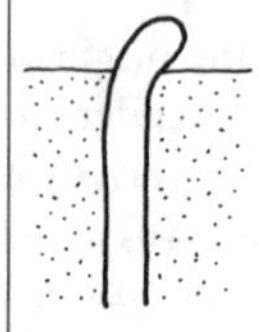 Bends towards light	Confirms that the tip perceives the stimulus
Boysen-Jensen	1913	 Mica — Unilateral light	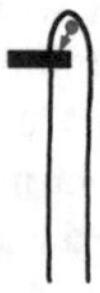 No response	It is not an electrical or nervous 'message' from the tip that induces elongation behind the tip since mica allows such 'messages' to pass. The 'message' must be a chemical passing from the tip down the shaded side. Mica prevents passage; therefore no response
		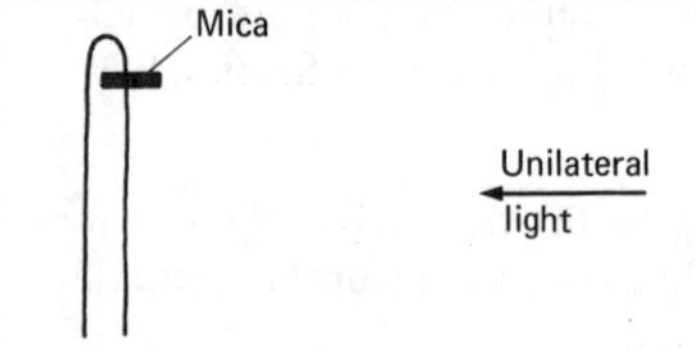 Mica — Unilateral light	 Bends towards light	When mica is on light side passage is unaffected and coleoptile bends. Reinforces the view that chemical causing cell elongation passes from tip down shaded side
		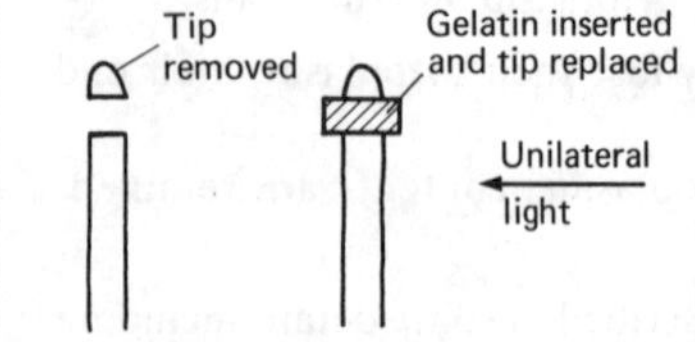 Tip removed / Gelatin inserted and tip replaced — Unilateral light	 Bends towards light	The gelatin allows chemicals to pass through it but not nervous or electrical impulses. Bending occurs; therefore the growth stimulus must be chemical and must accumulate on the shaded side

Fig. 5.17 Historical review of plant hormone experiments

response brought about by nerves. In the absence of a nervous system, the control of growth in plants must be hormonal.

Points of perspective

It would be an advantage for a candidate to have carried out a number of experiments involving tropic and tactic responses and to have observed the effects of plant hormones on growth.

Auxin concentrations can be measured by collecting the hormone in a gelatin block then placing it on one side of a decapitated coleoptile. The degree of curvature it causes under fixed conditions is a measure of the auxin concentration. This is an example of the important technique of bioassay, and an understanding of the principles underlying the process would be of value.

A range of practical applications for plant growth hormones is important, e.g. hormone weed killers and defoliants; artificial ripening of fruit by ethene; fruit development in the absence of pollination by auxins.

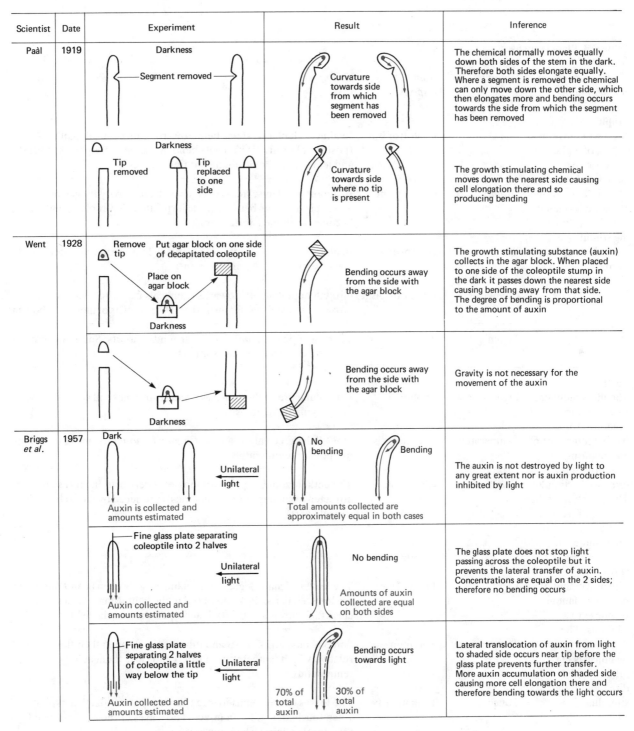

Fig. 5.17 Historical review of plant hormone experiments

Likewise a knowledge of the medical application of animal hormones and the effects of over-production and under-production would be helpful. The use of growth hormones in factory farming could also be considered. Finally candidates may consider the consequences of uncontrolled growth, in particular the production of tumours (cancers) and the problems associated with their cure.

Essential information

Control of growth in plants Plant growth substances, or hormones, are produced in very small quantities in one part of the plant and transported to another part where they promote, inhibit or in some way modify growth.

There are five main classes of plant growth hormones, which also affect other activities of the plant:

1 Auxins   4 Inhibitors
2 Gibberellins   5 Ethene
3 Cytokinins

Table 5.3 Plant responses

Type of movement and definition	Stimulus	Name of response	Examples
Tropic A growth movement of part of a plant in response to a directional stimulus. The direction of the response is related to the direction of the stimulus, e.g. towards it, away from it	Light	Phototropism	In almost all plants, stems bend towards a directional light source (positive phototropism), roots bend away (negative phototropism) and leaves become positioned at right angles
	Gravity	Geotropism	In almost all plants, stems bend away from gravity (negative geotropism), roots bend towards it (positive geotropism) and leaves become positioned at right angles
	Water	Hydrotropism	In almost all plants, roots are positively hydrotropic and stems and leaves show no directional response
	Chemicals	Chemotropism	Growth of pollen tube towards chemicals from the micropyle. Growth of fungal hyphae away from the products of their metabolism
	Touch	Thigmotropism	Twining of pea (*Pisum*) tendrils around supports. Spiralling of bean (*Phaseolus*) shoots around supports
Tactic The movement of a freely motile organism (or a freely motile part of an organism) in response to a directional stimulus. The direction of the response is related to the direction of the stimulus	Light	Phototaxis	Unicellular green algae such as *Chlamydomonas* will move to regions of optimum light intensity
	Temperature	Thermotaxis	Unicellular green algae such as *Chlamydomonas* will move to regions of optimum temperature
	Chemicals	Chemotaxis	The antherozooids (sperm) of mosses, liverworts and ferns are attracted to chemicals (e.g. malic acid) produced by the archegonium (female part)
Nastic The movement of part of a plant in response to a stimulus. The direction of the response is **not** related to the direction of the stimulus	Light	Photonasty	The leaves of many leguminous plants (e.g. French bean *Phaseolus*) are lowered in the dark and raised in the light. Many daisies (*Oxalis*) close their flowers in the dark and open them in the light
	Temperature	Thermonasty	Some plants, e.g. *Crocus* and tulip (*Tulipa*), open their flowers at relatively high temperatures (16°C) and close them at lower temperatures
	Touch	Thigmonasty	The leaves of the Venus flytrap (*Dionaea*) close together rapidly when touched, e.g. by an insect. The leaves of the sensitive plant (*Mimosa*) collapse when touched

Table 5.4 Plant hormones

Hormones	Effects		Examples
Auxins Indolyl acetic acid	Cause cell elongation	Phototropism	Oat coleoptiles bend towards light
		Geotropism	Roots grow downwards into the soil
	Cause cell division	Stimulate cambial activity	Development of wound tissue (calluses)
		Initiate root development	Root powders initiate root growth from cuttings
Synthetic auxins include 2-4-di-chlorophenoxy-acetic acid (2-4-D) and 2-4-5-trichloro-phenoxyacetic acid (2-4-5-T)		Stimulate fruit growth and parthenocarpic fruit development	Some crops, e.g. apples, are sprayed with synthetic auxins to cause fruit development without fertilization
	Maintain the structure of cell walls	Inhibit leaf abscission	Leaves do not fall when auxin from leaf exceeds that from stem
		Inhibit fruit abscission	Fruits do not fall when auxin from fruit exceeds that from stem
	Inhibit growth in high concentrations	Inhibit development of lateral buds	The dominance of apical buds is due to the auxin they produce inhibiting lateral ones; removal of apical buds therefore leads to branching
		Kill plants by disrupting growth	Synthetic auxins are used as selective weedkillers
Gibberellins Related to gibberellic acid which is a metabolic by-product of the fungus *Gibberella fujikuroi.* There are a number of different gibberellins. They affect cell elongation in stems, may increase the leaf area of some plants but have no effect on roots	Promote cell elongation		Cause elongation of plant stems
	Reverse some types of genetic dwarfism		Dwarf varieties of many plants, e.g. peas, *Chrysanthemum,* can be made to grow to normal size when gibberellins are applied
	Promote germination of seeds		Gibberellins promote germination of many seeds, e.g. oats (*Avena*)
	End dormancy of buds		The natural dormancy of many buds, e.g. birch, is broken when gibberellins are applied
	Affect leaf expansion and shape		*Eucalyptus* leaves are transformed from juvenile to mature shape when gibberellins are present. Reverse occurs in ivy
	Aid setting of fruit after fertilization		Some species of cherry, apricot and peach (*Prunus*) readily set fruit after treatment with gibberellins
	Remove the need for cold treatment in vernalization		Carrots (*Daucus*) normally flower only after a period of exposure to cold. They can be made to flower without this by application of gibberellins
	Affect flowering		The application of gibberellins to henbane (*Hyoscyamus*) kept in short day conditions will induce flowering
Cytokinins These are derived from the purine, adenine. Interact with auxins to promote cell division in cultures. Example is kinetin	Increase rate of cell division		Aid the growth of many plants, e.g. sunflower (*Helianthus*)
	Stimulate bud development		Cause buds to develop, e.g. on leaf cuttings of African violet (*Saintpaulia*)
	Increase rate of cell elongation in leaves		The size of fronds of duckweed (*Lemna*) increases in the dark when cytokinins are added

Table 5.4 Plant hormones (cont.)

Hormones	Effects	Examples
Inhibitors These are a group of substances that inhibit growth. Example is abscisic acid	Retard growth	Inhibit the growth of many plant parts, e.g. hypocotyls, radicles and leaves
	Induce dormancy in buds	The growing apex of some plants, e.g. birch (*Betula*) can be transformed into a dormant bud by addition of abscisic acid
	Inhibit germination	Some seeds have their germination inhibited by abscisic acid, e.g. rose (*Rosa*)
Ethene Ethene is a product of plant metabolism	Involved in many auxin-induced responses	Ethene production is frequently stimulated by auxin
	Causes leaf senescence and abscission	The leaves of some species, e.g. *Euonymus japonica*, die earlier and drop from the plant when treated with ethene
	Ripens fruits	Many fruits, e.g. oranges and lemons, ripen much more rapidly in the presence of ethene

Phytochromes

A number of plant growth reponses are influenced differently by light of different wavelengths. For light to have an effect it must be absorbed by a photoreceptor substance. In about 1960 the pigment phytochrome was isolated; this exists in two interconvertible forms: one which absorbs red light with a peak at about 660 nm (P_{660}) and the other which absorbs far-red light with a peak at about 730 nm (P_{730}).

$$P_{660} \xrightarrow{\text{red light (daylight)}} P_{730}$$

far-red light = rapid conversion
dark = slow conversion

Phytochrome is present in the leaves. After absorbing the appropriate wavelength of light, it causes the conversion of a hormone precursor to a hormone which then affects growth.

Photoperiodism

Flowering is regulated by daylength. Three basic groups of plants exist although all intermediates between them may be found:

1 Long-day plants, e.g. clover, barley, radish, petunia. These only flower when the light period in a 24-h cycle exceeds about 10 h. In temperate regions these plants flower in the summer.

2 Short-day plants, e.g. tobacco, cocklebur, poinsettia, chrysanthemum. These only flower when the light period is shorter than about 14 h. In temperate regions they generally flower in the spring or autumn.

3 Day-neutral plants, e.g. carrot, violet, begonia, cucumber. These are indifferent to daylength.

Long-day plants are thought to flower when the presence of red light, or a long period of sunlight, causes a sufficient accumulation of P_{730} and thus a low level of P_{660}.

Short-day plants flower when far-red light, or a long period of darkness, causes a sufficient accumulation of P_{660} and thus a low level of P_{730}.

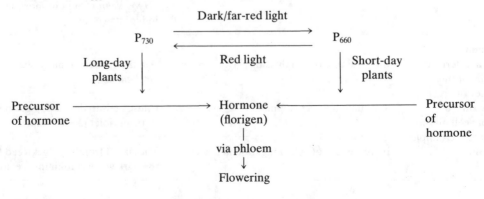

Other effects of red and far-red light (through the phytochrome system):

1 Germination: some seeds, e.g. *Lactuca,* germinate better in far-red than red light.
2 Formation of some plant pigments: e.g. red light induces the formation of anthocyanins.
3 Elongation of internodes: in many plants the internodes lengthen in far-red light and elongation is inhibited by red light.
4 Expansion of leaves: leaf area increases in response to red light.

Control of growth in mammals Hormones play a major role in regulating growth in mammals and they are co-ordinated by the activity of the hypothalamus.

$$\text{Hypothalamus} \xrightarrow{\frac{\text{1st order}}{\text{hormones}}} \underset{\text{gland}}{\text{Pituitary}} \xrightarrow{\frac{\text{2nd order}}{\text{hormones}}} \underset{\text{glands}}{\text{Endocrine}} \xrightarrow{\frac{\text{3rd order}}{\text{hormones}}} \underset{\text{effects}}{\text{Widespread}}$$

The main hormones influencing growth in mammals are:

1 Somatotrophin   4 Sex hormones
2 Thyroxine   5 Cortisol.
3 Insulin

Table 5.5 Mammalian growth hormones

Hormone	Produced by	Functions (concerned with growth)	Effect of imbalance
Somatotrophin (growth hormone)	Pituitary	Influences rate at which tissues grow; does not modify mechanism of tissue development. Stimulates synthesis of new proteins (in presence of insulin); raises levels of glucose and fatty acids in blood	Excess in adolescence causes gigantism Excess in adults causes acromegaly Deficiency causes dwarfism
Thyroxine	Thyroid	Influences way in which tissues (especially nervous tissue) develop. Increases rate at which cells oxidize glucose. N.B. Metamorphosis in Amphibia is controlled by variations in the amount of thyroxine present	Deficiency at birth causes cretinism Deficiency in adult causes myxoedema Excess causes Graves' disease
Insulin	Islets of Langerhans (in pancreas)	Presence required for growth hormone to stimulate protein synthesis	Deficiency causes diabetes (high blood sugar level has an inhibitory effect on growth hormone)
Gonadotrophins	Pituitary	Increase the production of:	
Testosterone	Testes	Development of secondary sex characteristics; stimulates bone growth before epiphyses fuse	Deficiency causes sexual development to be retarded and excessive growth of the long bones
Progesterone	Ovary	Development of secondary sex characteristics	Deficiency causes sexual development to be retarded
Oestrogen	Ovary	Development of secondary sex characteristics; stimulates bone growth before epiphyses fuse (because oestrogen matures long bone faster than testosterone, females are generally smaller than males)	Deficiency causes sexual development to be retarded and excessive growth of the long bones
Cortisol	Adrenal cortex	Opposes growth by breaking down proteins to amino acids as an energy source. Production increased in times of stress	Deficiency results in an inability to face physiological stress. Excess causes Cushing's syndrome

Hormonal control of insect metamorphosis Insect metamorphosis is controlled by three main hormones:

1 Brain hormone – produced by neurosecretory cells in the brain and stored in the corpus cardiacum.
2 Ecdysone (moulting hormone) – produced by the prothoracic gland when stimulated by brain hormone.
3 Juvenile hormone – produced by the corpus allatum.
   All moults require ecdysone. The type of moult is controlled by the concentration of juvenile hormone.
   High concentrations of juvenile hormone cause larval moults.
   Low concentrations of juvenile hormone cause pupal ecdysis and a pupa is formed.
   Absence of juvenile hormone causes imaginal ecdysis and an imago is formed.

Link topics

Section 5.4 Growth and development
Section 10.1 Hormones and homeostasis
Section 10.2 Nervous system and behaviour

Suggested further reading

Hill, T. A., *Endogenous Plant Growth Substances,* Studies in Biology No. 40, 2nd ed. (Arnold 1980)
Kendrick, R. E. and Frankland, B., *Phytochrome and Plant Growth,* Studies in Biology No. 68, 2nd ed. (Arnold 1983)
Lee, J. and Knowles, F. G. W., *Animal Hormones* (Hutchinson 1965)

QUESTION ANALYSIS

18 When a young plant shoot is illuminated from one side, there is
    A Growth only on the illuminated side
    B Growth only on the dark side
    C More rapid cell division on the dark side
    D More cell elongation on the dark side
    Mark allocation 1/40   Time allowed 1½ minutes   *In the style of the Cambridge Board*

This is a multiple choice question testing understanding of the phototropic response in shoots. As always with multiple choice questions the candidate should consider every option and not stop as soon as he feels he has found the correct answer. The candidate should be satisfied that there are valid reasons for rejecting an option rather than rejecting it because it 'does not seem right'. Consider each option in turn discounting first those that are obviously wrong. In this case A would probably be rejected initially since 'growth only on the illuminated side' would cause bending away from the light, whereas shoots are positively phototropic and bend towards the light. The remaining three options all cause bending in the correct direction and the candidate must now decide which best describes the mechanism of bending. C can be rejected because cell division occurs at the apex, whereas the bending is seen to occur behind the apex in the region of cell elongation. B can be rejected because since the overall length of the shoot increases during periods of unilateral illumination some growth must be occurring on both sides. Option D is therefore left as the correct answer, and reading the initial statement followed by option D should make accurate biological sense. The candidate should always adopt this habit as a means of checking the answer.

19 (a) Briefly define the following terms and give one example of each
     (i) phototaxis
     (ii) photoperiodism
     (iii) vernalization.

    (b) In a study of the growth of oat coleoptiles experiments were conducted as shown below.

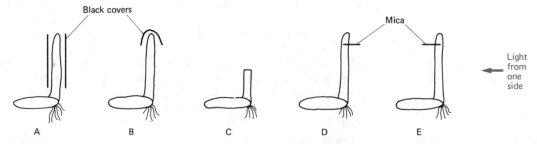

Fig. 5.18 Experiment on phototropism

For each coleoptile state the result which would be expected after three hours and give one reason for each statement.
Mark allocation 16/100   Time allowed 20 minutes

Associated Examining Board June 1980, Paper I, No. 10

This is a structured question examining knowledge of plant responses especially phototropism. While testing mostly factual recall there is some application of knowledge of plant hormones and their action in part (b).

In (a) the candidate should keep the definitions simple but at the same time clear and unambiguous. There are problems here because many terms, e.g. phototropism and phototaxis, differ only in one small, but important, respect. Candidates should therefore avoid giving definitions that are vague and could cover more than one term. In (a) (i) for instance, 'phototaxis' should be defined as the movement of **the whole** of a freely motile organism towards or away from a directional light stimulus. The fact that the whole organism moves distinguishes it from phototropism where only a part does. The examples chosen should be limited to ones clearly illustrating the definition. In most aspects of biology there are organisms which fit exactly into one category but many others that fit more than one or lie on the borders between them. In this type of answer it is not clever, but foolhardy, to give unusual or original examples; an example should be chosen that leaves the examiner in no doubt that the candidate understands the definition, for example in (a) (i) the movement of *Euglena* towards light of a suitable intensity.

For (a) (ii) a definition of photoperiodism could be 'the response of an organism to the relative duration of day and night' and the example could be the flowering of clover in response to long days.

In (a) (iii) 'vernalization' is the necessity for a period of cold exposure by plants or their seeds, in order to induce flowering, or at least earlier flowering. An example is the early flowering of winter varieties of wheat.

In (b) the candidate should understand that the hormones that induce bending of an oat coleoptile towards unilateral light are produced at the tip, where the stimulus is also perceived. The hormones move down the shoot and become more concentrated on the darker side, thereby causing greater cell elongation which results in bending towards the light (positive phototropism). The candidate may have learned the historical experiments given in the essential information and therefore rely on memory. This approach has the major disadvantage of being inflexible. If an original situation is encountered such a candidate, having had no previous experience to learn from, will be left to guess the answer. If the candidate understands the mechanism of hormone action in plants the knowledge may be applied to any situation with satisfactory results. This second method also involves the memorization of fewer facts, always a major advantage.

The question says 'after 3 hours' which is sufficient time for a phototropic response to occur where it is possible. In A the candidate should deduce that bending towards the light occurs because the stimulus is perceived at the tip and since this is exposed to the light the response is normal. In B there is no response since the tip cannot perceive the stimulus due to the black cover. Likewise in C the removal of the tip means that the stimulus cannot be perceived nor can the hormones be produced and so there is no response. In D there is bending towards the light because the hormones causing cell elongation move down the dark side, unimpeded by the mica. In E however the mica, being impervious to the hormones, prevents downward movement to the region of cell elongation and so there is no bending of the coleoptile.

20 The effect of increasing time of dry storage on the germination of a species of seed is shown in graph 1. All the seeds were kept at 15°C. Each curve represents a germination test on a seed sample stored for the number of weeks indicated.

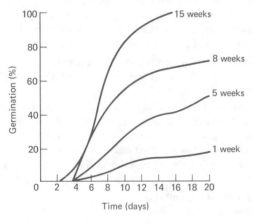

Fig. 5.19 (a) Graph I   **Fig. 5.19 (b)** Graph II

Germination of seeds of birch under long-day illumination cycles (20 h light: 4 h dark) and short-day cycles (4 h light: 20 h dark) produced the results shown in graph 2.

The percentage of seeds of a single species of plant which germinated during an eight-day period was recorded. Before starting the investigation, some of the seeds and some of the intact fruits were treated as indicated on the table.

Table 1

Treatment	Percentage germination by day			
	2	4	6	8
Intact fruits in air	0	6	10	10
Fruits with pericarp cut	0	12	28	38
Naked seeds in air	0	14	30	42
Seeds with testa pricked	14	45	53	53
Seeds in oxygen	40	56	72	78
Pricked seeds in oxygen	25	62	70	84

(a) Using the information given in the graphs and table list those factors which appear to influence the onset of germination and suggest a probable mechanism by which each of the factors operates.

(b) Suggest another factor, not indicated in the graphs and table, which might influence the onset of germination and devise an experiment to determine the extent of its influence.
Mark allocation 20/60   Time allowed 60 minutes

Associated Examining Board June 1979, Paper III, No. B4

The first part of this question is concerned with the comprehension of data in both graphical and tabular form and the ability to interpret its meaning. In addition the examiners require the candidate to select from his learned knowledge certain material that could explain the relationships shown by the data. The second part of the question requires the application of knowledge to devise an experiment which in all probability will not have been previously carried out by the candidate. Before attempting any answers the candidate should read through the whole question fully and then read part (a) as many times as it takes to understand exactly what is required. The candidate should note that only 'information given in the graphs and table' should be used and that factors that 'influence the onset of germination' are required. The word 'suggest' means that you are not expected to give the correct answer (if there is one) but simply offer a reasonable biological explanation. The words 'probable mechanism' reinforce the view that you should give only a plausible explanation and there is not necessarily just one correct answer. Do not despair if you have not read or heard an answer to these problems, the purpose of the question is to see how far a candidate is able to apply his basic biological knowledge to explain original problems. Take each set of data in turn and analyse it separately. Note the question says 'list the factors', this implies that no explanation of your answer is needed.

Graph 1 The candidate's interpretation of the graph may take the form: The temperature is constant ($15°C$) – assume, in the absence of other data, all other conditions except the period of dry storage are also constant. The longer the period of dry storage the more rapid is the germination (steeper gradient) and the greater the final percentage germination. Dry storage influences the onset of germination.

Mechanism: Germination is induced when the food reserves of a seed are mobilized by making them soluble through enzymatic breakdown. The production of these enzymes is controlled from the DNA in the nucleus via mRNA and tRNA (Section 3.3). Some substance produced in response to a long dry period must therefore increase transcription of DNA or possibly increase the permeability of the nuclear membrane to mRNA. This substance is probably a plant hormone, e.g. gibberellic acid (addition of GA to seeds can induce germination without the necessary period of dry storage).

Graph 2 Long periods of light (20 h) influence the onset of germination.

Mechanism: The reasoning in favour of gibberellic acid inducing germination is the same as that for graph 1. However a probable mechanism of how light induces GA production could be offered by an intelligent candidate who recalls that many light influencing activities in plants (e.g. flowering) are controlled by the interaction of two phytochrome pigments P_{660} and P_{730} (Section 5.5). During long light periods P_{660} is converted to P_{730} which induces the production of plant hormones including GA.

Table 1 The candidate should deduce that increasing exposure of the embryo to air induces more rapid germination, and that where pure oxygen is substituted for air the rate is further increased. Oxygen is therefore the factor influencing the onset of germination.

Mechanism: Germination involves growth, which requires energy. Energy is released by respiration which yields considerably more energy in the presence of oxygen. The influence of oxygen is therefore directly to increase respiration and hence energy yields for growth, thus stimulating germination.

In answering (b) the candidate would be well advised to draw up a list of other factors affecting germination and choose carefully the one that most lends itself to being tested experimentally. Other factors include:

1 Permeability of the testa to water   3 Immaturity of the embryo
2 Mechanical resistance to growth   4 The need of a cold period.

It would clearly be difficult to obtain quantitative data to test 2 and 3 and since the question says 'the extent of its influence' the need for some quantitative data is implied; therefore 1 or 4 should be chosen. To take the need for a cold period as an example, batches of 100 seeds should be subjected to a temperature of $5°C$ for varying lengths of time (do not freeze them as this can have complicating effects due to the physical damage freezing may cause). During this chilling period all other conditions necessary for germination of the seeds should be provided so that it cannot be argued that their absence is adversely influencing germination. These conditions, e.g. light, humidity, oxygen, will vary according to the seed being used. At weekly intervals one batch of seeds should be subjected to the optimum temperature for germination of that species, and the percentage germination recorded at daily intervals for about two weeks. This process should continue for about 15 weeks. From the data obtained a graph similar to graph 1 in the question would be obtained demonstrating the extent to which a period of chilling affects germination.

Answer plan

(a)
Graph 1
Influence: dry storage
Mechanism: dry storage induces GA production which influences DNA control of enzyme production, hence mobilization of food reserves, growth and germination.

Graph 2

Influence: light

Mechanism: light changes P_{660} to P_{730} which induces GA production, DNA control of enzyme production, mobilization of food reserves, growth and germination.

Table 1

Influence: oxygen

Mechanism: oxygen allows aerobic respiration; energy released for growth and germination.

(b) Cold period requirement: batches of 100 seeds subjected to various periods at 5°C (all other conditions favouring germination). Each week one batch is given optimum germination temperature and the percentage germination is measured. Plot percentage germination against time for each batch.

21 (a) Explain the difference between nastic, tropic and tactic responses in plants. (6)
  (b) (i) Describe an experiment to show that auxins are involved in the process of plant stems growing towards the light. Your answer should include details of the method to be used and the expected results. (8)
   (ii) What other effects do auxins have on the growth of plants? (6)
  Mark allocation 20/100 Time allowed 30 minutes *London Board June 1985, Paper 2, No. 5*

This structured essay illustrates the importance of revising practical work for theory examinations. It cannot be assumed that experiments studied or carried out in the practical part of a course will not appear on theory papers. In particular, experiments like the one referred to in part (b) (i) of the question may take days before results are obtained. Knowledge of these can only be examined on theory papers.

The key word in part (a) is 'differences' and the answer should attempt to highlight these. The definitions of nastic, tropic and tactic responses are given in Table 5.3. In tropic and tactic responses the direction of the response is directly related to the direction of the stimulus, whereas in nastic responses the stimulus is non-directional or diffuse and the direction of the response cannot therefore be related to it. While nastic and tropic responses involve the movement of part of a plant, tactic responses involve the movement of the whole organism (or a freely motile part of one). To complete part (a) the answer should be illustrated with at least one example of each type of response, chosen from Table 5.3.

The experiment in part (b) (i) could take a number of forms but whichever method is chosen the answer should include the use of many plants, a suitable control and ensuring all samples are left under exactly the same conditions for the same length of time. A typical experiment would be to set up three groups of coleoptiles (10–20 plants in each). All three should be exposed to light from one side only (unilateral light). One group should be left untreated, whereas the other two should have a mica strip inserted, in one case on the illuminated side and in the other the shaded side (see Boysen–Jensen's experiments in Fig. 5.17 and include similar diagrams in your answer). Bending towards the light (positive phototropism) occurs, except where the mica is placed on the shaded side of the coleoptile–here the downward translocation of auxin is prevented on this side. Another acceptable experiment involves using all round illumination and collecting auxin in agar blocks. These blocks are then put onto decapitated coleoptiles, displaced to one side. The results and explanations are given in the first of Went's experiments in Fig. 5.17.

The other effects of auxins required in (b) (ii) are given in Table 5.4 in the essential information. Do not confine your answer to a list but illustrate each effect with an example as shown in the table.

Answer Plan

(a) Nastic–movement of part of plant–non-directional stimulus–e.g. *Crocus* flowers opening at temperatures above 16°C (=thermonasty)

Tropic–movement of part of plant–direction of response is related to direction of stimulus–e.g. twining of pea *Pisum* tendrils around supports (=thigmonasty)

Tactic–movement of whole organism–direction of response is related to direction of stimulus–e.g. *Chlamydomonas* moves to regions of optimum light intensity (=phototaxis)

(b) Expose three groups (10–20 plants) of coleoptiles to unilateral light.

  1 untreated

  2 mica strip inserted on illuminated side

  3 mica strip inserted on shaded side

(Mica strips to be nearer tip than zone of elongation.)

Leave all samples for the same period of time (e.g. 24 hours).

Annotated diagrams to illustrate method.

Results–coleoptiles subjected to treatments 1 and 2 bend towards light; those treated as in 3, do not.

(c) Geotropism due to cell elongation, stimulation of cambial activity, initiation of root development, stimulation of fruit development, inhibition of leaf and fruit abscission and lateral bud development, maintenance of apical dominance, differentiation of cambium, leaf expansion, influence on other hormones.

22 (a) What are the essential features of complete metamorphosis in a named insect? (6)

(b) Explain briefly how the life cycle of an insect with incomplete metamorphosis differs from that of an insect with complete metamorphosis. (6)

(c) Give an account of the ways in which insect metamorphosis is controlled. (8)

Mark allocation 20/100   Time allowed 35 minutes   *London Board June 1979, Paper I, No. 7*

This is a structured essay testing recall of information and the ability to contrast the two main types of insect metamorphosis.

In part (a) the account should be a simple outline of the life cycle of a 'named' insect showing complete metamorphosis. The barest of outlines is given in Figure 5.2f of Section 5.1 for the cabbage white butterfly. If diagrams are used to illustrate this cycle they must be basic outlines since the time allowance for this part is only 10 minutes. The key words are 'essential features' and the candidate must therefore stress the features of the cycle that are typical of complete metamorphosis. These should include the presence of distinct developmental stages, each different in appearance, behaviour and habitat. These stages are the egg, larva, pupa and imago. The larva is active and undergoes major growth (usually of cell size rather than number). The pupa is an outwardly dormant stage in which the tissues are largely broken down and reorganized to give the imago (adult). The imago is the only sexually mature phase and its role is hence one of reproduction and dispersal.

In (b) the candidate must make clear differences between the two types and not simply describe each in turn. A table of differences could be used but since some expansion of the points is needed it is probably best to use one only if time is short. The question does not require a named insect and so generalized differences between the two types should be used. The question says 'briefly' and so the points should be expanded only enough to make the differences clear; superfluous detail must be avoided. The differences should relate to the same feature. For example, complete metamorphosis in insects frequently involves the larva and pupa using completely different food sources, whereas in incomplete metamorphosis the same type of food is consumed throughout. Candidates often compare different features: e.g. complete metamorphosis involves changes in diet whereas incomplete metamorphosis involves only growth in size. Clearly such statements bring no marks. The major difference is the distinct developmental stages that occur in complete metamorphosis compared with the gradual change from a 'miniature' adult to a full grown one in incomplete metamorphosis. Most of the other differences stem from this and are listed in the answer plan.

Part (c) has the slightly larger mark allowance and therefore warrants a little more time and detail. A summary is given in the essential information. This should be fully explained and, if time allowed, could include details of the experimental evidence for the process (i.e. removal of the corpus allatum, corpus cardiacum or prothoracic gland and the consequent effect on metamorphosis).

Answer plan

(a) Name the insect clearly.

Essential features will depend on the actual example chosen but should include definite developmental stages of:

Egg

Larva – active growing stage

Pupa – outwardly inactive, reorganizational stage

Imago – sexually mature, reproductive and dispersal stage.

Stages are distinct in all aspects, e.g. morphology, behaviour, habitat.

(b) Differences between complete and incomplete metamorphosis in insects

Complete	*Incomplete*
Distinct stages	Stages similar except in size
Sudden developmental changes	Gradual developmental changes
Changes in form	Changes in size
Juveniles and adults have different diets	Diets similar
Juveniles and adults differ in habitat and behaviour	Similar habitats and behaviour
Dormant/resistant stage is usually the pupa	Dormant/resistant stage is usually the egg

(c) See essential information in this section (5.5).

Expand the information slightly, possibly to include experimental evidence for the process.

6 Nutrition

6.1 AUTOTROPHIC NUTRITION (PHOTOSYNTHESIS)

Assumed previous knowledge Simple equation for photosynthesis.

Starch test in leaves and experimental evidence for necessity of chlorophyll, light and carbon dioxide.

Experiment to show that oxygen is produced during photosynthesis.

Underlying principles

One method of distinguishing living material from non-living is to compare the way they relate to the Second Law of Thermodynamics. This law states that all matter tends to high entropy (i.e. it tends to become disordered). This is true of non-living material which tends towards a state where it possesses the minimum of energy and becomes disordered. Living organisms on the other hand are highly ordered systems possessing much stored energy. In fact even living organisms tend to lose energy and become disordered. Where they really differ from non-living material is in their ability to replace the lost energy from outside themselves. In animals and the fungi this energy is obtained from food, which comes ultimately from green plants. The green plants obtain energy from the earth's major source, the sun, by using light to produce complex organic molecules from simple inorganic ones in the process of photosynthesis. All life is directly or indirectly dependent on this the most fundamental process in living organisms.

Points of perspective

A good candidate would have a knowledge of alternative mechanisms and pathways in auto-trophic organisms. Examples include the C_4 pathway and a range of chemosynthetic mechanisms where inorganic substances are used to provide energy rather than light, e.g. oxidation of ammonium compounds to nitrites by *Nitrosomonas*.

It would be useful to have some appreciation that photosynthesis can occur in some prokaryotic cells without chlorophyll, e.g. in the purple sulphur bacteria.

Some information on the inter-relationship between the photosynthetic pathway and other biochemical pathways in a plant cell would be helpful, in particular how the many other substances in a cell are formed from the products of photosynthesis.

Essential information

Light Visible light represents that part of the electromagnetic radiation spectrum which lies between 380 nm (violet) and 750 nm (far red). Three properties of light that are of importance to organisms are:

1 spectral quality **2** intensity **3** duration.

The Particle Theory proposed by Einstein in 1905 states that light is composed of particles of energy called photons. The energy of a photon is not the same for all kinds of light but is inversely proportional to the wavelength – the longer the wavelength, the lower the energy. To be of use the light must be converted to chemical energy.

Pigments The main photosynthetic pigments are: **1** chlorophylls **2** carotenoids.

Chlorophyll This name covers a group of closely related substances, the most important of which are chlorophylls *a* and *b*. They occur in all higher plants in the approximate ratio of 2 : 1.

Chlorophyll *a* $C_{55}H_{72}O_5N_4Mg$ Chlorophyll *b* $C_{55}H_{70}O_6N_4Mg$

Chlorophyll is usually contained within chloroplasts (see Section 2.1). The total weight of chlorophyll in green leaves varies between 0.55 – 0.20% of the fresh weight.

Carotenoids These are a large group of hydrocarbon pigments found throughout the plant. About 100 different carotenoids have been recognized. There are two major groups: the

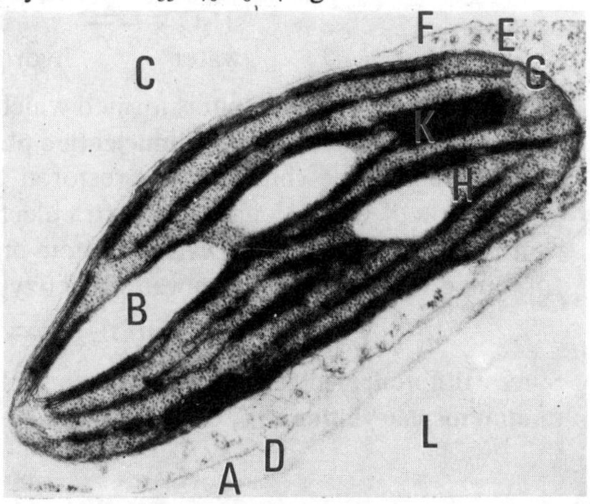

Fig. 6.1 Electronmicrograph of a chloroplast

Key
Chloroplast (in a leaf cell of *Agrostis*)
Magnification 50 000 ×

A Cell wall
B Starch grain
C Vacuole
D Plasmalemma
E Cytoplasmic ribosomes
F Tonoplast
G Chloroplast envelope
H Chloroplast stroma with ribosome
K Granum
L Intercellular space

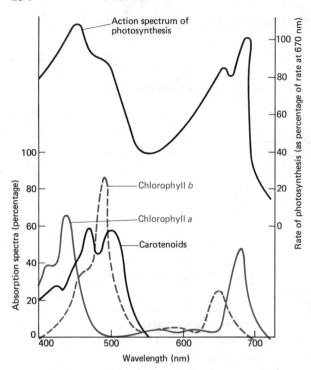

Fig. 6.1 (a) Graph to show relationship between action spectrum for photosynthesis and absorption spectra of major photosynthetic pigments

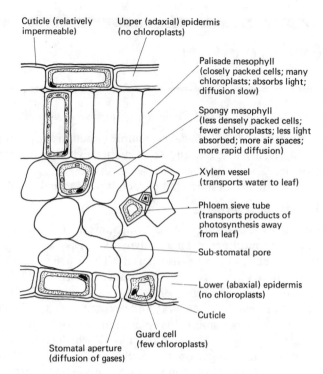

Fig. 6.1 (b) VS part of leaf to show structure in relation to photosynthesis

carotenes and xanthophylls. They are important pigments in photosynthesis, especially in some groups of algae, e.g. fucoxanthin in the brown algae (Phaeophyceae). They strongly absorb light in the violet end of the spectrum.

Mechanism of photosynthesis In 1932 Emerson and Arnold exposed plants to flashes of light lasting 10^{-4} seconds and varied the period of dark between the flashes. They found that the yield of carbohydrate increased as the length of the dark periods increased up to a maximum dark period of 20 milliseconds (0.02 s) at 25°C. If the temperature was reduced so was the yield of carbohydrate, although this could be compensated for by increasing the period of darkness. Increasing the duration of the flashes of light had no effect. These experiments suggest that photosynthesis has two stages:

1 A light stage (photochemical stage) which requires light, but is unaffected by temperature.
2 A dark stage (chemical stage) which does not require light and is affected by temperature.

The light stage When light is received by a chlorophyll molecule one of its electrons is lost, leaving the chlorophyll molecule positively charged and chemically less stable. This electron may return to the chlorophyll molecule, via the carrier system, thus stabilizing it again. In doing so some of the energy it loses is used to combine adenosine diphosphate and inorganic phosphate into adenosine triphosphate (phosphorylation). As further light can again raise the electron's energy and the process be repeated, it is termed **cyclic photophosphorylation.**

The electron may however not return directly to the chlorophyll molecule. It may combine with hydrogen ions (H^+) that result from the natural dissociation of water.

$$H_2O \rightleftharpoons H^+ + OH^-$$
water   hydrogen ion   hydroxyl ion

In doing so a hydrogen atom is formed which is immediately taken up by a hydrogen acceptor such as **nicotinamide adenine dinucleotide phosphate** (NADP) which enters the dark reaction. The stability of the chlorophyll is restored in this case by the hydroxyl ion (OH^-) from the dissociation of water donating its extra electron to the chlorophyll molecule. The electron is transported via an electron carrier system and ATP is yielded as before. The resulting OH is combined with others to form water and oxygen, the latter being evolved as oxygen gas.

$$4(OH) \rightleftharpoons 2H_2O + O_2 \uparrow$$

Since a different electron is returned to the chlorophyll molecule the process is termed **non-cyclic photophosphorylation.**

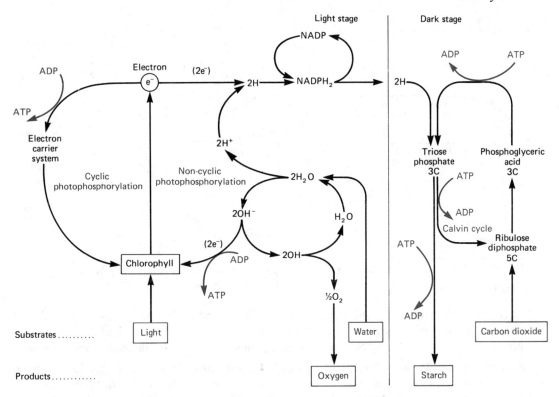

Fig. 6.2 Chemical reactions in photosynthesis

The dark stage (Calvin Cycle) The reduced nicotinamide adenine dinucleotide phosphate ($NADPH_2$) from the light reaction is used to reduce carbon dioxide using the ATP formed in the light reaction as the source of energy. Carbon dioxide absorbed by the plant is combined with a five carbon substance **ribulose diphospate** and the resulting unstable intermediate immediately splits to give two molecules of the three carbon substance **phosphoglyceric acid** (PGA). Most of this PGA is used to reform ribulose diphosphate but some is reduced by the $NADPH_2$ to give triose phosphate, which in turn is built into glucose phosphate which polymerizes into starch by condensation.

C_4 photosynthetic pathway
Plants such as sugar cane and maize have evolved an alternative mechanism for the fixation of carbon dioxide that involves the use of **phosphoenol pyruvic acid** (PEP) instead of ribulose diphosphate. The first formed product of this process is the four carbon oxaloacetic acid instead of PGA. It provides a means of obtaining carbon dioxide and storing it chemically for later conversion to PGA, which is of particular advantage in hot tropical climates where carbon dioxide levels may be low during the day.

Importance of photosynthesis

1 Releases oxygen required by animals and plants for aerobic respiration.
2 Provides a store of useful chemical energy by converting inorganic substances to stable organic substances of high potential chemical energy. It has been estimated that plants annually fix 35×10^{15} kg of carbon.

Synthesis of nitrogenous compounds The nitrogen of ammonia is brought into organic combination mainly by the formation of glutamic acid

$$\alpha \text{ ketoglutaric acid} + NH_3 + CoI - H_2 \rightleftharpoons \text{glutamic acid} + CoI$$
$$\text{ammonia} \quad \text{reduced} \quad \text{glutamic}$$
$$\text{coenzyme I} \quad \text{dehydrogenase}$$

Once glutamic acid is available other amino acids may be formed by the transfer of its amino group to another carbon skeleton, a process known as **transamination**.

Factors affecting photosynthesis Blackmans Law of Limiting Factors states that when the speed of a process is conditioned by a number of separate factors, the rate of the process is limited by the pace of the slowest factor. The main factors affecting the rate of photosynthesis are intensity and duration of light, carbon dioxide concentration and temperature. When temperature and the concentration of carbon dioxide are kept constant the rate of photosynthesis increases with

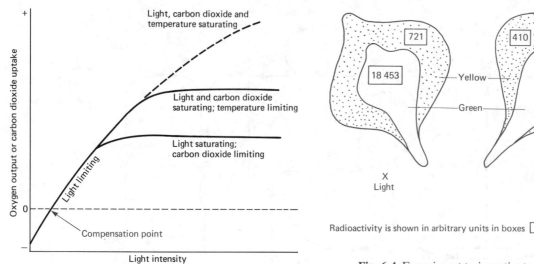

Fig. 6.3 Limiting factors in photosynthesis

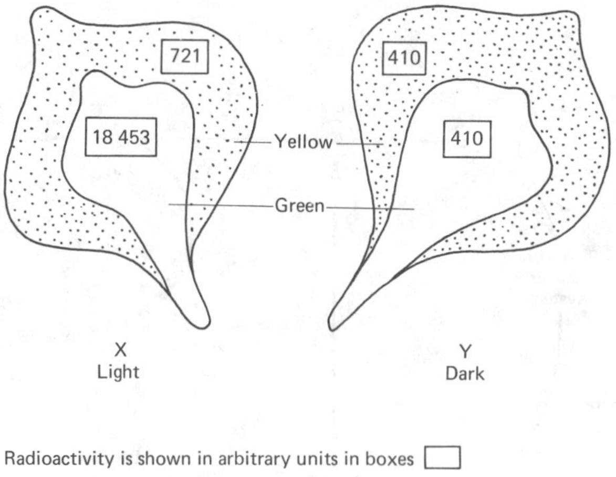

Fig. 6.4 Experiment to investigate photosynthesis in variegated leaves

an increase in light intensity and then levels off. When the carbon dioxide concentration is increased the rate of photosynthesis increases to a maximum before levelling out. If the temperature is then increased the rate of photosynthesis steadily rises instead of levelling off.

Compensation point This is the point at which the rate of carbon dioxide production from respiration is compensated exactly by the rate of carbon dioxide uptake in photosynthesis. Below the compensation point the rate of photosynthesis is less than the rate of respiration and so carbon dioxide is evolved. Above the compensation point the rate of photosynthesis exceeds the rate of respiration and so oxygen is evolved. At the compensation point there is no net exchange of gases.

Link topics

Section 3.2 Enzymes (including biochemical techniques)
Section 6.2. Heterotrophic nutrition (holozoic)
Section 7.1 Cellular respiration
Section 11.1 Ecology

Suggested further reading

Fogg, G. E., *The Growth of Plants*, 2nd ed. (Pelican 1970) (Chapter 4)
Hall, D. O. and Rao, K. K., *Photosynthesis*, New Studies in Biology, 4th ed. (Arnold 1987)
Roberts, M. B. V., *Biology, A Functional Approach*, 3rd ed. (Nelson 1982) (Chapter 10)
Tribe, M. and Whittaker, P. A., *Chloroplasts and Mitochondria*, Studies in Biology No. 31, 2nd ed. (Arnold 1982)
Whatley, J. M. and F. R., *Light and Plant Life* Studies in Biology No. 124 (Arnold 1980)
Whittingham, C. P., *Photosynthesis*, Oxford/Carolina Biology Reader No. 9, 2nd ed. (Packard 1977)

QUESTION ANALYSIS

1 A plant with variegated leaves was supplied with radioactive carbon dioxide ($^{14}CO_2$) during an experiment. Leaf Y was kept in the dark and leaf X was illuminated. The radioactivity in the leaves was measured at the end of the experiment and found to be as shown in Fig. 6.4. The most likely explanation for the level of radioactivity found in the yellow zone of X is:

A Photosynthetic products diffuse into the yellow zone.
B Photosynthesis takes place in this zone but no storage of starch occurs.
C Some photosynthesis occurs in the yellow zone, but in the absence of chlorophyll the amount is small.
D Radioactive carbon dioxide diffuses into the leaf and accumulates in this zone.
Mark allocation 1/40   Time allowed 1½ minutes   In the style of the Cambridge Board

This is a multiple choice question involving application of the candidate's knowledge of photosynthesis to an experimental situation and ability to make accurate deductions from the results obtained. The candidate needs to appreciate that the level of radioactivity is an indication of the amount of ^{14}C present. Since carbon dioxide is utilized by plants during photosynthesis the ^{14}C need not be in the form of $^{14}CO_2$ but could be part of any molecule derived from it during photosynthesis.

The two sets of figures for X and Y should then be compared. In Y where no light is available the amount of radioactivity, and hence photosynthesis, is the same regardless of the colour of the leaf. Because no photosynthesis has taken place, this level of radioactivity must be due to the $^{14}CO_2$ that has diffused into the air spaces of the leaf. This is the 'control' level against which the radioactivity in X can be compared. In X the green area has more than 45 times the radioactivity of the green area of Y indicating considerable incorporation of radioactive ^{14}C into the plants, presumably as starch during photosynthesis. The yellow area of X, which must lack chlorophyll, might be expected to show no increase in ^{14}C because no photosynthesis should have occurred. However there is a slight increase and it is this increase that the examiner requires the candidate to explain. Note that 'the most likely explanation' is required which implies that all might be possible, but one fits both the facts given and known details of photosynthesis better than the others. It is therefore all the more important to consider all possibilities and not stop as soon as a feasible explanation is found.

In A, the diffusion of the products of photosynthesis from the green area to the yellow must be considered seriously. Soluble precursors of starch diffuse into the yellow area where they are built up into starch and account for the small increase of ^{14}C in that region.

Knowledge of photosynthesis would allow the candidate to dismiss B on the grounds that carotenes (yellow pigments) cannot carry out photosynthesis in the absence of chlorophyll. The same reasoning can be used to discount C.

If D were correct, and it was diffusion of $^{14}CO_2$ into the leaf that accounted for the level of radioactivity in the yellow area of X, it would be difficult to explain why a similar level of radioactivity was not found in the yellow area of Y. $^{14}CO_2$ diffusion is not affected by the presence or absence of light and so levels in X and Y would be very similar if D were correct.

The answer is therefore A.

2 (a) What is meant by a 'limiting factor'? (2)
(b) Discuss how variations in (i) light intensity and (ii) temperature, influence the rate of photosynthesis. (8)
(c) Outline the sequence of events by which a molecule of carbon dioxide in the atmosphere might be converted into a storage carbohydrate by a green plant. (Exclude details of the light reaction in photosynthesis.) (10)
Mark allocation 20/100   Time allowed 30 minutes   *London Board 1985, Paper II, No. 1*

For part (a) a short definition should suffice. A limiting factor is one whose presence is essential to a process; if present below a certain minimal level the process will not occur, even though all other factors may be present in excess. In other words, the rate of a given process is determined by the factor in shortest supply. This factor is called the limiting factor.

In part (b), section (i) warrants slightly more time than section (ii). Light is needed in the reaction where ATP and NADPH$_2$ are formed. An increase in light intensity therefore increases the rate of photosynthesis. The precise relationship between the two is illustrated in Fig. 6.3. The effect of light intensity on the compensation point and the opening and closing of stomata (which affects the availability of carbon dioxide) are worthy of inclusion. In (b)(ii) it is essential that the effect of temperature on the rate of enzyme action is related to the fact that the dark reaction of photosynthesis is enzymatically controlled. The rate of enzyme-controlled reactions will double for a 10°C rise in temperature up to an optimum temperature above which the enzyme becomes denatured and the rate decreases.

In part (c) one key word is 'atmosphere'. Many candidates throw away a considerable number of marks by ignoring the important sequence of events by which the molecule of carbon dioxide reaches the stroma of the chloroplast. Carbon dioxide enters the plant from the atmosphere by diffusing through stomata. The absence of key words such as 'diffusion' or 'stomata' could lose marks. Other essential points are: movement along a concentration gradient, dissolving into the wet surface of a palisade cell and diffusion across the cytoplasm into the stroma of the chloroplast. The answer should give the outline biochemical processes of the dark reaction (Calvin Cycle) as given in the essential information of Section 6.1, including the relevant part of Fig. 6.2. A common and costly error is to give the hydrogen carrier as NADH$_2$ rather than NADPH$_2$.

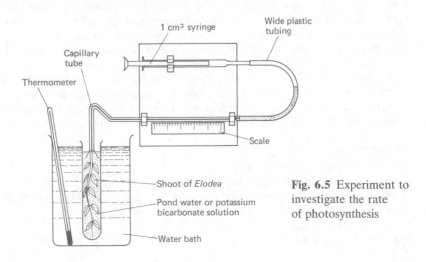

Fig. 6.5 Experiment to investigate the rate of photosynthesis

3 The apparatus shown above can be used to indicate the rate of photosynthesis of an aquatic plant.

(a) What data can be obtained with this apparatus? (2)

(b) Explain how you would use the apparatus to investigate the effect of changes in light intensity on the rate of photosynthesis. (7)

(c) What relationship between light intensity and rate of photosynthesis would you expect the apparatus to show? Explain your answer. (4)

(d) Describe two possible sources of inaccuracy in the experiment and explain how you might overcome them. (4)

(e) When the gas given off by an aquatic plant is collected and analysed the percentage of oxygen present is frequently as low as 40% even under optimum-photosynthesizing conditions. Give reasons for this phenomenon. (3)

Mark allocation 20/140   Time allowed 25 minutes

Associated Examining Board November 1980, Paper II, No. 5 (Alternative I)

This is a longer type structured essay question testing knowledge of the use of experimental apparatus to measure photosynthesis and the ability to predict the results and criticize their accuracy. It is not essential for the candidate to have carried out the experiment, although to have done so would be a considerable advantage. The apparatus, called a photosynthometer, should be studied carefully for, while the principle is the same in most types, the actual design varies considerably.

In (a) the amount of gas produced by the plant in a given time can be obtained. If the scale is calibrated in mm^3 or similar units the volume may be read directly. If it is a linear scale (mm) the volume can be calculated using the formula $\pi r^2 h$, where r is the radius of the capillary tube and h is the distance along the tube occupied by the gas given off.

In (b) the candidate should give full experimental details. Minor details such as depressing the plunger of the syringe fully before starting the experiment are no less important for successful results than major ones. There should be a means of adjusting the intensity of the light source or, failing this, a lamp of fixed intensity should be used and its distance from the apparatus varied. In either case light meter readings close to the apparatus should be taken to determine the actual light intensity each time. The whole experiment should be conducted in a dark room. The temperature should be measured on the thermometer and it should be kept as constant as possible during the experiment by the addition of hot or cold water. The light should be switched on for a fixed period of time, say 2 h, after which the gas that has collected at the cut end of the shoot of *Elodea* should be drawn along the capillary tube by withdrawing the plunger of the syringe. The capillary tube is full of pond water and so the column of gas can be easily seen. It should be drawn alongside the scale and its length measured. It is then drawn away from the capillary tube by further withdrawal of the syringe plunger until it enters the upper part of the wide plastic tubing. The plunger is depressed fully again, a different light intensity is used and the experiment repeated. A series of results should be obtained and the graph drawn of volume of oxygen produced against light intensity.

The relationship required in part (c) is shown in Figure 6.3 with the graph tailing off as shown by the lowest line 'light saturating; carbon dioxide limiting'. The graph is the simplest description of the relationship which can be explained as follows: while the output is limited by light intensity an increase in light intensity will produce more oxygen; when the amount of carbon dioxide or some other factor is in short supply, it begins to limit photosynthesis and an increase in light intensity ceases to affect the oxygen evolved and so the graph tails off.

One possible inaccuracy, as required in part (d), is variation in temperature during the experiment, especially when the light source is near the apparatus. To overcome this a transparent heated water bath (e.g. aquarium) with a sensitive thermostat control could be used. Another would be fluctuations in light intensity due to voltage changes during the experiment, especially if the supply is from a battery. Where low voltages are used a zener diode can be introduced to rectify this and produce a constant voltage and hence a constant light intensity. A more likely inaccuracy is fluctuation in daylight around the apparatus, hence the need to carry out the experiment in a dark room. Yet another inaccuracy is the trapping of gas bubbles on and under the leaves. Tapping or swirling the shoot may dislodge many of these but complete success is unlikely. One reason for finding only 40% oxygen in the gas collected is the release from solution in the pond water of dissolved nitrogen and possibly other gases during the experiment. These dilute the oxygen produced. Furthermore, if the plant has been kept in the dark prior to the start of the experiment, the carbon dioxide from respiration will have accumulated in spaces in the leaves and stem. The oxygen produced by photosynthesis when the light source is switched on will firstly displace this carbon dioxide so that this gas rather than oxygen will be collected initially.

Answer plan

(a) Volume of gas evolved per unit time.

(b) Use various light intensities and measure oxygen evolved in a set period at each intensity. Use syringe to move gas produced alongside the scale. Expel gas before taking another reading. Use constant temperature and carry out experiment in a dark room.

(c) See Figure 6.3.

(d) Choose any 2 from:

Inaccuracy	Remedy
Temperature fluctuation	Thermostatically controlled waterbath
Light intensity fluctuation	Use zener diode to give constant voltage and hence light intensity
External light fluctuation	Carry out in dark room
Trapped gas bubbles	Tap/swirl *Elodea* to release them

(e) Dissolved nitrogen or other gases released from solution and accumulated carbon dioxide from respiration being displaced initially; both dilute the oxygen.

4 Five small discs cut from spinach leaves were floated on a small volume of buffered bicarbonate solution in a flask attached to a respirometer. The discs were first exposed to bright light, then to dim light and finally left in the dark. Oxygen release was recorded as positive values and the oxygen uptake as negative values.

The results obtained from the experiment are given below

Light intensity	Time interval in minutes	Oxygen uptake or release for each 3 minute interval (mm³)
Bright light	0 – 3	+57
	3 – 6	+64
	6 – 9	+58
	9–12	+60
Dim light	12–15	+16
	15–18	+ 3
Dark	18–21	−16
	21–24	−12
	24–27	−15
	27–30	−14

(a) Present the data in suitable graphical form (5)
(b) Calculate the mean rate of oxygen release in bright light (2)
(c) Explain the significance of the results obtained from the experiment (8)
Mark allocation 15/175   Time allowed 11 minutes   *London Board June 1980, Paper I, No. 9*

This is a structured question testing the ability to present data in a suitable graphical form, carry out a simple mathematical calculation and interpret the data provided.

The graph required for (a) could either be in the form of a histogram or a point graph. The candidate should always choose axes carefully to make maximum use of the paper and not leave most of it empty. The scale should be chosen to make the plotting of points simple. To represent one minute by seven squares or some equally unusual scale will only lead to mistakes when plotting the points. Label each axis clearly and state the units. Put a title on the graph. The finished graphs should resemble either Figure 6.6a or 6.6b.

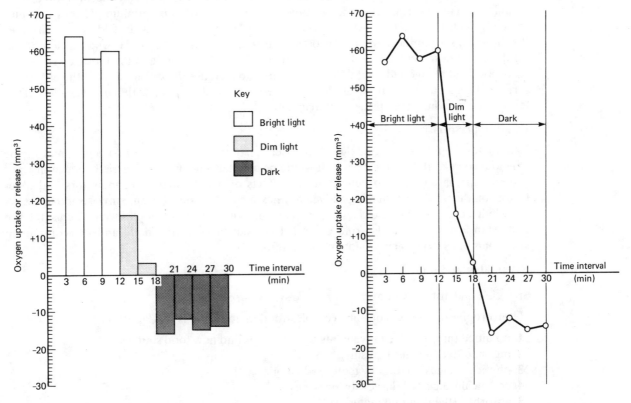

Fig. 6.6 (a) and (b) Oxygen uptake or release of spinach leaf discs in different light intensities

In (b) the period of bright light referred to is from time 0 to 12 min. The most likely error is that candidates will take the final figure for 9–12 min of 60 mm^3 and divide this by 12 to give the answer 5 mm^3/min. The fact that the division produces a whole number answer may even reinforce the candidate's view that he has the correct answer. The good candidate will appreciate that the amount of oxygen released is given for each separate 3 min period and so the total volume evolved in the first 12 min is the sum total of that evolved in each of the first four 3 min periods, i.e. $57 + 64 + 58 + 60$ making a total of 239 mm^3. Since the total time is 12 min the mean rate of oxygen release is

$$239/12 = 19.9 \text{ mm}^3/\text{min}$$

Make certain that the units are clearly stated.

In explaining the significance of the results (part c), the candidate should first make it clear that the rate of oxygen release is a measure of the rate of photosynthesis. The rate of oxygen uptake or release is fairly constant in both bright light and in the dark. In dim light where light intensity is limiting the rate of photosynthesis, the gradient of the graph is very steep, indicating that a change from bright to dim light produces a rapid decrease in the rate of photosynthesis. In bright light much photosynthesis occurs and therefore much oxygen is released; the reverse is true in the dark where no photosynthesis occurs and where oxygen is taken up during respiration.

It is important that the candidate refers to the fact that the oxygen released in bright light is not truly a measure of the oxygen produced in photosynthesis since some is immediately used in respiration. A good candidate may make reference to the point where the line crosses the x axis and oxygen is neither released or taken up, i.e. the compensation point.

6.2   HETEROTROPHIC NUTRITION (HOLOZOIC)

Assumed previous knowledge The nature of a balanced diet.
The nature of food substances (Section 3.1).
The nature and properties of enzymes (Section 3.2).
Deficiency diseases related to the lack of particular food requirements.
Structure of teeth; structure of mammalian alimentary canal.

Underlying principles

The evolution of autotrophic organisms inevitably led to the development of organisms that obtained their energy not by photosynthesis but by consuming the autotrophs. These were the herbivores. They developed various mechanisms for ingesting the food and digesting the cellulose walls that surrounded the plant cells. In turn there evolved organisms that obtained energy by consuming the herbivores. These were the carnivores. They developed a variety of mechanisms for capturing prey organisms as well as for ingesting them. Unlike autotrophs the energy source of heterotrophs is in complex form and contains some unwanted materials. The food must therefore be digested, the required materials absorbed and the unwanted ones eliminated. This digestion is carried out by enzymes each with an optimum pH. For enzymes to be efficient the food they act upon must have a large surface area. In addition to enzymes the digestive system therefore also produces substances to adjust the pH, water as a medium for the enzymes to act in and substances to increase the surface area of the food. To speed the absorption of the digested food there are methods to increase the surface area of the absorbing surface. To maintain a diffusion gradient this surface is well supplied with blood vessels and the food is kept moving. The surface is thin and moist.

Points of perspective

Candidates should have carried out a full dissection of the digestive system of a mammal and be able to recognize all the major parts. It would be beneficial to also have carried out a number of experiments on enzymes to determine the effects of pH, temperature, inhibitors and other factors on the rate of action of digestive enzymes such as pepsin, trypsin, amylase and lipase. A knowledge of a wide range of feeding mechanisms with specific examples would be useful. An understanding of the precise role of each food substance in the body and the consequent symptoms of its deficiency would also be helpful.

Essential information

Structure and histology of human digestive system – see Fig. 6.7.

Heterotrophic organisms consume complex food material. This food must be

1 obtained (may involve movements to capture or find new food sources)
2 ingested (feeding mechanisms)
3 physically digested (e.g. by teeth, radula, gizzard)
4 chemically digested (largely by enzymes)
5 absorbed (the required material)
6 eliminated (the unwanted material).

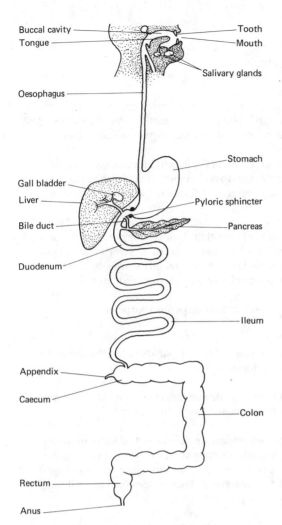

Buccal cavity
Tongue
Tooth
Mouth
Salivary glands
Oesophagus
Stomach
Gall bladder
Liver
Pyloric sphincter
Bile duct
Pancreas
Duodenum
Ileum
Appendix
Caecum
Colon
Rectum
Anus

Fig. 6.7 (a) Human digestive system

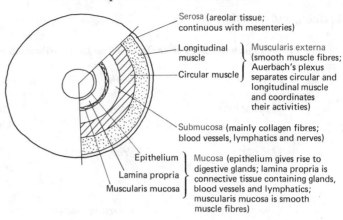

Serosa (areolar tissue; continuous with mesenteries)

Longitudinal muscle
Circular muscle
} Muscularis externa (smooth muscle fibres; Auerbach's plexus separates circular and longitudinal muscle and coordinates their activities)

Submucosa (mainly collagen fibres; blood vessels, lymphatics and nerves)

Epithelium
Lamina propria
Muscularis mucosa
} Mucosa (epithelium gives rise to digestive glands; lamina propria is connective tissue containing glands, blood vessels and lymphatics; muscularis mucosa is smooth muscle fibres)

Fig. 6.7 (b) General plan of the alimentary canal

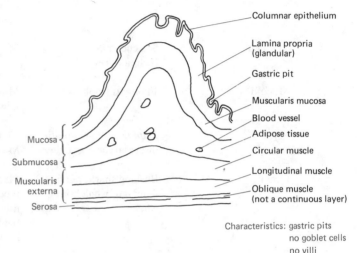

Columnar epithelium
Lamina propria (glandular)
Gastric pit
Muscularis mucosa
Blood vessel
Adipose tissue
Circular muscle
Longitudinal muscle
Oblique muscle (not a continuous layer)

Mucosa
Submucosa
Muscularis externa
Serosa

Characteristics: gastric pits
no goblet cells
no villi

Fig. 6.7 (c) LS stomach

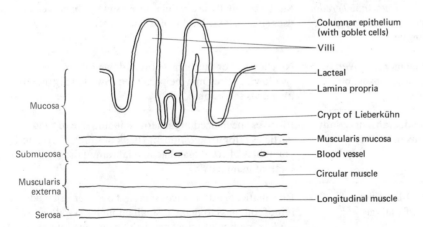

Columnar epithelium (with goblet cells)
Villi
Lacteal
Lamina propria
Crypt of Lieberkühn
Muscularis mucosa
Blood vessel
Circular muscle
Longitudinal muscle

Mucosa
Submucosa
Muscularis externa
Serosa

Characteristics of duodenum and ileum:

villi
goblet cells
crypts

Differences between

Duodenum	Ileum
Many villi	Few villi
Villi leaf-like	Villi finger-like
Brunner's glands present	Brunner's glands absent

Fig. 6.7 (d) LS ileum

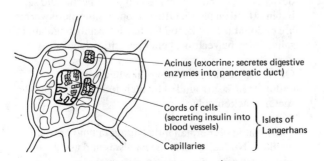

Acinus (exocrine; secretes digestive enzymes into pancreatic duct)

Cords of cells (secreting insulin into blood vessels)
Capillaries
} Islets of Langerhans

Fig. 6.7 (e) TS pancreas

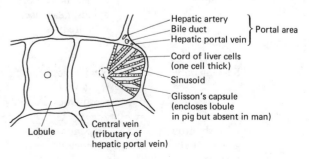

Hepatic artery
Bile duct
Hepatic portal vein
} Portal area

Cord of liver cells (one cell thick)
Sinusoid
Glisson's capsule (encloses lobule in pig but absent in man)

Lobule
Central vein (tributary of hepatic portal vein)

Fig. 6.7 (f) TS liver

Obtaining and digesting

Table 6.1 Feeding mechanisms in animals (based on the classification by Yonge)

Food type	Mechanism	Examples	Notes
Small particles	Pseudopodial	*Amoeba*	Often associated with locomotion. Pseudopodia enclose food in a vacuole in which digestion takes place
	Flagellate	*Euglena,* sponges	Beating flagellum (or flagella) directs microscopic food particles to region specialized for their ingestion. In flagellate protozoa, associated with locomotion
	Ciliary	*Paramecium, Sabella* (polychaete), *Mytilus* (bivalve), *Amphioxus*	Cilia create currents carrying microscopic food particles to region of ingestion. Continuous process. Mucus used to trap food. Found in almost all invertebrate phyla, except Arthropods. Frequently marine because sea contains much microscopic organic material
	Tentacular	Sea cucumber (echinoderm)	Mucus on tentacles traps food particles
	Mucoid	*Vermetus* (mollusc)	Mucus forms veil to trap particles, later swallowed and new veil formed
	Setous	*Daphnia,* some aquatic insect larvae e.g. *Culex*	Chitinous setae on appendages trap small food particles. May be associated with locomotion
	Other	Basking shark, herring, flamingo, whalebone whales	Planktonic organisms filtered out of water by various mechanisms. Basking sharks and herrings use gills to filter particles out of respiratory currents; flamingo uses beak for blue-green algae; whales feed on krill using keratinous plates
Large particles	Swallowing inactive food	*Lumbricus, Arenicola*	Non-selective swallowing of mud, silt, sand etc., the organic content supplying nourishment
	Scraping and boring	*Helix, Littorina, Teredo* (molluscs), caterpillars, termites	Molluscs usually use radula, insects mandibles. *Teredo* uses shell valves. Herbivores and carnivores
	Seizing and swallowing	Coelenterates, *Nereis,* dogfish, snakes, birds, seals	No chemical or mechanical breakdown before swallowing. Prey may be paralysed. Swallowing aided by mucus secretions
	Seize and masticate before swallowing	Squid, crabs, dragonfly, many mammals	Involves specialized organs for reducing size of food particles prior to digestion e.g. beak of squid, chelae and mandibles of crab; maxillae and labium of dragonfly, teeth of mammals
	Seize and digest externally before swallowing	Starfish, spiders, blowfly larvae	Enzymatic digestive juices secreted over, or injected into, food mass to make it small enough to swallow. In spiders, pedipalps inject proteolytic enzyme into prey
Fluids or soft tissues	Piercing and sucking	Leeches, hookworms, female mosquitoes, aphids, fleas, lice, vampire bats	Mouth parts modified for piercing and also sucking devices developed. Often feed on plant nectar and sap but may feed on blood. Often ectoparasites (hookworms are endoparasites). Blood feeders have anticoagulant in saliva. May be vectors of endoparasitic organisms
	Sucking only	Nematodes, lepidoptera, male mosquitoes, housefly	Often live on plant secretions like nectar or as endoparasites surrounded by their food source. Simple suctorial action only
	Absorb through general body surface	*Trypanosoma, Monocystis,* cestoda	Only in animals permanently bathed in fluid nutrient medium. No real feeding mechanisms and usually no alimentary canal. Generally endoparasitic

Physical digestion

To make food manageable for movement along the alimentary canal and give it a larger surface area for enzymes to act on, different mechanisms for physically breaking up the food have evolved. The structures involved include a gizzard (earthworms, birds), a radula (gastropod molluscs), a gastric mill (crayfish), mandibles (insects) and teeth (vertebrates). The teeth in particular are well adapted to the diet. Indeed, not only teeth, but the whole digestive system has become modified according to whether the organism is herbivorous or carnivorous.

Table 6.2 The main differences between herbivorous and carnivorous mammals

Herbivores	*Carnivores*
Open pulp cavity in teeth	Closed pulp cavity in teeth
Incisors for cutting or gnawing food or absent e.g. upper jaw of sheep	Incisors pointed for nipping and biting
Canines small or absent leaving gap called diastema e.g. in rabbit	Canines very well developed for piercing and tearing
Carnassial teeth absent	Carnassial teeth present for shearing flesh
Cheek teeth flattened with ridges of enamel and grooves of dentine. Adapted for grinding	Cheek teeth pointed (cusps) and adapted for shearing
Upper teeth meet lower teeth to facilitate grinding	Upper jaw wider than lower to facilitate shearing action
Articulation of lower jaw permits lateral movement	Articulation of lower jaw prevents lateral movement and therefore reduces the danger of dislocation
Skull smooth without well-developed processes for muscle attachment	Coronoid process, zygomatic arch and sagittal crests well developed for the attachment of large powerful muscles
Cellulose digestion carried out by micro-organisms in rumen and reticulum (ruminants) or caecum and appendix	No cellulose in diet therefore no rumen; caecum and appendix are small
Relatively long alimentary canal to digest large volumes of vegetation	Relatively short alimentary canal because diet rich in protein with relatively little indigestible material

Chemical digestion

Chemical digestion is carried out primarily by enzymes with the aid of many other substances that help to break up the food (e.g. bile salts) and adjust the pH to the optimum for particular enzymes.

Table 6.3 Summary of digestion

Organ	*Secretion*	*pH*	*Site of action*	*Production induced by*	*Secretion (including enzyme)*	*Effect*
Salivary glands	Saliva	Slightly alkaline	Mouth	Visual/olfactory expectation; reflex stimulation	Salivary amylase	Starch → maltose via dextrin
					Mucin	Sticks bolus
					Salts	Correct pH
Gastric glands of stomach	Gastric juice	Very acid	Stomach	Gastrin (hormone) Reflex stimulation	Hydrochloric acid	Provides correct pH
					Pepsin	Protein → polypeptides
					Rennin (chymase)	Caseinogen → casein
					Mucus	Lubrication; prevents autolysis
Liver	Bile	Alkaline	Duodenum	Cholecystokinin induces release of bile. Reflex action. Secretin causes production of bile	Bile salts	Emulsify fats
					Bile pigments	Excretory products of haemoglobin breakdown
					Sodium hydrogen carbonate	Correct pH
					Cholesterol	Excretory

Table 6.3 (cont.)

Pancreas	Pancreatic juice	Alkaline	Duodenum	Secretin induces production of pancreatic juice Pancreozymin induces release of pancreatic juice	Trypsin	Protein → peptides and amino acids
					Carboxypep-tidase	Peptides → amino acids
					Chymotrypsin	Casein → peptides + amino acids
					Amylase	Starch → maltose
					Lipase	Fats → fatty acids and glycerol
					Nucleases	Nucleic acids → nucleotides
Small intestine wall	Intestinal juice	Alkaline	Small intestine	Mechanical stimulation of intestinal lining	Enterokinase	Trypsinogen → trypsin
					Aminopeptidase	Peptides → amino acids
					Amylase	Starch → maltose
					Maltase	Maltose → glucose
					Sucrase	Sucrose → glucose and fructose
					Lactase	Lactose → glucose and galactose
					Nucleotidases	Nucleotides → organic bases, pentose, sugar and phosphoric acid

Absorption

The soluble products of digestion are absorbed both by diffusion and active transport. To aid these processes an absorbing surface such as the villi of the small intestine has the following features:

1 It is thin (usually a single-celled epithelial layer).
2 It is moist.
3 It is well supplied with blood vessels (to maintain a diffusion gradient by removing the absorbed material).
4 It has a large surface area (for more rapid absorption).

The absorbed materials are transported to various organs of the body and either utilized immediately or stored for later use.

Table 6.4 Absorption and fate of major digestive products

End product of digestion	Form in which absorbed	Where absorbed	Mechanism of absorption	Fate of products	Storage	Use
Monosaccharides	Possibly as a complex e.g. phosphate sugar	Capillaries of villi	Diffusion and active transport	Most remain in general circulation. Excess converted to glycogen in the liver	Glycogen in liver	Respiratory substrate
Fatty acids and glycerol	Fatty acids and glycerol; some as droplets of neutral fat (chylomicrons)	Mainly lacteals in villi; some into capillaries	Diffusion and active transport	Fat enters blood stream from lymphatic system (in neck)	Fat under skin, in mesenteries etc.	Respiratory substrate; insulation; protection; phospholipids (structural)
Amino acids	Amino acids	Capillaries of villi	Diffusion and active transport	Deamination in liver: nitrogenous residues converted to urea; carbon residues mainly in general circulation but some used for carbohydrate or fat synthesis	After deamination as fat or carbo-hydrate	Protein synthesis; deaminated proteins may become part of carbohydrates or fats

Elimination

Unwanted food is eliminated by periodic egestion from the organism, a process known as defaecation. It is important to note that, as such material has never crossed an epithelial boundary or been involved in the organism's metabolism, the process is elimination (egestion) and not excretion.

Interconversion

Although each absorbed product of digestion has a particular role in the body, a temporary shortage of one product can be compensated for by conversion of other readily available products.

Table 6.5 To show interconversions of absorbed food products

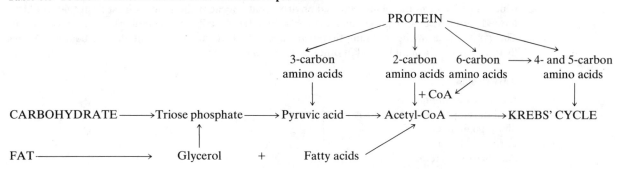

Link topics

Section 3.1 Carbohydrates, lipids and proteins
Section 3.2 Enzymes (including biochemical techniques)
Section 6.1 Autotrophic nutrition (photosynthesis)

Suggested further reading

Jennings, J. B., *Feeding, Digestion and Assimilation in Animals*, 2nd ed. (Macmillan 1972)
Sanford, P., *Digestive System Physiology* (Arnold 1982)
Swindler, D. R., *The Teeth of Primates*, Oxford/Carolina Biology Reader No. 97 (Packard 1978)

QUESTION ANALYSIS

5 Give an account of the adaptations of mammals to a carnivorous mode of life.   (10)
  Contrast the teeth of a herbivore with those of a carnivore.   (5)
  Briefly discuss the expression 'carnivorous plants'.   (5)
  Mark allocation 20/100   Time allowed 30 minutes   In the style of the Oxford and Cambridge Board

This is a structured essay question testing candidates' knowledge of herbivores and carnivores in a way that demands both a broad knowledge of the adaptations to a particular diet and the ability to contrast the two types of feeding.

In the first part where the candidate is required to give an account of the adaptations of carnivorous mammals, it is important not to restrict such an account to teeth. Methods of locomotion, adaptations for catching prey and modifications of the alimentary canal should all be included. The second part likely to escape the notice of many candidates is that carnivorous mammals are not restricted to the order Carnivora. The account should also include whales, ant-eaters and insectivores such as the hedgehog. Even within the Carnivora, fish-eating mammmals such as seals are likely to be ignored.

A logical approach would be to consider the following in turn, for a full range of mammals (examples are given in the answer plan):

1 Adaptations for finding and catching prey
2 Adaptations for ingesting the food
3 Adaptations for digesting the food.

A good account will include less obvious examples such as the extremely elongated canines of a walrus for digging up buried crustaceans and molluscs and their flattened molars for crushing them; the long, sticky, protrusible tongue of ant-eaters and the baleen of certain whales. Other important features include claws in the cat family, the spike-like teeth of seals and the overall shorter gut in most carnivores. The candidate should not forget some of the most common adaptations of carnivores, including well developed canines and incisors, carnassial teeth, absence of diastema, closed roots, powerful jaw muscles and the up and down jaw action to avoid dislocation. The important point, however, is not to limit the essay to these features alone. A common failing of candidates is to list adaptations as a series of features without ever indicating how the feature is related to a carnivorous mode of nutrition. The contrast between the teeth of a herbivore and those of a carnivore, required in the second part, are summarized in the early part of the table in the essential information. With only five marks available the points should be brief. The key words are 'contrast' and 'teeth'. The account should therefore only include features where the two types of teeth differ. There is always the possibility that, after the wide ranging adaptations required in the first part, the candidate will broaden the discussion to include contrasting features concerned with jaw motion, etc., rather than restricting it to teeth.

The final part of this question is rather open-ended. The word 'discuss' suggests that there is room for opinion and argument and that marks will probably be obtained for any relevant well-reasoned and supported answer. 'Carnivorous' comes from two Latin words, *carno* meaning 'flesh' and *vorare* meaning 'to devour'. The question clearly refers to insectivorous plants such as the Venus fly trap, sundew, pitcher plant and bladderwort. In various ways such plants capture animals, usually insects, which they digest in order to obtain nutrients. As all such plants are photosynthetic the prey is clearly not a primary source of food, indeed it is never a direct one. The prey is simply a source of some valuable element which is deficient in the plant's normal environment. Because the plants digest flesh they could be termed 'carnivorous' though whether 'devour' adequately describes their feeding behaviour is doubtful. The word devour implies consumption in an eager or greedy manner. While this is true of the Venus fly trap, the same cannot be said of others such as sundew.

Answer plan

Carnivorous adaptations
Well-developed canines and incisors
Sharp pointed teeth
Carnassial teeth
Absence of diastema
Closed roots to teeth   These are features of most carnivores,
Powerful jaw muscles   particularly the dog and cat family
Up-and-down jaw action only
Less developed caecum and colon
Fast moving
Well developed sense of smell

Other features specific to particular groups:
Claws – cat family
Spiked teeth – Piscivores, e.g. seal
Long, sticky tongue – e.g. ant-eater
Short elongated snout – Insectivores, e.g. hedgehog
Baleen – baleen whales
Extremely long canines to dig up prey – e.g. walrus
Flat molars to crush prey – e.g. walrus

Contrast of herbivore and carnivore teeth:
These are outlined in the table in the essential information.

Carnivorous plants:
'carno' = prey; 'vorare' = to devour.
Capture prey but slowly digest rather than 'devour' it. Prey is not a direct energy supply but a source of some necessary mineral, e.g. nitrogen.

6 (a) What roles do the liver and pancreas play in
   (i) food digestion
   (ii) metabolism of absorbed products?
   (b) How can a diet of raw liver prevent the disease pernicious anaemia?
   Mark allocation 20/100   Time allowed 40 minutes   In the style of the Joint Matriculation Board

This is an essay question which tests recall of factual information on much of heterotrophic nutrition. The main difficulty with this question is to ensure that all the required information is included. It is possible for a candidate to write an essay covering many important points and yet only include half those on the examiner's answer scheme.

In (a) the four key words are 'liver', 'pancreas', 'digestion' and 'metabolism'; each of the organs needs to be related to each of the processes. A useful method of planning such an essay is a matrix as shown in the final answer plan. Although no mark distribution is given for the parts of the question it should be obvious that part (b) has a very short answer and therefore much of the time allowed, say 80%, should be devoted to part (a). This underlines the need to read the whole question and to plan carefully before starting an answer. Provided revision has been adequate such planning should produce many roles for both liver and pancreas and so indicate that there is little, if any, time for a detailed account of these roles; a brief explanation must suffice. One danger is that candidates will give a detailed account of two or three roles and completely ignore the remainder. If only half the roles are covered only a maximum of half the marks can be obtained regardless of how well they are done. A common failing of candidates is to give much detail on one role and very little on others. While some roles are clearly more important and may warrant a little more detail, this must not be hopelessly out of proportion to the detail given to others. Three sides of writing on one role followed by a list of ten others is not what is required here.

In dealing with the roles of the pancreas and liver in digestion, it is probably best to begin by listing all constituents of the pancreatic and bile juice and then discount the few substances such as bile pigments and cholesterol that are excretory and play no role in digestion. The remainder can then be included with a short account of the function they perform (see answer plan). Care should be taken to include all digestive substances and not just enzymes. For instance mention should be made of the role of bile salts in fat emulsification thereby increasing the surface area for lipase action. In dealing with metabolism it is important to limit the answer to 'absorbed products' as the question asks. The unwary candidate may waste much time dealing with all functions of the liver. Perhaps the most important aspect of answering questions well is the selection of relevant facts and the dismissal of superfluous ones. Blanket recall of all learned facts on a topic rarely brings good marks at this level.

Although part (b) clearly requires a much shorter answer than part (a) it needs more than a statement such as 'the liver contains vitamin B_{12}'. The candidate should give details of the precise mechanism by which pernicious anaemia is prevented by vitamin B_{12}. The question says 'raw liver' and a good candidate may deduce that cooked liver is not such a good source of vitamin B_{12}, since it can be destroyed by prolonged high temperatures. Vitamin B_{12} is stored in the livers of animals which are therefore a rich source of it. It is essential for the formation of blood in the red bone marrow and, as it cannot be synthesized by the body, a lack of it in the diet leads to a drastic, even fatal, reduction in the number of erythrocytes. The patient becomes pale and weak, rapidly worsening without vitamin B_{12}.

Answer plan

(a)	(i) *Digestion*	(ii) *Metabolism*
Pancreas	Trypsin secreted in inactive form – protein to polypeptides and amino acids Polypeptidase – polypeptides to amino acids Chymotrypsin secreted in inactive form – casein to polypeptide Amylase – starch to maltose Lipase – fats to fatty acids and glycerol Nuclease – nucleic acids to nucleotides Mineral salts – neutralize acid chyme from stomach	Insulin – lowers blood sugar level converting it to glycogen, fat or metabolizing it to carbon dioxide and water Glucagon (hormone) – raises blood sugar level by facilitating the breakdown of glycogen into glucose
Liver	Bile salts – emulsifying fats Mineral salts – neutralize acid chyme from stomach	Regulation of blood sugar level – stores glycogen and is the site of the interconversions of glucose and glycogen Regulation of blood lipid level – controls inter-conversion of blood lipids and fat deposits Breaks down lipids Deamination of amino acids – breaks down excess amino acids to carbohydrate and urea Stores vitamins – stores excess vitamins A, D and B_{12} Detoxification – removes or renders harmless toxic materials such as alcohol, cholesterol and hydrogen peroxide

(b) Raw liver contains B_{12} – needed for formation of erythrocytes – lack of it leads to paleness, listlessness and death (pernicious anaemia).

7 The figure below is a diagramatic representation of part of a transverse section through the duodenal wall of a mammal.

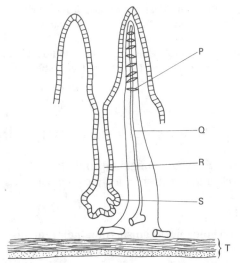

Fig. 6.8 Section through a part of the alimentary canal ×80

(a) Identify the parts labelled P Q R S and T.   (5)

(b) State four features, visible in the diagram, which enable the duodenum to carry out its functions of digestion and absorption. Relate each of the four structural features to the part which it plays in digestion or absorption.   (4)

(c) Comment on the pH conditions which prevail from the stomach to the ileum and the effect of these varying conditions on enzyme activity.   (5)

(d) How is autodigestion of the intestine wall prevented in the presence of proteolytic enzymes?   (2)

(e) The villi of the intestine possess muscles which enable them to move. Suggest a possible advantage of this movement.   (2)

(f) Name two hormones which stimulate the secretion of digestive juices into the gut.   (2)

Mark allocation 20/140   Time allowance 25 minutes

Associated Examining Board November 1979, Paper II, No. 1

This is a longer structured question testing recall of factual information and the ability to relate the structure of the duodenum to its function.

In part (a) the candidate should look exactly where the label line goes to before committing himself to an answer. Although P has the shape of smooth muscle cells and these are present in the walls of the villi, these cells are not arranged longitudinally. In any case P connects the arteriole and venule and is almost certainly part of the capillary plexus. Q is a lacteal (lymph vessel) and R is a crypt of Lieberkühn. S is more of a problem; it could be a Paneth cell that lines the crypt of Lieberkühn, although its overall shape suggests S is meant to represent the Brunner's gland. T represents the longitudinal and circular muscle layers.

In selecting the features in (b) the candidate should choose those most easily related to their role in digestion or absorption. Probably the most obvious are the villi with their large surface area to increase the rate of absorption. The capillary plexus is used to remove absorbed food and so maintain a gradient of diffusing molecules. Capillary structure is related to this role as they are numerous, thin and close to the villi epithelium. The muscle layers help maintain a diffusion gradient by moving food along the duodenum; furthermore they churn food within it to ensure efficient mixing of food and enzymes. Their structure is related to this role in that longitudinal and circular layers allow the duodenum to be made narrower and wider respectively. The epithelium may be related to the function of absorption because it comprises a single layer of cells and hence is thin.

The pH conditions referred to in (c) are very acid in the stomach (as low as pH 2) but become progressively more neutral or even slightly alkaline (pH 8) in the ileum. In considering the effect of pH on enzyme activity candidates should not limit their comments to how a particular pH stimulates enzyme activity by providing an optimum pH, but also discuss how other enzyme activities are prevented by an unfavourable pH. In the stomach, for instance, acid conditions favour the action of pepsin on proteins but prevent further action of salivary amylase on starch. In the same way the more neutral conditions in the duodenum favour the action of pancreatic amylase, trypsin and lipase but halt further action of pepsin. Enzymes such as maltase, sucrase, lactase and peptidase all work well in the neutral or slightly alkaline pH of the ileum.

In a question such as (d) the candidate should give as many answers as possible and not settle for one. Indeed the mark allocation of 2 may reasonably suggest there are two means of preventing auto-digestion, at least. The main methods are the production by the intestinal wall of enzymes in an inactive form which are only activated when they are in the lumen of the intestine, e.g. pepsin is produced as pepsinogen and only activated in the presence of dilute hydrochloric acid. The other method is the production of mucus by the wall of the intestine which acts as a protective barrier against the proteolytic enzymes. The key word in (e) is 'villi' and a common mistake is to discuss peristalsis which is performed not by muscles in the villi but by the muscle layers of the muscularis externa. The villi muscles are to allow the villi to move, helping to churn food which prevents accumulation next to the villus of a layer of food from which all soluble products have been absorbed. The advantage of this is that diffusion is considerably speeded and the efficiency of absorption increased.

Part (f) is straightforward recall and the candidate should choose any two from gastrin, cholecystokinin, pancreozymin and secretin (see table summarizing digestion in the essential information).

6.3   Heterotrophic nutrition (saprotrophs and parasites)

Assumed previous knowledge An appreciation of the term heterotrophic nutrition (see Section 6.2).

Structure and feeding of a typical fungal saprophyte, e.g. *Mucor*.

Details of the life cycle and nutrition of a typical parasite, e.g. *Taenia, Schistosoma*.

Underlying principles

When autotrophic and holozoic organisms die, their remains comprise complex chemicals which may be broken down to simpler forms. The energy released in this process is utilized by groups of organisms called saprotrophs, which speed up the breakdown of the dead and decaying remains. The energy released is used to build up their own complex structures. In addition saprotrophs play a vital role in releasing valuable elements such as carbon, nitrogen and phosphorus in a form in which they can be absorbed by autotrophs and so be recycled. It is possible that such organisms arose when an autotrophic organism became unable to photo-synthesize, e.g. lost the ability to make chlorophyll. If it were to survive it would need to feed heterotrophically. With no means of obtaining and ingesting complex food it would need to live on dead organisms and digest them externally, absorbing the soluble products.

In a similar way any organism that lost, by mutation, the ability to obtain or utilize a particular substance, would need to obtain it from some external source. One way would be to obtain it from another living organism. This is the basis of parasitism. In due course the parasite becomes increasingly dependent on its host for other nutrients. It is not in the parasite's interest to kill the host, indeed the less obvious it makes its presence the more likely it is to survive. The host and parasite may develop such a close relationship that they become mutually beneficial. This is symbiosis.

Points of perspective

It would be an advantage to have a wide knowledge of the economic importance of saprophytes and parasites. An understanding of the various methods of controlling important human parasites would be beneficial. A good candidate will understand the problems of defining terms such as parasitism, saprotrophism, symbiosis, mutualism and commensalism, and appreciate that the divisions between the categories are not clear.

Essential information

There are three types of heterotrophic nutrition:

1 holozoic
2 saprotrophic ('sapros' = rotten; 'trophe' = nourishment)
3 parasitic.

Saprotrophic nutrition Saprotrophs are organisms obtaining their energy from the dead remains of other organisms. Saprotrophic plants are known as **saprophytes** and animals as **saprozoites.** The importance of saprotrophs lies in their role in breaking down complex organic material and so releasing the valuable elements such as nitrogen, carbon and sulphur, usually in the form of an oxide, e.g. nitrate, carbon dioxide, sulphur dioxide, or in a reduced state, e.g. ammonia, methane. Many bacteria and fungi are saprotrophs, e.g. *Myxococcus, Mucor.*

Parasites Parasitism is an association between two organisms whereby one (the parasite) derives some benefit and the other (the host) does not, and may in fact be harmed. However, the most well-adapted parasites inflict little damage on their hosts.

Plants are parasitized mainly by fungi and bacteria; they are poor hosts for invertebrates because:

1 They remain in one position and so it is more difficult for the parasite to reach a new host.
2 Their cellulose cell walls are difficult for animals to digest.
3 They have few internal cavities suitable for parasites.

The most successful invertebrate parasites of plants are nematodes. In animals the habitats most used are:

1 The body surface and its infoldings such as the buccal cavity, lungs, external gills and nostrils.
2 The alimentary canal and its associated organs such as the liver and bile duct.
3 Internal tissues such as blood and muscle.

Parasites using the body surface as a habitat are termed **ectoparasites** (*ektos*=outside) and those using internal habitats are **endoparasites** (*endon*=within).

Most parasites are endoparasites because the inside of the body provides an environment which is rich in nutrients and not subject to extremes of environmental conditions.

Transmission of parasites The transmission of ectoparasites is relatively simple, e.g. by jumping in the flea *Pulex*. Since ectoparasites are subjected to varying conditions existence away from the host can be tolerated temporarily. The transmission of endoparasites however presents three main problems:

1 leaving the host
2 living away from the host
3 entering a new host.

Table 6.6 Methods used by parasites to escape from their hosts

Habitat	Example	Method
Lungs	*Bacillus tubercle*	Through the air
Intestines	*Ascaris*	In the faeces
Intestines (larvae encysted in muscle)	*Trichinella*	Disintegration of host tissues after death
Blood	*Schistosoma*	Urine (eggs in wall of bladder)
Blood	*Plasmodium*	Intermediate blood sucking organism
Under skin	*Dracunculus*	Discharge larvae through skin of host

Living away from the host
There may be:

(a) dormant stages, e.g. *Taenia*
(b) free-living stages, e.g. *Haemonchus*
(c) intermediate hosts, e.g. *Fasciola*.

Entering a new host
The parasite may:

(a) be eaten, e.g. *Taenia*
(b) penetrate the skin or body surface of a new host after leaving an intermediate host, e.g. *Schistosoma*
(c) be injected by an intermediate host, e.g. transmission of *Plasmodium* by the mosquito.

The main features of parasites

1 A means of attachment to help the parasite maintain its position in/on the host, e.g. hooks, suckers.
2 Degeneration of certain unnecessary organ systems, e.g. the digestive system of gut parasites is reduced or absent because the parasite is surrounded by a large supply of predigested food. There are often special mechanisms developed to move a parasite from one host to another and so there is no need for a well developed locomotory system.
3 Protective devices are often developed to protect the parasite from the host's defences, e.g. schistosomes absorb glycolipids from the host on to their surface to imitate the host tissue and thus forestall immunological attack.
4 Penetrative agents such as cellulases in fungal plant parasites.
5 Vector or intermediate host, e.g. *Fasciola* (liver fluke of sheep), requires an intermediate host, the snail, for the completion of its life cycle. The blood-sucking female mosquito *Anopheles* is the intermediate host for *Plasmodium* (malarial parasite) and is thus the vector of the disease malaria.
6 Production of many eggs because of the difficulty of finding a new host, e.g. *Diphyllobothrium* (fish tapeworm) produces 36 million eggs a day.
7 Dormant or resistant phases to overcome the period spent away from the host, e.g. eggs in *Ascaris*, encysted cercariae in *Fasciola*.

Symbiosis This term is most commonly applied to a relationship in which both organisms benefit, e.g. *Zoochlorella* inhabiting the endodermal cells of *Hydra viridissima*. In this relationship *Zoochlorella* obtains protection, shelter, carbon dioxide (for photosynthesis) and nitrogen (from excretory wastes) from *Hydra; Hydra* obtains oxygen and carbohydrates from the photosynthesis of *Zoochlorella*. Lichens are a symbiotic relationship between an alga and a fungus in which the alga gains water, mineral salts and protection from desiccation and the fungus obtains oxygen and carbohydrates. The hermit crab *Pagurus* gains protection and camouflage from the sea anemone *Adamsia palliata* which lives on its shell. The sea anemone obtains food dropped by the hermit crab.

Alternative classification of relationships In some schemes symbiosis is taken to mean 'living together' and this is subdivided as follows:

1 Mutualism: a relationship in which both organisms benefit.
2 Commensalism: a relationship in which one organism benefits and the other neither benefits nor suffers.
3 Parasitism: a relationship in which one organism benefits and the other is harmed.

Link topics

Section 6.1 Autotrophic nutrition (photosynthesis)
Section 6.2 Heterotrophic nutrition (holozoic)
Section 11.1 Ecology

Suggested further reading

Baker, J. R., *The Biology of Parasitic Protozoa*, Studies in Biology No. 138 (Arnold 1982)
Jackson, R. M. and Mason, P., *Mycorrhiza*, Studies in Biology No. 159 (Arnold 1984)
Lyons, K. M., *The Biology of Helminth Parasites*, Studies in Biology No. 102 (Arnold 1978)
Whitfield, P. J., *The Biology of Parasitism, an introduction to the study of associating organisms* (Arnold 1979)
Wilson, R. A., *An Introduction to Parasitology*, Studies in Biology No. 4, 2nd ed. (Arnold 1979)

8 When legumes are grown on sterile soil they do not develop fully. The addition of certain living, nitrogen-fixing bacteria to the soil improves the growth of the legume. Nitrogen-fixing bacteria can be isolated from nodules that form on the roots and can be shown to utilize carbohydrates formed from the legumes. The relationship between the organisms is:

A – commensalism   C – parasitism

B – symbiosis   D – saprophytism

Mark allocation 1/40   Time allowed 1½ minutes   *In the style of the Cambridge Board*

This is a multiple choice question testing application of knowledge on modes of nutrition. There are three main descriptive sentences before the actual question. Each sentence can be used to narrow progressively the options to the correct one. The word 'legumes' in the first sentence may be a clue to candidates, many of whom will immediately think of nitrogen fixation by symbiotic bacteria such as *Rhizobium* that live in nodules on the roots of some leguminous plants. It would however be grossly unwise to decide upon the answer on the basis of the word 'legume'. In effect this first sentence states 'no bacteria – poor growth'. The second sentence states that nitrogen-fixing bacteria improve growth. This eliminates C, because the plant would not gain advantage if the bacteria were parasitic. D is also eliminated since the legume is clearly alive if its growth is improving. The possibility of the legume living saprophytically on the bacteria can also be discounted as all legumes are autotrophic. The remaining choices both involve benefit by one organism at least. For B to be correct however, both must benefit. The third sentence confirms that bacteria can utilize carbohydrate from the legume, and presumably benefit from it. B is therefore the correct response.

9 (a) Explain the term 'saprophyte'.

  (b) Compare and contrast saprophytes with parasites.

  (c) In what ways are saprophytes important to humans?

Mark allocation 28/140   Time allowed 35 minutes   *In the style of the Oxford Board*

This essay question, largely testing recall of knowledge about saprophytes, also requires candidates to have knowledge of parasites adequate to make comparisons with saprophytes.

In defining a saprophyte in (a) (see essential information) it is useful to make it apparent that the definition is not absolute and the division between saprophytes and parasites is at times unclear.

In part (b), however, it is essential to avoid any of the problem border-line cases and keep rigidly to examples which clearly belong to one group or the other. The key words here are 'compare and contrast' indicating that both similarities and differences are sought. The similarities between saprophytes and parasites are that they are both heterotrophic, both absorb soluble food and where a digestive system exists it is relatively simple. Both types exhibit sexual and asexual phases of re-production often involving resistant phases and the production of large numbers of offspring.

The most obvious difference between the two is that while saprophytes obtain their energy from dead organisms, parasites obtain it from living ones. Further differences may not be apparent to some candidates, but clearly more are needed for an adequate A-level answer. Many candidates at O-level study *Mucor* (pin mould) as a saprophyte example and *Taenia* (tapeworm) as a parasite. There is therefore a temptation to compare and contrast these two examples specifically rather than saprophytes and parasites in general. In a number of ways both *Mucor* and *Taenia* are not typical of the types they represent. For instance most parasites, especially ectoparasites, are aerobic; rarely are they completely anaerobic. Saprophytes are often anaerobes. *Taenia* and *Mucor* are exceptions to this general trend. Another difference is that parasites are very specific in the species from which they obtain energy whereas saprophytes are not. Parasites have some representatives in almost all plant and animal groups, but saprophytes are mainly restricted to the bacteria and fungi. Parasites are more highly adapted nutritionally than saprophytes. In a similar way parasites, e.g. *Plasmodium, Puccinia,* have complex life cycles often with different stages. Saprophytes have far simpler life cycles without various intermediate stages.

A list of the importance of saprophytes to humans is given in the answer plan. It should be expanded to provide additional detail in so far as time allows.

Answer plan

(a) Saprophytes are plants which obtain energy from the dead remains of other organisms.

(b)(i) Similarities

Both

1 are heterotrophic

2 absorb soluble food

3 have simple digestive systems where they are present

4 have sexual and asexual phases in reproduction, often involving resistant stages.

5 produce large numbers of offspring.

(ii) Differences

Parasites	**Saprophytes**
Energy derived from living organisms	Energy derived from dead organisms
Many stages in life cycle	Usually a single adult stage + spores
Very specific to their hosts	Use a variety of food sources

Parasites	Saprophytes
Nutritionally highly adapted	Simple methods of nutrition
Most plant and animal groups have representatives	Almost totally bacteria and fungi
Most aerobic	Anaerobic and aerobic

(c) Importance of saprophytes

1 Recycling of materials, e.g. carbon, nitrogen, phosphorus

2 Brewing and baking, e.g. *Saccharomyces* (yeast)

3 Making antibiotics, e.g. Penicillin

4 Decomposition of wastes such as sewage

5 Production of yoghurt and cheese

6 Food source, e.g. mushrooms

7 Industrial processes, e.g. tanning of leather, production of vitamins

10 (a) Discuss the main advantages of a parasitic mode of life.   (10)

(b) Give an account of the principle adaptations shown by parasites, using examples from both plant and animal kingdoms.   (10)

Mark allocation 20/60   Time allowed 35 minutes

Welsh Joint Education Committee June 1979, Paper A1B, No. 14

This is an essay question testing knowledge of parasitism. It requires not only a detailed and broad knowledge of the topic but also an understanding adequate to discuss the advantage of this mode of life.

In part (a) the key word is 'discuss'. The candidate must therefore put forward reasoned arguments to support the views expressed. As always 'discuss' implies that there are a variety of views; a number of different answers may all obtain high marks. The question says 'main advantages' so a few should be discussed in detail rather than many superficially. Parasitism is essentially a nutritional relationship; thus the most important advantage for the parasite is that it has a source of food, or at least one requirement of its diet, that it would otherwise not be able to obtain or manufacture. Not only is food available but it is often in a readily usable form. This is particularly true of blood and gut parasites whose food has been digested and needs only to be absorbed. An additional advantage to the parasite is that it need not maintain a complex digestive system to break down its food. As the parasite has a relatively constant supply of food, mechanisms of storage are unnecessary except during dispersal. Candidates should try to support such comments with actual examples and these should be drawn as far as possible from both plant and animal parasites and hosts. In this case the tapeworm *Taenia* is an obvious example of an animal with a degenerate digestive system.

The second major advantage is that the parasite is buffered to some degree from the external environment and lives in a relatively constant environment. This is especially true of parasites of mammals and birds which live in a homeostatically controlled environment, but not generally true of the parasites of plants. Other hosts, however, avoid harsh conditions, either through their behaviour or by other means. Provided the parasite has a physiology that functions well in the environment of its host, it need not develop specialized mechanisms for overcoming harsh conditions. The main disadvantage of this is that it may require such mechanisms during its transfer to a new host.

Another advantage is that the parasite is afforded protection from predators. This is particularly true of endoparasites. The host's mechanisms for avoiding predators are just as valuable to the parasite. In addition the sheer size of the host normally makes it impossible for a predator to attack a parasite. The parasite can often survive with a degenerate sensory system since detection of predators, adverse conditions and food becomes unnecessary. This is not true of ectoparasites such as fleas which are not protected within the host.

Overall the advantage may be summarized by saying that many of the physiological processes of the parasite are carried out for it by the host. The parasite can therefore allow some of its own systems to degenerate and acquire a more simple anatomy and physiology. The notable exception is the reproductive system which is often especially well developed. The simplification of structure can present problems to the parasite during its existence away from the host, e.g. during its dispersal.

Part (b) of the question stresses the need for examples from both the plant and the animal kingdoms. Despite such emphasis candidates frequently give few examples, use the same repeatedly, or restrict them to either plants or animals (usually the latter). Plan in advance which examples you propose to use and then check to see they are about equally divided between the plant and animal kingdoms. If possible use examples from a number of groups within a kingdom rather than restrict them to fungi and protozoa. Alter the planned examples until the balance is correct before commencing the essay. Do not start writing in the hope that a good balance will develop of its own accord. The question asks for the 'principle adaptations' and so again the major ones should be dealt with in detail and those shown by only a few types and those of little significance may be ignored. Keep to adaptations of parasites in general rather than specific examples. A range of such adaptations is given both in the essential information and the answer plan. It is important not just to list these adaptations but to show exactly how each one is useful in a parasitic mode of life.

Answer plan

(a) Advantages of parasitic mode of life:
1 Constant supply of food
2 No need to store food
3 Relatively constant environment – protected from climatic extremes (especially endoparasites of homeotherms)
4 Protection from predation (especially endoparasites)
5 Degeneration of many systems and simplified form means less energy required to maintain themselves, and allows rapid development.

(b)
1 Protective devices to prevent destruction by the host's defence mechanisms, e.g. *Schistosoma*
2 A means of attachment, e.g. hooks on legs of *Pulex* (flea)
3 A means of penetrating host, e.g. digestive enzymes produced by *Erysiphe* (powdery mildews)
4 Degeneration of parasite systems, e.g. simplified locomotory, digestive, excretory and sensory systems in *Taenia* (tapeworm)
5 Use of a vector or intermediate host either as a means of entering the primary host or surviving when the primary host is not available, e.g. *Puccinia* (wheat rust/fungus) uses the barberry bush to survive during periods when wheat is not available (e.g. winter)
6 Production of large numbers of offspring, e.g. 36 million eggs a day produced by the fish tapeworm *Diphyllobothrium*
7 Dormant or resistant phases to overcome periods away from host, e.g. dormant asci in *Venturia* (apple scab fungus).

7 Respiration

7.1 CELLULAR RESPIRATION

Assumed previous knowledge The simplified overall equation for the oxidation of glucose. Mechanism of enzyme action, properties of enzymes and biochemical techniques (Section 3.2). Structure of a mitochondrion (Section 2.1).

Underlying principles

All energy exchanges whether in living or non-living systems are governed by the laws of thermodynamics. The first of these states that energy cannot be created or destroyed, only changed from one form to another. The second states that natural processes proceed when there is an increase in the entropy (randomness) of a system. In effect this means that systems tend to lose their energy and assume their lowest possible energy level. Living organisms are highly ordered (have low entropy) which they achieve by the breakdown of glucose in the process called respiration. In doing so they are increasing the entropy of their environment.

The breakdown of glucose occurs in three main stages in aerobic respiration and one in anaerobic respiration. In both cases the first stage is the splitting of the 6-carbon glucose molecule into two 3-carbon molecules (glycolysis). In anaerobic respiration this is followed by one of a variety of reactions. In aerobic respiration the other stages are the tricarboxylic acid (Krebs') cycle and the electron carrier pathway. All processes have evolved to allow the controlled release of the energy which is temporarily stored in chemical form as adenosine triphosphate (ATP).

Points of perspective

It would be an advantage to understand the basic chemistry of oxidation and reduction reactions. Although examination boards do not require knowledge of the chemical formulae of intermediates in the biochemical processes of respiration, a good candidate will at least know the names of the major intermediates. It would be helpful to know the inter-relationships between the respiratory pathway and other metabolic pathways, e.g. photosynthesis. In particular, it would be useful to know where fatty acids, glycerol and amino acids can be fed into the respiratory pathway.

Essential information

The energy essential to maintain the highly structured state of living organisms is derived from the food they manufacture (autotrophs) or consume (heterotrophs). Whatever form the food initially takes it is converted into carbohydrate, usually the hexose sugar glucose, before being respired. Respiration is basically the oxidation of the glucose to carbon dioxide and water with the release of energy. It can conveniently be divided into two parts:

1 Cellular (internal or tissue) respiration: the biochemical processes which take place within living cells that release the energy from glucose.
2 Gaseous exchange (external respiration): the processes involved in obtaining the oxygen needed for respiration and removing the gaseous wastes such as carbon dioxide.

Cellular respiration is a complex metabolic process of over seventy reactions; these may be conveniently divided into three main stages:

1 Glycolysis
2 Krebs' (tricarboxylic acid) cycle
3 Electron transfer system.

Electron (hydrogen) carriers Before discussing these stages, it is necessary to mention a group of substances that are employed throughout cellular respiration. These are the electron (hydrogen) carrier or electron (hydrogen) acceptor molecules. Many reactions in cellular respiration involve the release of electrons at a high energy level. These electrons are collected by electron carrier molecules which pass them to electron carriers at lower energy levels. The energy released is used to form ATP from ADP. The electrons are initially released as part of a hydrogen atom which later splits into a proton and an electron. This is why the carriers have the alternative name hydrogen carriers. They act as coenzymes because they are essential to the functioning of the dehydrogenase enzymes that catalyse the removal of the hydrogen atoms. One of the most important is nicotinamide adenine dinucleotide (NAD) and others include nicotinamide adenine dinucleotide phosphate (NADP), flavoprotein (FP) and cytochromes.

Glycolysis The first stage in the breakdown of glucose, glycolysis, is common to all organisms. In anaerobic organisms it is the only stage in respiration. It occurs in the cytoplasm of the cell. At this level it is not necessary to know all the intermediate compounds or enzymes but an overall appreciation of the major features is required.

Stages of glycolysis

Stage 1
The glucose molecule has a phosphate group added to make it more reactive (= activation). The phosphate group is donated by ATP

Stage 2
The glucose phosphate is reorganized into a fructose phosphate molecule

Stage 3
The fructose phosphate is further activated by the donation of a second phosphate group by an ATP molecule

Stage 4
The 6-carbon fructose diphosphate is split into two 3-carbon triose phosphate molecules

Stage 5
Hydrogen atoms are removed from the triose phosphate molecules and taken up by NAD. Inorganic phosphate is added to further activate the triose phosphate

Stage 6
A phosphate molecule is lost and ATP is regenerated

Stage 7
A phosphate molecule is lost and ATP is formed; a water molecule is also lost, for each triose phosphate

Glucose (6C)
│ ⤷ ATP
│ ⤷ ADP
▼
Glucose phosphate (6C)
│
▼
Fructose phosphate (6C)
│ ⤷ ATP
│ ⤷ ADP
▼
Fructose diphosphate (6C)
│
Triose ⟵——— Triose
phosphate (3C)   phosphate (3C)
│ ⤷ Inorganic phosphate
│ ⤷ 2 NAD (oxidized)
│ ⤷ 2 NAD (reduced)
▼
2 × Triose diphosphate (3C)
│ ⤷ 2 ADP
│ ⤷ 2 ATP
▼
2 × Triose phosphate (3C)
│ ⤷ 2H$_2$O
│ ⤷ 2 ADP
│ ⤷ 2 ATP
▼
2 × Pyruvic acid (3C)

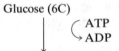

The process of glycolysis produces two molecules of pyruvic acid for each molecule of glucose degraded. Effectively every substance after Stage 4 is doubled in quantity for each glucose molecule. The total energy yield is therefore two molecules of ATP directly and six molecules of ATP produced from the two reduced NAD molecules. A total of eight ATP molecules.

Krebs' (tricarboxylic acid) cycle The pyruvic acid formed as a result of glycolysis may in the absence of oxygen be converted to a variety of substances to yield a little energy. These processes constitute the anaerobic pathways. In the presence of oxygen the pyruvic acid enters the Krebs' cycle. Before entering the actual cycle one of the three carbon atoms of pyruvic acid is oxidized to carbon dioxide and a molecule of NAD is reduced by the addition of two hydrogen atoms (the NAD yields a further three ATPs). This leaves an acetyl group (CH_3CO) which is readily accepted by a coenzyme called coenzyme A. The resulting substance is acetyl coenzyme A. The 2-carbon acetyl group of this compound combines with a 4-carbon substance called oxaloacetic acid to give the 6-carbon molecule citric acid. In a series of reactions two carbon dioxide molecules are produced and the 4-carbon oxaloacetic acid molecule is regenerated in readiness to receive another 2-carbon acetyl group from acetyl coenzyme A. Other products include a total of eight hydrogen atoms which are used to reduce three molecules of NAD and one molecule of FP. These reduced electron (hydrogen) carriers eventually pass on the hydrogen atoms to oxygen yielding 11 more ATP molecules for each pyruvic acid. In addition a further ATP molecule is yielded directly during the cycle to give a total of 12 ATPs per pyruvic acid molecule.

Differential centrifugation (see Section 3.2) may be used to separate parts of a cell. By testing each part it can be shown that the Krebs' cycle takes place in the mitochondria.

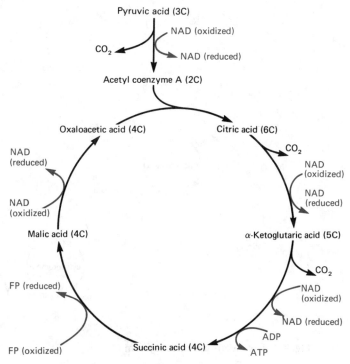

Fig. 7.1 Summary of Krebs' cycle

Importance of Krebs' cycle

1 It provides hydrogen atoms which ultimately yield the major part of the energy derived from the oxidation of a glucose molecule.
2 It is a valuable source of intermediates which are used to manufacture other substances, e.g. fatty acids, amino acids, carotenoids.

Electron transfer system Although the glucose molecule has been completely oxidized by the end of the Krebs' cycle, much of the energy is in the form of hydrogen atoms which are attached to the hydrogen carriers NAD and FP. These atoms are passed along a series of carriers at progressively lower energy levels. As they lose their energy it is harnessed to produce ATP molecules, three for each molecule of NAD and two for each one of FP. The other carriers in the system are iron-containing proteins called cytochromes. The hydrogen atoms split into their protons and electrons during the pathway and the electrons are carried by the cytochromes. They recombine with their protons before the final stage where the newly reformed hydrogen atoms combine with oxygen to form water. Although oxygen, so closely connected with aerobic respiration, only plays a role at this final stage, it is nevertheless vital since it drives the whole process. Except for the ability of oxygen to act as the final hydrogen acceptor the hydrogens would accumulate and the process of aerobic respiration would cease. The whole process whereby oxygen effectively allows the production of ATP from ADP is called **oxidative phosphorylation.** The electron transfer system occurs in the mitochondria. At each stage of the system the hydrogen atoms (or electrons) are passed from the reduced carrier to the oxidized carrier further along the pathway. Thus the reduced carrier becomes oxidized and able to accept more hydrogen atoms (or electrons) and the oxidized one becomes reduced.

e.g. for NAD and FP

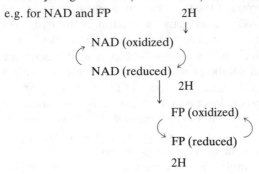

The full pathway can be summarized.

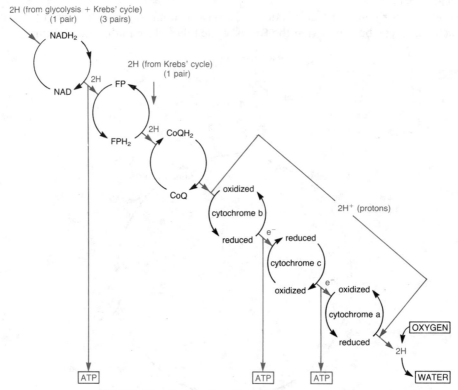

Fig. 7.2 The electron transfer system

Anaerobic pathways If no oxygen is available the pyruvic acid molecules formed at the end of glycolysis do not enter the Krebs' cycle but follow one of a number of anaerobic pathways. The anaerobic pathways are often referred to as fermentation. The pathways often use up 2H and so regenerate oxidized NAD from reduced NAD, allowing it to be used again in glycolysis.

Table 7.1 Summary of three major anaerobic pathways

Acetaldehyde fermentation	*Alcoholic fermentation*	*Lactic acid fermentation*
$CH_3COCOOH$ pyruvic acid	$CH_3COCOOH$ pyruvic acid $2H \rightarrow$	$CH_3COCOOH$ pyruvic acid $2H \rightarrow$
$\rightarrow CO_2$	$\rightarrow CO_2$	
CH_3CHO acetaldehyde, e.g. certain anaerobic bacteria	CH_3CH_2OH ethanol, e.g. yeasts and other plants. This process forms the basis of brewing and baking	$CH_3CHOHCOOH$ lactic acid, e.g. higher animals, especially in the muscles when oxygen usage exceeds supply, i.e. an oxygen debt is incurred

The role of the vitamin B complex in cellular respiration The group of vitamins collectively called vitamin B play a major role in cellular respiration, particularly by acting as coenzymes. Some of the individual vitamins and their roles in respiration are tabulated below:

Table 7.2 The roles of some B vitamins

Vitamin		Role in cellular respiration
B_1	Thiamine	Involved in formation of some Krebs' cycle enzymes. Forms part of acetyl coenzyme A
B_2	Riboflavin	Forms part of the hydrogen carrier flavoprotein (FP)
B_3	Niacin (nicotinic acid)	Forms part of the coenzymes NAD and NADP. Forms part of acetyl coenzyme A
B_5	Pantothenic acid	Forms part of acetyl coenzyme A

ATP and energy yields Adenosine triphosphate (ATP) is the form in which energy from the breakdown of glucose is temporarily stored. ATP consists of the organic base adenine, the 5-carbon sugar ribose and three phosphate groups.

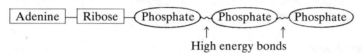

High energy bonds

The removal of the final phosphate to form adenosine diphosphate (ADP) releases about 34 kJ/mole of energy.

$$ATP + H_2O \longrightarrow ADP + phosphate + 34\ kJ$$

Although the terminal bonds are referred to as 'high energy bonds' the term is misleading because the energy is not within these bonds but spread throughout the molecule. In fact, energy is only released because there is a net lowering of energy when one bond is broken and another reformed. High energy bonds as such do not exist; they are a simplified way of expressing a more complex process. The amount of ATP produced during anaerobic respiration is small. For instance in alcoholic fermentation glycolysis yields two molecules of ATP directly and two hydrogen atoms. However the hydrogen atoms do not enter the electron transfer system as they are used to form the alcohol. The total yield is therefore only two ATPs or 68 kJ of energy from a potential of nearly 3000 kJ for the complete oxidation of a glucose molecule. The process is therefore only about 2% efficient.

In aerobic respiration the yield per glucose molecule is:

Glycolysis = 8 ATP (2 ATP molecules directly + 2 molecules of
reduced NAD which yields a further 6 ATP)
Pyruvic acid→Acetyl CoA = 6 ATP (2 molecules of reduced NAD
each yielding 3 ATP)
Krebs' cycle = 24 ATP (6 molecules of reduced NAD each yielding
3 ATP + 2 molecules of reduced FP yielding
2 ATP + 2 molecules of ATP formed
directly)
Total ATP yield = 38 molecules of ATP.

This gives a total energy yield of 38 × 30.6 kJ = 1162.8 kJ, which is about 40.3% efficient.

One problem arising from the totalling of these ATPs that causes confusion is that the yields refer to the number of ATPs from a single glucose molecule. Early in glycolysis the 6-carbon sugar is split into two 3-carbon sugars. The yields from this point onwards are therefore doubled to take account of this.

Respiratory quotients

A respiratory quotient (RQ) is a measure of the ratio of carbon dioxide evolved to the oxygen consumed.

$$RQ = \frac{CO_2\ evolved}{O_2\ consumed}$$

For a hexose sugar such as glucose it can be seen from the equation

$$C_6H_{12}O_6 + 6O_2 \longrightarrow 6CO_2 + 6H_2O$$

that the ratio is $\dfrac{6CO_2}{6O_2} = 1.0$

For a fat such as stearic acid however the equation for its complete oxidation is

$$C_{18}H_{36}O_2 + 26O_2 \rightarrow 18CO_2 + 18H_2O$$

The $RQ = \dfrac{18CO_2}{26O_2} = 0.7$

Other RQs are

> Malic Acid – 1.33
> Proteins   – 0.9 (Though this varies slightly according to the particular protein)

In practice organisms rarely oxidize one compound alone and therefore RQs may lie between these values, e.g. for humans the RQ is typically 0.85 indicating the oxidation of a mixture of carbohydrate (1.0) and fat (0.7) rather than protein which is normally only respired in extreme circumstances, e.g. starvation.

Link topics

Section 2.1 The cell (especially mitochondria)
Section 3.1 Carbohydrates, lipids and proteins
Section 3.2 Enzymes (including biochemical techniques)
Section 7.2 Gaseous exchange

Suggested further reading

Chappell, J. B., *Energetics of Mitochondria,* Oxford/Carolina Biology Reader No. 19, revised ed. (Packard 1978)
Hughes, G. M., *Vertebrate Respiration* (Heinemann 1963) (Chapter 9)
Nichols, P., *Cytochromes and Biological Oxidation,* Oxford/Carolina Biology Reader No. 66, 2nd ed. (Packard 1983)
Widdicombe, J. and Davies, A., *Respiratory Physiology* (Arnold 1983)
Wood, D. W., *Principles of Animal Physiology,* 3rd ed. (Arnold 1983) (Chapter 3)

QUESTION ANALYSIS

1 (a) There are three stages in the release of energy from a molecule of glucose: glycolysis, tricarboxylic acid (Krebs') cycle and the electron transfer system. What are the essential features of each of these processes?   (10)
   (b) In what circumstances would you expect anaerobic respiration of glucose to occur in
   (i) yeast?   (2)
   (ii) a flowering plant?   (4)
   (iii) a mammal?   (4)
Mark allocation 20/100   Time allowed 40 minutes   In the style of the Joint Matriculation Board

This is a structured essay involving knowledge of cellular respiration. It tests the ability to isolate the important features of the biochemical pathway of respiration and the circumstances in which certain of the pathways are followed.

To the candidate who has revised the three processes mentioned in (a) the answer should be straightforward. The important words are 'essential features': a full detailed account of every stage regardless of its importance is not required. The approach of many candidates is to write everything known about the pathways. Even if the candidate does not lose marks directly he will almost certainly penalize himself by wasting valuable time on detail which will not warrant marks. The main problem for a candidate will be to determine just what the essential features are and just how much detail should be included. In this, as in many other cases, the detail needed is determined by the mark allowance. For each of the three stages there are only about three marks and so three or four features for each stage would seem reasonable. In glycolysis, for instance, the main stages are: phosphorylation of the glucose, the splitting of the hexose sugar into two triose molecules, the production of electrons which yield some ATP and the production of ATP directly to form pyruvic acid, the final product of glycolysis. Although there are other stages these are mostly reactions involving rearrangement of the molecules into different isomers. While such reactions are essential to the process they play only a supporting role to the four main stages listed above. The essential information gives a fuller account of all the stages.

For the tricarboxylic acid (Krebs') cycle the essential features are

1 The loss of a molecule of carbon dioxide from pyruvic acid to produce a 2-carbon molecule (acetyl coenzyme A) that enters the cycle.
2 Combination of this with a 4-carbon molecule (oxaloacetic acid) to give a 6-carbon molecule (citric acid).
3 The oxidation of two carbon atoms to carbon dioxide in order to regenerate the 4-carbon molecule and perpetuate the cycle.
4 Production of energy, directly as ATP (one molecule) or indirectly through the production of electrons which later form a further 11 molecules of ATP for each sequence of the cycle.

The essential feature of electron transfer is the progressive transfer of electrons from one carrier to another. Each carrier is at a lower energy level than the preceding one and the energy lost in moving down this energy 'hill' is used to generate ATP from ADP. The final electron acceptor is oxygen which becomes reduced to water.

Part (b) requires much skill and a broad knowledge of the ecology and physiology of the three examples mentioned. The ability to think widely about environmental and internal conditions that restrict the oxygen supply to organisms is extremely valuable. It is important to remember that, with the exception of yeast, the other two examples are not wholly anaerobic. The 'circumstances' will therefore only affect specific parts of the organisms. No account of the anaerobic pathways is required. Precise details of the 'circumstances' should be given. To say 'when there is no oxygen available' is clearly an inadequate answer at this level.

For (i) 'yeast' the mark allowance is half that for (ii) and (iii). This is reflected in the relatively few circumstances in which anaerobic respiration occurs. The obvious answer is in stagnant water where dissolved oxygen is quickly used up and the lack of turbulence in the solution makes dissolving of more very slow. This is the basis of brewing. Other more natural situations include the centre of rotting fruits or in other decomposing organic matter. In (ii) ('a flowering plant') the circumstances occur either where oxygen is used faster than it is being delivered or where it is deficient in the environment. Examples of the former include germinating seeds and the centre of large fruits or herbaceous stems. Examples of the latter include roots in very compacted or waterlogged soils and aquatic plants in stagnant ponds (in the dark).

In (iii) ('mammals') it is best to consider the circumstances in two groups. Firstly there are situations where the supply of oxygen to the cells is reduced for one reason or another. These include failure of the lungs (e.g. disease), failure of the heart (e.g. inadequate output), coronary thrombosis, failure of the capillary network (e.g. inadequate for the tissues needs), failure of the blood (e.g. haemorrhage, anaemia, bone marrow disease), or a restricted supply (e.g. pressure on an artery). Secondly there are circumstances where the supply is adequate for normal situations but a temporary excessive demand occurs, e.g. during strenuous exercise or pregnancy. Incidental occurrences of anaerobic respiration in mammals may be mentioned, e.g. in hibernating animals, in a sperm swimming through the oviduct, in the muscles of unacclimatized mammals exercising at high altitude. The circumstances are therefore shown to be many and various and the answer not quite as simple as it first appeared. Part (b) does after all carry half the total marks and therefore something more than a list of one or two circumstances for each is needed.

Answer plan

(a) Glycolysis
1 Phosphorylation of glucose
2 Splitting of hexose to two triose molecules
3 Production of ATP directly
4 Production of ATP indirectly – via the electron transfer system
Krebs' cycle
1 Loss of carbon dioxide to give a 2-carbon molecule
2 Combination of 2-carbon and 4-carbon to give a 6-carbon molecule
3 Oxidation of two carbon atoms to carbon dioxide to regenerate the 4-carbon molecule
4 Production of ATP directly and indirectly (via electron transfer system)
Electron transfer system
1 Progressive transfer of electrons to carriers at lower energy levels
2 The use of the energy associated with the electrons to generate ATP from ADP
3 The reduction of oxygen to water
(b) (i) Yeast
1 In stagnant solutions
2 Centre of decomposing fruits and other organic matter
  (ii) Flowering plants
1 Young seeds, centre of fruits or large stems
2 In roots in compacted or waterlogged soils
3 In aquatic plants growing in stagnant ponds.
  (iii) Mammals
1 Inefficiency of lungs, e.g. emphysema
2 Reduction in blood supply, e.g. haemorrhage, pressure on artery
3 Low oxygen-carrying capacity of blood, e.g. anaemia, bone marrow disease
4 Low cardiac output, e.g. slow heart rate, coronary thrombosis
5 Capillary network inadequate, e.g. angina
6 High oxygen demands, e.g. strenuous exercise, pregnancy
7 Others, e.g. hibernation, sperm in oviduct, high altitude

2 Compare the biochemical changes, beginning with pyruvic acid, which take place in aerobic and anaerobic respiration. What is the difference in energy output?
Mark allocation 10/100   Time allowed 15 minutes

Southern Universities Joint Board June 1980, Paper I No. 10

This is a shorter type essay question testing the ability to compare aerobic and anaerobic respiration. The answer requires recall of factual information but it needs to be presented in a way that brings out the similarities and differences between the biochemistry of aerobic and anaerobic respiration. All too frequently candidates with adequate biological knowledge to pass an A-level examination fail to do so simply because they do not answer the question asked. In this case the key words are 'compare', 'biochemical changes' and 'beginning with pyruvic acid'. This may seem a strange statement when the question is concerned with aerobic and anaerobic respiration but almost every candidate reading the question will appreciate this fact immediately. What they will do is not 'compare' but simply describe both processes. They will discuss the types of organisms involved and even methods of gas exchange rather than limit themselves to biochemical changes. They will include glycolysis when the question says 'beginning with pyruvic acid'. It is all too easy to read the question superficially, observing only the obvious words and not paying adequate attention to the other, equally important words.

In the first part of the answer candidates should compare the first stages of the biochemical pathways beginning with pyruvic acid. In anaerobic respiration there are a number of routes according to whether carbon dioxide is removed (acetaldehyde fermentation) or two hydrogen atoms are added (lactic acid fermentation) or both (alcoholic fermentation). In aerobic respiration there is a single pathway, via acetyl coenzyme A to the Krebs' cycle. In anaerobic respiration the pathway is short but the energy yield very low. In aerobic respiration the pathway is long and much more energy is obtained. In anaerobic respiration no oxygen is needed whereas in aerobic respiration it is the final acceptor of hydrogen at the end of the electron transfer system and therefore maintains the whole pathway. Aerobic respiration involves some cycling of substances (Krebs' cycle); anaerobic respiration does not. The end products of anaerobic respiration can be toxic if they accumulate but those of aerobic respiration are less so. However, the products of anaerobic respiration are capable of being further oxidized to yield more energy; those of aerobic respiration cannot be further broken down by living organisms. If time allows a brief summary of the various pathways giving the major intermediate stages would be a useful way of concluding this part of the answer. Some candidates may find it better to start with such a summary and use it as a type of plan from which to make the necessary comparisons. Such an arrangement has much to recommend it provided the necessary comparisons are made rather than leaving the summary as the entire answer.

Although no separate mark allocation is given for the second part, it clearly requires a much shorter answer than the first part. Ideally the candidate should show the working that leads to the answer and the figures are better expressed in kJ/mole than as a percentage efficiency. Under 'ATP and energy yields' in the essential information the yields are compared, including the methods by which they are calculated. The theoretical energy yield for 1 mole of glucose is 2957 kJ, that of anaerobic respiration is 68 kJ and that for aerobic respiration is 1292 kJ. The difference is therefore 1224 kJ. In practice estimates of both the number of ATPs evolved in aerobic respiration and the yield of energy for each ATP vary slightly giving different total yields. Provided the figures you use for the basic values are reasonable and clearly stated, some credit, if not all, should be allowed. The main disagreement lies in whether or not the two hydrogen atoms evolved during glycolysis are transferred to the mitochondria in order to enter the electron transfer system and yield energy or if they are used in non-energy producing processes in the cytoplasm.

Answer plan

Anaerobic	Aerobic
Various metabolic pathways from pyruvic acid	Single metabolic pathway from pyruvic acid
Low energy yield	Higher energy yield
Short pathway of reactions, none of which are cyclic	Longer pathway in which some reactions are cyclic
No oxygen used, no Krebs' cycle and no electron transfer system	Oxygen used, Krebs' cycle and electron transfer system involved
Toxic products which can be further oxidized	Products not toxic and oxidation is complete

Summary of the various pathways
Energy yields
Anaerobic – 2 ATP each yielding 34 kJ =   68 kJ
Aerobic   – 38 ATP each yielding 34 kJ = 1292 kJ
                                Difference = 1224 kJ

3 When an athlete is running in a 200-metre sprint, his rate of oxygen consumption rises above the resting level. After the race his oxygen consumption does not fall to the resting level for some time. A blood sample taken at the end of the race shows a significant increase in lactic acid. The diagram on page 161 summarises the metabolic processes involved in the release of energy in the muscles during and after exercise.

(a)   (i) Explain why, although oxygen consumption rises during the race, more lactic acid is present in the blood at the end of the race.   (4)

        (ii) Why is the athlete's oxygen consumption still higher than normal some time after the race?   (1)

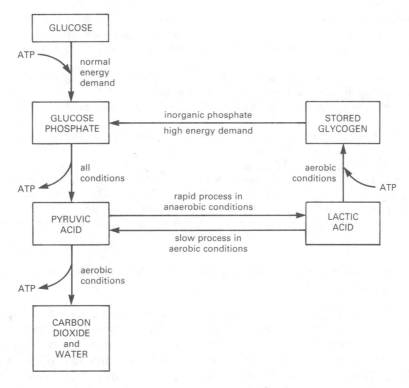

(iii) After the race ATP is used to convert most lactic acid to glycogen. A small amount of lactic acid is fully oxidised with the production of ATP. With the help of information in the diagram, explain why this is an efficient use of lactic acid. (4)

(iv) For every molecule of glucose 38 molecules of ATP are produced during aerobic respiration but only 2 molecules of ATP are produced during anaerobic respiration. Explain how this difference occurs. (3)

 (v) With the help of information in the diagram, explain why it is advantageous to respire glycogen rather than glucose during strenuous exercise. (2)

(b) Athletes devote much time to training. Suggest **three** ways in which exercise may improve an athlete's subsequent performance. (6)

Mark allocation 20/120   Time allowed 30 minutes

Associated Examining Board, June 1984, Paper 2, No. 2

This question is designed to test knowledge and understanding of cellular respiration and how this relates to the physiology of an athlete running a race. Always study carefully any information provided. Even though you may be conversant with the topic you can often gain vital new information from the facts provided in the question.

In answering part (a) (i) the candidate should be aware that in a 200 metre sprint, although the energy expenditure is large it is over a relatively short period, say 30 seconds. There is a limit to the total amount of oxygen which can be inspired in so short a time. While, therefore, the oxygen consumption rises, it is insufficient to meet demand. Some anaerobic respiration must hence take place resulting in a rise in the level of lactic acid in the blood. The oxygen debt built up during the race must be repaid once it has been completed. This explains why the athlete's oxygen consumption remains higher than normal after the race.

Part (a) (iii) requires the candidate to interpret parts of the diagram provided. One key word in the question is 'efficient'. In this case it can be taken to mean making the best possible use of the available lactic acid. The diagram shows just two alternative ways in which the lactic acid can be removed; either into stored glycogen or into pyruvic acid which is then oxidised. The conversion to stored glycogen uses up energy in the form of ATP and might at first appear wasteful. However, the diagram shows that glycogen can later be converted into pyruvic acid with an equivalent gain in the amount of ATP. Therefore either route yields the same quantity of energy. Why then is most lactic acid converted to stored glycogen? Mainly because it represents a store of energy for later use. Having completed the race, and presumably resting, the athlete has no need of an immediate source of ATP. To completely oxidise all the lactic acid would produce ATP which could not be utilised and would be wasted. Therefore only enough is oxidised to meet the body's present needs. The majority is converted to stored glycogen. This can be later used to meet any future energy demands. This is more efficient than producing ATP at a time when it is not required.

In (a) (iv) the explanation for the much lower energy yield in anaerobic respiration is because in the absence of oxygen Krebs' cycle and the electron transport system cannot operate and so glucose is only partially broken down. As it is the electron transport system which generates most ATP, in its absence the energy yield is very low. A full explanation of the differences in energy yield between aerobic and anaerobic respiration is given in the essential information on page 157.

The answer to (a) (v) lies in the fact that in order to convert glucose to glucose phosphate, ATP is used up; this reduces the total energy produced. When glycogen is converted to glucose phosphate no energy is used up and therefore more is available to an organism, a distinct advantage during strenuous exercise.

The ways in which exercise may improve an athlete's performance (part (b)) include:

1 The build up and improved functioning of the muscles.

2 Improvements in the circulation including more efficient cardiac output, i.e. the heart can pump more blood in a given time.

3 Improved oxygen carrying capacity of the body, e.g. through a greater concentration of red blood cells with more haemoglobin in each one.

4 Greater lung capacity permitting more air, and therefore oxygen, to be inspired in a given time.

Candidates often leave the answer here rather than linking these points to an improved performance as asked. The answer should therefore be concluded by pointing out that all these changes result in more oxygen being carried more quickly to the muscles so allowing more aerobic respiration to occur. This means a greater energy yield and a consequent improved performance.

7.2   GASEOUS EXCHANGE

Assumed previous knowledge An outline of cellular respiration (Section 7.1)

Underlying principles

Much of the energy utilized by organisms for osmotic, mechanical, electrical and chemical work is derived from ATP formed in the mitochondria during oxidative phosphorylation. This process requires oxygen which the organism must obtain from its environment of air or water. Regardless of source, the oxygen enters the organism by diffusion. In order to obtain it in adequate quantities and remove the carbon dioxide produced, a relatively large, moist surface must be exposed to a source of oxygen. The distance between this oxygen source and the cells requiring or transporting it must be small and the diffusion gradient must be maintained if the process is to satisfy the needs of the organism. In small organisms such as protozoa and unicellular algae the surface area is sufficiently large compared with volume so that diffusion over the whole body surface satisfies their needs.

As organisms became multicellular and grew in size they were only able to utilize the whole body surface for obtaining oxygen if their energy demands were small and their shape became flattened (e.g. platyhelminthes) or the centre contained no respiring material (e.g. coelenterates). Any further increase in size or metabolic rate required some specialized gaseous exchange system to compensate for a lower surface area/volume ratio and greater oxygen demand. Such systems included tubular ingrowths to carry air directly to the cells (e.g. insects) or organs of gaseous exchange situated in one part of the organism and supplied with a means of transporting gases from it, to and from the other cells. In water this second system takes the form of gills and on land lungs are used. Whichever system is used the respiratory surface must have a constant supply of the oxygen-carrying medium flowing over it, to maintain the diffusion gradient.

Points of perspective

It would be useful for candidates to have seen mammalian lungs and fish gills, either by dissection or demonstration. As this topic lends itself quite well to data type questions, it would be beneficial for candidates to have interpreted such data, or even to learn some figures, e.g. oxygen consumption of different animals and under different levels of activity. The effects of certain activities on the human lungs would be useful to know, e.g. effects of smoking, other types of atmospheric pollution, as well as the effect on lungs and breathing of diseases such as cancer, bronchitis, tuberculosis, pneumoconiosis, silicosis, emphysema and asbestosis. Knowledge of a wide range of respiratory surfaces in organisms would be especially helpful, e.g. lung books in spiders, lungs of pulmonate snails, cloacal respiration in holothurians.

Essential information

Air and water as respiratory media

Table 7.3 Comparison of water and air as respiratory media

Property	Water	Air
Oxygen content	$0.04 - 9.0 \text{ cm}^3/\text{l}$	$105 - 130 \text{ cm}^3/\text{l}$
Oxygen diffusion rate	Low	High
Density	Relative density of water about $1000 \times$ air	
Viscosity	Water about $100 \times$ air	

Since the volume of oxygen in a given volume of air is much greater than in the same volume of water, an aquatic animal must pass a greater volume of the medium over its respiratory surface in order to obtain the oxygen necessary for cellular respiration. Water is also far denser and has a greater viscosity than air at the same temperature and so it is more difficult for the organism to extract the necessary oxygen from it. It would, therefore, seem easier for animals to obtain oxygen from air than from water but this is not the case since terrestrial animals must avoid desiccation and therefore have relatively impermeable skins. Terrestrial animals and the more active aquatic ones do not rely on the general body surface for respiratory exchange but have developed specialized respiratory surfaces. The simplest of these surfaces is gills which may be external, e.g. annelida, amphibian tadpoles, or internal, e.g. fish. Insects have a tracheal system. Terrestrial vertebrates and some invertebrates, e.g. pulmonate snails, have developed lungs. For maximum efficiency internal respiratory surfaces such as internal gills and lungs need to be ventilated.

General properties of respiratory surfaces
1 Large surface area/volume ratio: the body surface is adequate in very small organisms; infoldings of a restricted part of the body provide a large respiratory surface, e.g. lungs and gills, in larger organisms.
2 Moist: surfaces are moist so that diffusion may occur in solution.
3 Thin: diffusion is only efficient over short distances because the rate of diffusion is inversely proportional to the distance between the concentrations on the two sides of the respiratory surface.
4 Transport: this must occur in order to maintain the diffusion gradient. In large metazoans an efficient vascular system is required.

Mechanisms of gaseous exchange

Terrestrial flowering plants
By virtue of their autotrophic mode of nutrition the supply of gaseous oxygen to the tissues of flowering plants is facilitated for three reasons:

1 When photosynthesizing oxygen is produced as a by-product and may be used directly for respiration.
2 Photosynthesis requires specialized structures – the stomata – in the leaves to allow diffusion of carbon dioxide and these can be utilized to obtain oxygen from the atmosphere when no photosynthesis is taking place.
3 The roots have a thin surface and large surface area to facilitate the collection of water by osmosis for photosynthesis. This feature allows oxygen to diffuse into the root easily from the soil spaces.

The stems of plants obtain their oxygen either through stomata (herbaceous flowering plants) or lenticels (woody flowering plants). The central tissue of woody stems and roots is composed of dead xylem cells. Therefore no living cell of a plant is ever far from a source of oxygen and no specialized respiratory surface is required as diffusion over the whole surface will suffice.

Insects – tracheal system
The respiratory system of an insect comprises a series of tubes, tracheae, with paired openings to the exterior called spiracles. The tracheae branch to form terminal tracheoles, often less than 1.0 μm in diameter which reach more or less every part of the insect's body. In insects blood does not carry oxygen and carbon dioxide. Except at their very ends the tracheae and tracheoles are filled with air; oxygen and carbon dioxide are transported primarily by diffusion. Ventilation of the larger tracheae occurs through movements of the muscles or exoskeleton. The spiracles are surrounded by hairs to prevent the entry of dust and parasites and they may be opened and closed by valves. The larger tracheae are lined by cuticle which must be shed at each moult and this, together with the rate of diffusion in narrow tubules, is the main factor limiting the size of insects. See Fig. 7.4.

During exercise lactic acid accumulates in the muscle, raising its osmotic pressure. Water passes from the tracheole into the muscle by osmosis thus drawing air further down the tracheole.

Fish – gills
Although fundamentally similar, there are some differences in structure, and therefore in mechanisms of gaseous exchange, between cartilagenous and bony fish. See Fig. 7.5.

Mammal (e.g. man) – lungs
Ventilation, which is clearly essential for a respiratory surface so deep inside the body, is

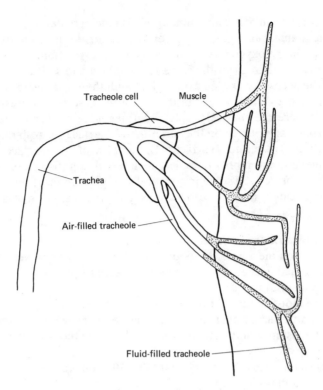

Fig. 7.4 Tracheoles in an insect

brought about by the diaphragm and the ribs. The diaphragm is a fibrous sheet of tissue whose muscular edges are attached to the thoracic wall. Contraction of these muscles flattens the diaphragm and increases the volume of the thoracic cavity. Between the ribs are the intercostal muscles whose contractions cause the ribs to move. The external intercostal muscles contract to move the ribs upwards and outwards, at the same time raising the sternum. This also causes the volume of the thoracic cavity to increase. As the volume increases the pressure decreases to below that of the atmosphere and air rushes in. This is inspiration (breathing in). Expiration is brought about when the diaphragm muscle relaxes and the internal intercostal muscles contract decreasing the volume of the thoracic cavity and forcing air out.

The respiratory epithelia across which gaseous exchange takes place are the alveoli, each of which is less than 0.5 μm thick and only 100 μm across. Each lung of man may contain 350 million alveoli and so the total internal surface of the respiratory epithelium is enormous (an estimated $80-100$ m^2).

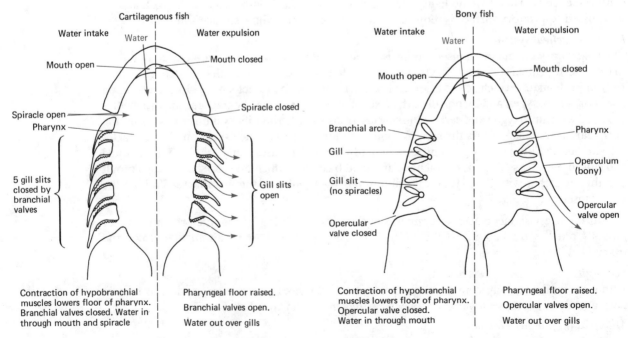

Fig. 7.5 (a) Gaseous exchange in fishes

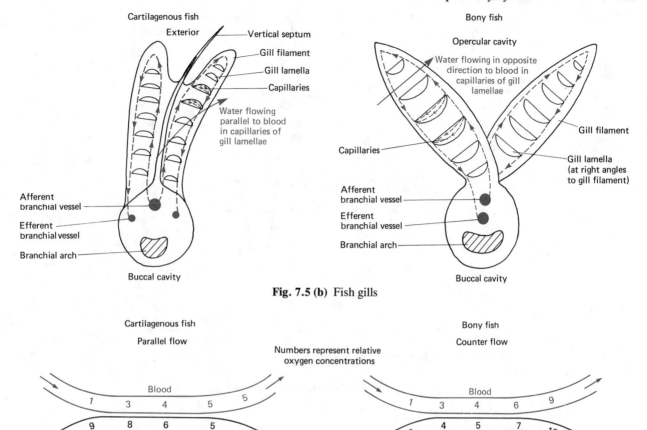

Fig. 7.5 (b) Fish gills

Fig. 7.5 (c) Comparison of parallel and counter flow systems

Cartilagenous fish

Parallel flow

Numbers represent relative oxygen concentrations

Water and blood flow side by side in the same direction. Oxygen is exchanged by diffusion. The oxygen concentration in the blood leaving the gills can never exceed that of the water leaving the gills. Equilibrium is soon reached.

Bony fish

Counter flow

Blood flows in the opposite direction to water. At each point of contact, the water has a higher oxygen concentration than the blood. Therefore diffusion can occur over the whole region and much higher blood oxygen concentrations can be reached than with parallel flow

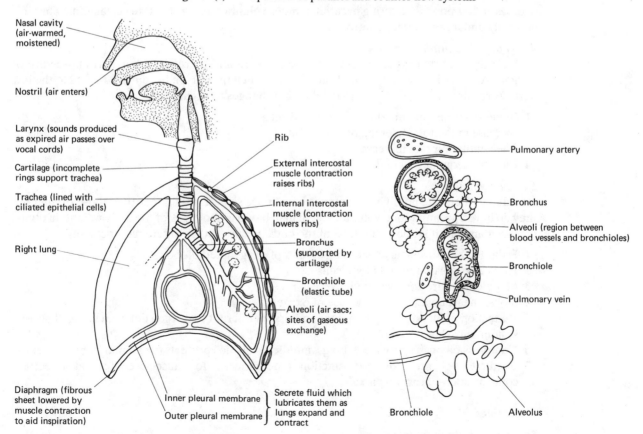

Fig. 7.6 (a) Human respiratory system

Fig. 7.6 (b) LS part of mammalian lung

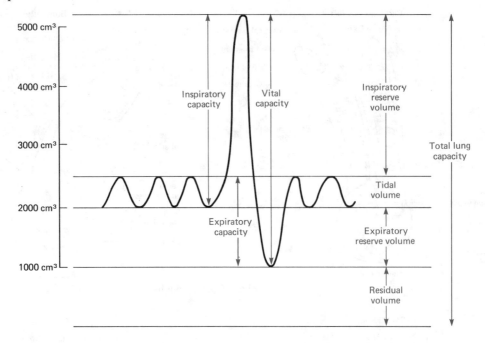

Ventilation rate = Tidal volume x Frequency of inspiration

Fig. 7.7 Graph to illustrate lung capacities

Control of respiration

The respiratory centre, which is in the medulla oblongata at the base of the brain, is particularly sensitive to changes in the concentration of carbon dioxide in the blood; a slight increase in the level of carbon dioxide causes deeper, faster breathing until the carbon dioxide concentration returns to normal. Changes in breathing are brought about by signals from the respiratory centre to the spinal nerves controlling the intercostal muscles and the muscles of the diaphragm. The respiratory centre is also sensitive to signals from other parts of the body, e.g. chemoreceptor cells in the aorta and carotid artery which detect when the oxygen concentration in the blood decreases. It is possible to bring breathing under voluntary control within certain limits but it is normally under involuntary control.

Effect of high altitude

At high altitudes there is a low atmospheric pressure and therefore a low partial pressure of oxygen. Acclimatization, a period of time during which the person becomes adjusted to the low partial pressure of oxygen, causes the following changes to occur:

1 Increase in the total number of red blood cells
2 Increase in the total haemoglobin content
3 The ventilation rate increases
4 Cardiac frequency increases.

Diving mammals

Relative to their size no diving mammals have lungs significantly larger than those of man but they do have a greater blood volume, e.g. in man blood is about 7% of the body weight; in diving marine mammals it is about 10–15% of the body weight. Other differences include:

1 Enlarged blood vessels to act as reservoirs of oxygenated blood
2 Higher proportion of red blood cells
3 Myoglobin more concentrated
4 Slower heart beat (to conserve use of oxygen)
5 Reduction of blood supply to organs and tissues tolerant to oxygen deficiency, e.g. digestive system, muscles
6 Compression of air spaces, e.g. lungs, middle ear, which protects the animal from the 'bends'
7 Respiratory centres do not function automatically to cause breathing at a certain concentration of carbon dioxide.

Link topics

Section 7.1 Cellular respiration
Section 8.2 Blood and circulation

Suggested further reading

Hughes, G. M., *The Vertebrate Lung*, Oxford/Carolina Biology Reader No. 59, 2nd ed. (Packard 1979)
Hughes, G. M., *Vertebrate Respiration* (Heinemann 1963)
Wigglesworth, V., *Insect Respiration*, Oxford/Carolina Biology Reader No. 48 (Packard 1972)

QUESTION ANALYSIS

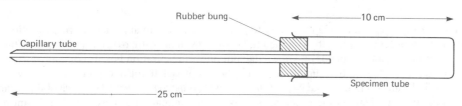

Fig. 7.8 A simple respirometer

4 The diagram above is of a simple form of respirometer for measuring oxygen consumption in small organisms.
   (a) What chemical is used in the specimen tube?
   (b) What function does this chemical perform?
   (c) How can it be kept in position in the tube and isolated from the organisms?
   (d) How would you use the apparatus to measure the volume of oxygen taken up by the organisms?
   (e) Give any two sources of error in using the apparatus.
   (f) Could the apparatus be used to measure anaerobic respiration?
   Mark allocation 9/100 Time allowed 15 minutes   *In the style of the Joint Matriculation Board*

This is a structured question testing the ability to use a piece of experimental apparatus and evaluate its usefulness. Candidates who have used such a piece of apparatus are clearly at some advantage, although even those that have not should still be able to apply their general biological knowledge to good effect. The device is simple and much more complex ones are commonly used by students in schools and colleges. In all cases the principle is the same. During aerobic respiration the volume of oxygen taken up is often equal to the volume of carbon dioxide evolved. There is therefore no net change in volume when an organism is left to respire in a closed vessel. If, however, the carbon dioxide evolved is absorbed chemically so that its volume is negligible, the volume will decrease by an amount equivalent to the amount of oxygen taken up. The chemical referred to in (a) must be one that absorbs carbon dioxide (b) and yet is not toxic to the organisms used in the specimen tube. Ten per cent solutions of either sodium or potassium hydroxide could be used provided a suitable means is found of isolating them from the organisms. Soda lime is a suitable solid, and less toxic, alternative. In (c) to isolate the chemical, a piece of zinc gauze should be positioned firmly about half way along the specimen tube after the chemical has been added. The animals may then be introduced to the left of the gauze before the bung and capillary tube are placed in position. The zinc gauze has holes to allow gases to diffuse through; clearly the holes must not be too large or the organisms will fall through. If the chemical is a liquid it must be added absorbed onto cotton wool.

For (d) the apparatus is set up as explained in (c) except that the end of the capillary tube should first be dipped in water to allow a drop to enter the tube. The point of this drop in the tube should be marked immediately and again after a certain period of time, say one hour – although this depends on the organisms used and the prevailing conditions. The distance between the marks (h) is measured in millimetres. The diameter of the capillary tube, if not known, can be found using a travelling microscope; division by two gives the radius (r). The volume of oxygen consumed in the time can then be calculated using the formula $\pi r^2 h$.

Sources of error for (e) include possible leakage of air in or out of the specimen tube, temperature changes during the experiment altering the volume of gas in the specimen tube and failure of the chemical to absorb all the carbon dioxide produced. The latter two are the best to quote since the first is relatively easy to overcome.

In (f) the answer is yes and no. If the organisms' anaerobic respiration yields carbon dioxide then the volume in the specimen tube will increase in proportion to the rate of respiration. The water drop must, however, be positioned initially at the specimen tube end of the capillary tube and the carbon dioxide absorbing chemical omitted. If the form of anaerobic respiration used by the organism does not yield carbon dioxide (e.g. lactic acid fermentation) then the apparatus is useless for measuring respiratory rate. In any case the problems of suitably housing the organisms, since many are aquatic, make it a rather unsatisfactory method in practice even where it is theoretically possible.

5 The oxygen consumption of six animals at rest is given on page 168. Study the table and explain the differences between the figures by reference to mechanisms of gas exchange, the systems of transport of oxygen, the body temperature and general level of activity.

Animal	Oxygen consumption (cm^3/kg/h)
Earthworm	75
Butterfly	500
Frog	120
Humming bird	11 000
Mouse	3000
Man	200

Mark allocation 20/100   Time allowed 30 minutes   In the style of the Oxford and Cambridge Board

This is an essay question testing the ability to interpret data and to apply a wide range of biochemical knowledge in order to explain it. Although the figures given measure oxygen consumption, the ability to answer this question depends on knowledge not only of respiration but also of many aspects of the physiology and ecology of the animals listed. The best approach to the question is to construct a matrix as shown in the answer plan so that each of the four aspects listed is reviewed for each of the six animals. Such a matrix will highlight the differences between the animals and these can then be more easily correlated to the differences in the figures given. One approach is to deal with one aspect, e.g. gas exchange, at a time for each of the animals in turn. This however is a rather long method and may fail to stress the differences. It is probably better to take each animal in turn and compare its oxygen consumption with that of the other animals. In some aspects there will be no difference between one animal and another and these can, therefore, be ignored as they cannot 'explain the differences'. The animals need not be taken in the order given; indeed it makes more sense to deal with them in ascending or descending order of oxygen consumption.

To begin with, the earthworm (75 cm^3/kg/h) has the lowest oxygen consumption because its method of gaseous exchange does not involve any special mechanism for increasing the surface area; as it has a fairly low surface area/volume ratio it cannot absorb oxygen rapidly. Being ectothermic (normally living within the range 5–15°C) it has a lower metabolic rate than any other animal on the list. Because it is not a very active animal its oxygen demands are relatively low. The frog (120 cm^3/kg/h) is larger but has a fairly low metabolic rate and is ectothermic, living around 5–15°C. Its oxygen demands are therefore low. Although it has lungs it rarely uses them when at rest; the oxygen required is diffused in via the skin and buccal cavity, but their relatively small surface area only allows a low oxygen consumption. The circulatory system mixes oxygenated and deoxygenated blood and this 'mixed' blood at the respiratory surface lowers the diffusion gradient and hence the oxygen consumption. Man (200 cm^3/kg/h) has perhaps a surprisingly low consumption in view of his high metabolic rate, high body temperature, large respiratory surface and efficient oxygen transport system. A good candidate will observe that in the table the mouse has similar features in every respect to those of man, and yet its oxygen consumption is fifteen times greater. The explanation lies in the relative sizes of the two organisms. When at rest a large proportion of the energy produced during respiration is used to maintain the body temperature. The larger surface area/volume ratio of a small animal such as a mouse causes more rapid heat loss at normal environmental temperatures than would occur in a large animal such as man. The extra oxygen consumed is used to release more energy to replace that lost as heat.

The butterfly (500 cm^3/kg/h) has an efficient gaseous exchange system for its size and one that removes the need for an oxygen transport system. It is ectothermic which reduces its metabolic rate and hence oxygen consumption, but it is a flying insect; flight is an energetic process and its metabolic rate is therefore not too low, even when at rest, as it must retain the capacity to produce energy quickly when the need arises. This means it has a fairly high level of oxygen consumption. The humming bird (11 000 cm^3/kg/h) has by far the greatest oxygen consumption of the six animals. This is because it combines many of the features of others that require large energy production and hence oxygen uptake. The humming bird is small, having an even greater surface area/volume ratio than the mouse; like the butterfly it needs a high metabolic rate for flight. It has, by virtue of its air sacs, an even more efficient gaseous exchange system than the mouse and man, and it has a much faster rate of circulation which increases the uptake of oxygen. It has the highest body temperature of the animals listed and this further increases the metabolic rate and need for energy to maintain the body temperature constant.

Answer plan

	Gas exchange mechanism	Oxygen transport system	Body temperature	Activity
Earthworm	Diffusion through moist body surface	Closed blood system with oxygen carrying pigments, e.g. haemoglobin	Ectothermic (poikilothermic); normally in the range 5–15°C	Slow moving; low metabolic rate
Butterfly	Tracheal system; oxygen delivered direct to cells	Open blood system; no respiratory pigments	↓	Active; high metabolic rate

Frog	Diffusion through skin, buccal cavity, lungs (not normally at rest)	Closed blood system; 3-chambered heart; haemoglobin present		Fairly low metabolic rate
Humming bird	Lungs–air sacs improve efficiency but do not themselves absorb oxygen	Closed blood system; 4-chambered heart; haemoglobin present	Endothermic; around 42°C	Very active; high metabolic rate
Mouse	Lungs		Endothermic 37°C	Active; high metabolic rate
Man	Lungs			

Give explanations on the basis of the comparisons in the table above.

6 (a) State precisely the respiratory surface(s) employed by the following organisms:
   (i) a protozoan   (ii) a flatworm   (iii) a tadpole
   (iv) an adult frog   (v) a fish   (vi) a lizard   (vii) a mammalian foetus
                                                                                (2)
   (b) What are the essential features of a good respiratory surface?   (2)
   (c) With the aid of a fully labelled diagram, describe the process of inspiration in a human being (5)
   (d) (i) What factors influence breathing?
       (ii) Which part of the brain controls breathing, and how is this control achieved?   (6)
   Mark allocation 15/70   Time allowed 40 minutes   Northern Ireland 1983, Paper 1, No. 2

This structured question tests a range of knowledge on respiration, both general and specific. Part (a) is straightforward, although candidates should ensure that answers are precise and not vague. In (i) for instance 'the body' or even 'the body surface' is not as suitable an answer as 'the whole of the plasma membrane which surrounds the organism'. Similarly in (ii) the answer should be 'the whole of the epidermis which covers the body'. In (iii) and (iv) more than one surface is used with the tadpole employing at various stages of development external and internal gills. The frog can use the whole epidermis, the lining of the buccal cavity and the lungs. A fish (v) uses gill filaments, a lizard (vi) uses the epithelial lining of the lungs and a mammalian foetus (vii) employs the villi of the placenta.

   With 5 minutes available for (b) the answer should be brief and yet be more than a list of three or four words. The features are listed under 'General properties of respiratory surfaces' in the essential information, and the detail is adequate for this question.

   Part (c) clearly requires a single diagram and Figure 7.6a would be suitable. The written explanation is given in the essential information under 'Mechanism of gaseous exchange – mammal (e.g. man)'. Candidates often fail to restrict their answers to the information required. The question says 'inspiration' and therefore detail on 'expiration' and 'diffusion across the respiratory epithelia' (at the end of the passage) must be excluded.

   The factors influencing breathing required in (d)(i) include all processes which affect the uptake of oxygen and production of carbon dioxide, e.g. the metabolic rate (this is affected by temperature in ectothermic animals); the extent of physical activity. .

   These factors affect the carbon dioxide concentration of the blood and thus its pH. It is this variation in pH that causes alteration in the breathing rate. Details for the answer to part (d)(ii) are given in the essential information under the heading 'control of respiration'.

7 (a) Graph 1 shows the percentage of haemoglobin associated with oxygen to form oxyhaemoglobin over a range of partial pressures of oxygen. Graph 2 shows the relationship between altitude and partial pressures of oxygen. (See Fig. 7.9.)
   Note that chemical details of glycolysis and Krebs' (TCA) cycle are not required in any part of this answer.
   (i) Using the information given on both graphs explain why most people who are not acclimatized to living at high altitudes will lose consciousness at altitudes between 6000 and 8000 metres.   (5)
   (ii) Permanent human habitations occur up to approximately 7000 metres and people who are acclimatized to high altitudes can survive for a few hours when breathing air at approximately 9000 metres. Suggest three adjustments which probably occur in the physiology of such acclimatized people.   (3)
   (iii) Explain the physiological reasons for each of the adjustments you have suggested.   (3)

   (b) The following data refer to Olympic Games held at the sites stated.

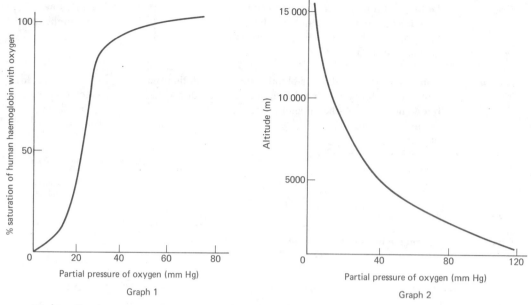

Fig. 7.9 Partial pressures of oxygen in relation to (a) haemoglobin saturation and (b) altitude

Results of the 10 000 m race

Tokyo 1964 (200 m above sea level)

1 M. Mills, USA
2 M. Gammoudi, Tunisia**
3 R. Clarke, Australia
4 M. Wolds, Ethiopia**
5 L. Ivanov, Russia
6 K. Tsuduroya, Japan
7 M. Halberg, New Zealand
8 A. Cook, Australia

Mexico 1968 (2242 m above sea level)

1 N. Tamu, Kenya**
2 M. Wolds, Ethiopia**
3 M. Gammoudi, Tunisia**
4 J. Martinez, Mexico**
5 N. Sviridov, Russia*
6 R. Clarke, Australia*
7 R. Hill, UK*
8 W. Masresha, Ethiopia**

**Indicates athletes who had lived most of their lives at high altitudes.
* Indicates athletes who trained at high altitudes for an extended period prior to the games.

Carefully explain why unacclimatized athletes were relatively unsuccessful during the 10 000 m race at the Mexico Olympic games. (4)

(c) Some unsuccessful athletes collapsed and were given oxygen. Clearly explain the role of this oxygen with specific reference to the athletes' livers. (4)

(d) By referring only to general principles explain the role of oxygen in energy release in mitochondria. (6)

(e) Describe a quantitative experiment which you have carried out to compare the composition of inspired and expired air in any named living organism. (5)

Mark allocation 30/140   Time allowed 40 minutes

Associated Examining Board June 1979, Paper II, No. 6

This is a longer style structured question testing knowledge of gaseous exchange and transport of oxygen. It involves the ability to analyse graphical and written data, draw conclusions and make explanations. It also tests knowledge of certain biochemical principles and the ability to describe experimental details.

Part (a) requires the candidate to study the two graphs. In (a) (i) note that the explanations should use 'the information given on both graphs'. Using a ruler carefully, the partial pressures of oxygen at 6000 and 8000 metres should be read from Graph 2. These will give figures for the partial pressure of oxygen of approximately 30 and 20 mm Hg, respectively. Then these figures should be used to determine the corresponding percentage saturation of haemoglobin with oxygen using Graph 1. The figures are approximately 80% and 40%, respectively. These levels are inadequate to supply oxygen to the cells of an unacclimatized person at a rate sufficient to fully maintain their functioning. The nerve cells, especially those of the brain, use large quantities of oxygen and are particularly sensitive to any lack of oxygen. They cease to function adequately and consciousness is slowly lost. In answering (a) (ii) the candidate should give any three of the four changes shown under 'living at high altitude' in the essential information. The physiological reasons required in (a) (iii) will depend on the changes chosen in (a) (ii), but involve an increase in the amount of oxygen carried by the blood (more red blood cells and more haemoglobin), an increase in the rate at which oxygen is absorbed (increased ventilation rate) and an increase in the rate at which the oxygen is delivered to the cells (increased cardiac frequency). In answering (b) the candidate may ignore the names of the athletes and concentrate on the asterisks. Basically two asterisks indicate a fully acclimatized athlete, one indicates at least partial acclimatization and no asterisk means the athlete is unacclimatized. It is clear that at the high altitude of Mexico only those athletes who were fully acclimatized did well, the unacclimatized ones not appearing in the first eight places. In addition to the features listed in (a) (ii), a lifetime at high altitude produces other changes such as an increase in the number of capillaries in the tissues, designed to increase the supply of oxygen and hence the efficiency of that tissue. The oxygen supply to the muscles of the unacclimatized athlete would therefore have been much less than to the muscle of the acclimatized athlete and the performance correspondingly lower.

In (c) the candidate should note that the 'role' of the oxygen is needed and with 'special reference' to 'the liver'. It would be worth mentioning that a lack of oxygen to vital organs such as the brain, precipitated the collapse, and the administering of oxygen supplied these organs with it more rapidly. The main point however is that during the race, lactic acid had accumulated and the athlete had incurred an oxygen debt. The debt had to be repaid and the oxygen applied did this more rapidly. The lactic acid was transported to the liver where it was largely reconverted to glucose and glycogen – a process requiring oxygen.

The answer to (d) is covered in the essential information to Section 7.1 (Cellular respiration) under 'electron transfer system'. The main point is that the oxygen acts as the final acceptor of the hydrogen atoms produced during glycolysis and Krebs' cycle. It therefore effectively drives the whole electron transfer system that releases energy as ATP.

The candidate's answer to (e) will depend on the precise experiment performed, but remember the candidate is expected to have carried it out and the use of radioactive labelling, mass spectrometry or other elaborate techniques are not likely to be acceptable. The organism should be clearly named and again this will depend on the candidate's experiences. The percentage composition of oxygen in inspired and expired air may be measured by absorption of oxygen using a suitable substance such as pyrogallol. Similarly the amount of carbon dioxide in the air can be measured by absorbing it in sodium or potassium hydroxide solutions. In both cases the total volume of the air should be measured initially and then the volume remaining after the gas in question has been absorbed. The percentage concentration is then calculated by:

$$\frac{\text{Initial volume} - \text{Final volume}}{\text{Initial volume}} \times 100$$

Actual experimental details should be given briefly and these will depend very much on the organism chosen.

8 Water relations and Transport

8.1 PROPERTIES AND IMPORTANCE OF WATER

Assumed previous knowledge None

Underlying principles

Life on earth originated in water, which remains essential to all forms of life. It is the most abundant liquid on earth and makes up more than half of every living organism, even as much as 95% of some. Probably more than any other substance it determines the distribution, anatomy and physiology of organisms. Abundant it may be, ordinary it certainly is not. It possesses some unusual chemical and physical properties owing to the hydrogen bonds formed between each water molecule. These special properties make it an ideal constituent of, and medium for, living things. Its roles are mainly ones of lubrication, a solvent and support.

Points of perspective

Questions exclusively about water are often concerned with its importance to living organisms and the functions it has within them. Candidates will benefit from being able to give not only a wide range of functions, but also a wide range of examples, supported by facts and figures. It would be useful to understand the physical chemistry of a water molecule including its atomic structure. Detailed knowledge of ionization, pH, buffers and hydrogen bonding would all help the overall understanding of the properties of water.

Essential information

The water molecule (hydrogen bonding and polarity) Naturally occurring water consists of 99.76% by weight of $^1H_2^{16}O$. The remainder consists of various isotopes, e.g. ^2H and ^{18}O. The commonest of these is ^2H (deuterium) which is most often found with the normal hydrogen atom as HDO, though occasionally as D_2O. Both are called 'heavy water' and have a deleterious effect on living organisms.

The molecule of water consists of the two hydrogen atoms bonded to the oxygen atom covalently (by the sharing of electrons). Although the molecule overall is neutral the oxygen atom retains a slight negative charge and the hydrogen atoms a slight positive one. Such molecules are termed **polar**.

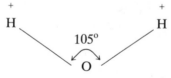

This polarity of molecules causes them to be attracted to each other by their opposite charges. These attractive forces form what are called **hydrogen bonds.** Although very weak and short-lived, such bonds collectively constitute an important force which holds water molecules together and makes water a much more stable substance than would otherwise be the case.

Properties of water

Cohesion and surface tension
Cohesion is the tendency of molecules of one substance to hold together by mutual attraction. The hydrogen bonding of water results in strong cohesive forces. One effect of this is that the surface of a drop of water will assume the smallest possible area, and the drop therefore forms a sphere. The water molecules at the surface are drawn in towards the body of the drop forming a skin-like layer of molecules at the surface. This force is called surface tension. Insects walking on the surface of water and the movement of water up plants are two biological processes than can occur as a result of the cohesive properties of water molecules.

Adhesion and capillarity
Adhesion is the attraction of molecules of different compounds to one another. That water clings readily to other molecules is responsible for the upward movement of water when a small bore tube is dipped in water. This phenomenon is called capillarity. Xylem vessels of diameter 0.02 mm can, in theory, support a column of water 1.5 m by capillarity forces. One of its main biological effects is the upward movement of water in the soil.

Thermal capacity (specific heat)
Another consequence of hydrogen bonding in water is that much heat is needed to cause increased molecular movement and hence gas (steam) formation. The heat energy must first be used to break the hydrogen bonds. For this reason the temperature of water rises only very slowly for a given amount of heat added, when compared with other substances. Similarly it cools more slowly. In all it is thermally stable and so biochemical reactions in a water medium are not subjected to large temperature fluctuations and can take place at a more constant rate. Were it not for hydrogen bonding, water would be a gas at normal environmental temperatures and life as we know it could not exist. For the same reasons much heat is needed to evaporate water and therefore even the evaporation of a small amount of water from the surface of an organism has a large cooling effect, e.g. sweating.

Density
Water has its maximum density at 4°C. Unlike most other substances it is less dense as a solid than as a liquid. It freezes from the top downwards and the ice that forms at the surface can insulate the warmer water below from the colder temperatures above it. This prevents large bodies of water from freezing solid and has contributed to the survival of aquatic organisms.

Dissociation (ionization), pH and buffers
There is a slight tendency for water molecules to dissociate into ions according to the equation:

$$2H_2O \rightleftharpoons H_3O^+ + OH^-$$

water molecule   hydronium ion   hydroxide
                                      (hydroxyl) ion

It is simpler however to consider the dissociation as:

$$H_2O \rightleftharpoons H^+ + OH^-$$

water molecule   hydrogen ion   hydroxide ion
                                     (hydroxyl)

In a litre of water this dissociation produces 1/10 000 000 (10^{-7}) mole of hydrogen ions. This is equivalent to a pH of 7 which is neutral. If the concentration of hydrogen ions was greater, say 1/1000 (10^{-3}) mole hydrogen ions per litre, the pH would be 3 and the solution would be acid. Any pH below 7 is acid, any above is basic. An acid is therefore a substance that donates hydrogen ions and a base is a hydrogen ion acceptor. Note that the pH scale is not linear but logarithmic.

A buffer solution is one that retains a constant pH despite the addition of small quantities of acids or bases. Buffers contain both hydrogen ion donors and acceptors. Bicarbonate ions may act as an acceptor.

$$HCO_3^- + H^+ \rightleftharpoons H_2CO_3$$

bicarbonate ion   hydrogen ion   carbonic acid

or a donor

$$HCO_3^- + OH^- \rightleftharpoons CO_3^{2-} + H_2O$$

The removal of OH^- allows more water to dissociate and so produce more H^+

$$H_2O \rightleftharpoons H^+ + OH^-$$

Bicarbonate salts and phosphate salts are responsible for the buffering of the blood, maintaining its pH at a constant 7.4. Apart from dissociating itself, water readily causes the dissociation of other substances placed in it. Thus it is an excellent solvent.

Colloids When two substances are added to each other they may separate out or form a mixture. If the mixture consists of one substance finely dispersed throughout the other it is termed a colloid. The substance that is dispersed is usually of a high molecular weight and does not readily diffuse through a semipermeable membrane. An example of a colloid is jelly, which comprises the protein gelatin finely dispersed in water.

Diffusion This is the net movement of a substance from a region where it is more highly concentrated to a region where it is at a lower concentration, due to the motion of its constituent particles. Generally a slow process, its rate is increased when it occurs over a large area, through a short distance and over a large concentration gradient.

Osmosis Although water moves freely through cell membranes, the substances dissolved or dispersed in it may pass through more slowly, or not at all. Membranes which act in this way are termed **semi-permeable.** If two equal volumes of water are separated by a semi-permeable membrane, water molecules will move from one to the other in both directions. The tendency of water molecules to move from one side to the other is called the diffusion pressure. The diffusion pressures are equal on each side of the semi-permeable membrane and an equilibrium is established where the movement from one side to the other is exactly counterbalanced by an equal movement in the reverse direction. If a substance which cannot permeate the membrane is added to one side only, it effectively impedes the movement of water molecules on that side and thereby reduces its diffusion pressure. The solution is said to have a **diffusion pressure deficit** (DPD). The greater the concentration of the solution, the greater the DPD. Water moves from pure water, which has a DPD equal to 0, into the solution by the process of **osmosis.** The pressure which must be applied to prevent this movement is called the **osmotic pressure** (OP), i.e. it is the pressure which must be applied to prevent osmosis into a solution, when separated from pure water by a semi-permeable membrane. In practice the term can apply to solvents other than water. Because this is a special situation and most solutions do not actually exert a pressure, the term **osmotic potential** is more accurate. While numerically equal, osmotic pressure is a positive value, whereas osmotic potential is a negative one.

When two solutions have the same OP they are said to be **isotonic.** When one solution has a higher OP (more concentrated) than another it is said to be **hypertonic** to it. The one with the lower OP (more dilute) is said to be **hypotonic.**

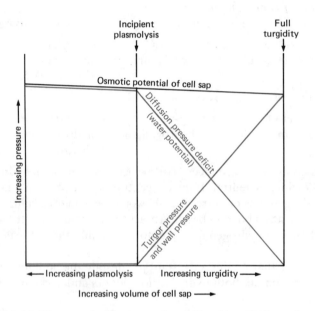

Fig. 8.1 (a) Diagram showing osmotic relationships within a plant cell

Osmosis and plant cells

For practical purposes a plant cell can be considered as a solution of salts and sugars in the vacuole surrounded by a semi-permeable membrane (tonoplast, cytoplasm and plasmalemma) and a slightly elastic but completely permeable cell wall. If a single plant cell were placed in pure water, the water would enter by osmosis; the vacuole would swell and push on the cell wall. This outwardly directed pressure is called the **turgor pressure.** The wall would expand a little, but would increasingly oppose the entry of water by pushing inward on the cell. Such inwardly directed pressure is called the **wall pressure** (WP). When the wall pressure equals the turgor pressure there will be no net entry of water. At this point the cell is fully turgid. If the same plant cell is placed in a hypertonic solution, water will leave it by osmosis. The wall pressure will fall to zero, at which point the cell is said to be at **incipient plasmolysis.** Further loss of water from the vacuoles will cause the protoplast to shrink so that the plasmalemma pulls away from the cell wall. The cell is said to be **plasmolysed.**

If the plant cell is returned to a hypotonic solution, water will once again enter. The force with which it enters is called the DPD or **water potential** (Ψ) (formerly suction pressure) and is proportional to the OP of the cell sap. Initially then, $\Psi = $ OP. As the cell becomes turgid the influx of water is resisted by the wall pressure, and hence after incipient plasmolysis $\Psi = $ OP + WP (Ψ and OP are normally negative and WP is positive). At full turgidity $\Psi = $ O and hence OP = WP. Water potential (Ψ) is most easily understood if it is thought of as a measure of the kinetic energy of water. The greatest kinetic energy of water occurs when there is no solute, i.e. in pure water. This is given the value O. In a concentrated solution the water's kinetic energy is reduced by the presence of solute molecules. Such a solution therefore has a lower water potential, i.e. it is more negative.

Link topics

Section 8.2 Blood and circulation
Section 8.3 Uptake, transport and loss in plants
Section 8.4 Osmoregulation and excretion
Section 9.1 Locomotion and support in plants and animals

Suggested further reading

Knight, R. O., *The Plant in Relation to Water,* 2nd ed. (Heinemann 1967) (Chapters 1 and 3)
Sutcliffe, J., *Plants and Water,* Studies in Biology No. 14, 2nd ed. (Arnold 1979)

QUESTION ANALYSIS

1 The sap of a plant cell has an osmotic pressure of 1 200 kPa and there is a wall pressure of 400 kPa. When this cell is placed in a solution with an osmotic pressure of 500 KPa, the force causing water to enter the cell is
A 900 kPa
B 800 kPa
C 500 kPa
D 300 kPa
Mark allocation 1/40   Time allowed 1½ minutes   *In the style of the Cambridge Board*

This is a multiple choice question testing knowledge of the relationship between various pressures relating to osmosis in plant cells, and the ability to make a simple mathematical calculation. The basic equation for this question is $\Psi = OP + WP$. Here Ψ is 'the force causing water to enter the cell'. However the calculation is not so straightforward as it would be if the cell were placed in pure water. If this were the case the Ψ would be equal to the cell sap osmotic potential (-1200 kPa) plus the wall pressure (400 kPa). Many candidates may fall into this trap and give the answer B (8 atmospheres). However the solution into which the cell is placed has its own osmotic potential of -5 atmospheres and hence Ψ is changed by this amount, compared with a solution of pure water. The true pressure with which water enters a cell is hence only (800–500) 300 kPa.

The calculation should be based on

$$\Psi \text{ (external)} + \Psi \text{ (internal)} = OP + WP$$
$$\text{i.e. } 500 + \Psi \text{ (internal)} = -1200 + 400$$
$$\text{Therefore } \Psi \text{ (internal)} = 300 \text{ kPa}$$

The correct response is therefore D.

2 (a) What is meant by the terms osmosis and osmotic pressure?   (6)
(b) Describe an experiment you have carried out to determine the osmotic pressure (potential) of a living tissue. Explain carefully the observations you made and how you calculated the result.   (7)
(c) Discuss the importance of osmosis in one process in either a named plant or a named animal. (7)
Mark allocation 20/100   Time allowed 30 minutes   In the style of the Oxford and Cambridge Board

This is a structured essay question testing recall of factual knowledge, the ability to relate an experiment carried out during the A-level course and application of knowledge of osmosis to the role it plays in an organism.

In part (a) the terms should be not only defined but also explained. 'What is meant by' requires some detail in addition to the basic definition. The mark distribution shows that up to one third of the time (10 minutes) can be spent on this part of the question and therefore some detail is essential. The two terms are discussed under 'Osmosis' in the essential information. In dealing with osmosis the candidate should discuss carefully the role of the solute in reducing the activity of solvent molecules and so allowing solvent from a more dilute solution (where the solvent molecules are less impeded) to enter through a membrane permeable only to the solvent and not the solute. Although osmosis is generally concerned with the solvent water, it may be applicable to other solvents.

The answer to part (b) should describe an experiment the candidate has carried out. It may be possible to answer such a question from information obtained in textbooks, but the answer is unlikely to be as thorough or as detailed as one relating a candidate's own experiments. In effect the osmotic pressure can only be measured by placing a solution next to its pure solvent and separated from it by a membrane permeable only to the solvent. The question however says 'living tissue' and the above procedure cannot be carried out effectively using 'living tissue'. It is, however, possible to measure the water potential (Ψ) of living tissues in a number of ways. If then, Ψ is calculated at incipient plasmolysis for a plant tissue, the OP must be numerically the same since at this point the WP is equal to zero.

$$\Psi = OP + WP$$
$$\text{where } WP = 0, \Psi = OP$$

To calculate Ψ a tissue such as a potato tuber should be taken, peeled, cut into small discs, dried and weighed. Samples of around 20 g should be placed into a range of salt or sugar molarities (1.0, 0.9, 0.8, 0.7 M, etc.). After a few hours (or even a day) they should be removed, dried and reweighed. The ratio of initial to final weight can then be calculated for each molarity (% weight loss or gain could equally be used). A graph of the weight ratio against molarity should be plotted. The molarity where the ratio is equal to one (or % loss or gain equals zero) should be read off the graph; this molarity is isotonic with the cells of the potato, because at this point there is no net movement of water in or out of the potato. The OP of this molarity of sucrose and hence the potato cells can then be found by separating it from pure water using a membrane permeable to water. In practice the OP values for solutions such as sucrose have been calculated and can be obtained from standard tables. This experiment can be carried out by measuring strips of potatoes, or observing onion epidermal cells for incipient plasmolysis through a microscope. Candidates should describe whichever method they are conversant with.

In answering part (c) the process chosen should clearly illustrate the importance of osmosis. A process which only involves osmosis as a minor aspect will not bring good marks. There are a number of processes to choose from such as support of herbaceous tissues in almost all plants, uptake of water, opening and closing of stomata in plants and the uptake of water in many animals such as *Amoeba*. Candidates should remember to name the organism they choose and make clear the importance of osmosis in the process.

Answer plan

(a) **Osmosis** – see essential information, but refer to 'solute' and 'solvent' and 'membrane permeable only to solvent'.

(b) **Osmotic pressure** – see essential information. Name tissue, e.g. potato tuber; state method fully: measurements taken such as weighings, materials used, e.g. sucrose, how OP is calculated, e.g. readings taken and graphs plotted.

(c) **Name organism** – answer depends very much on example chosen, but in all cases state the 'importance'.

3 'Water is essential for life'. Explain in what ways this statement is true for plants and animals.
   Mark allocation 20/100   Time allowed 35 minutes   London Board June 1980, Paper II, No. 7

This is an essay question testing the ability of candidates to review almost every aspect of biology in justifying the statement given. While it tests recall of factual information, it requires very careful selection of facts in order to effectively justify the statement. The most common error that candidates make is to simply catalogue the uses of water in much the same way as they have learned them. The key words are 'essential for life'. A wide range of plants and animals should be used to illustrate the statement. Examples drawn only from mammals and flowering plants will not bring as much credit as ones from the whole range of plant and animal phyla. It pays to have a mental checklist of all the major groups and to attempt to use at least one example from each. In this type of question, on Paper II of the London board, marks are given for the way in which the answer is presented as well as the actual content. For this reason candidates should carefully marshal their ideas by drawing up an answer plan first. Answers should follow a clearly defined course and not jump from point to point in a haphazard and disorganized way. Writing should be legible and care taken with spelling, especially of biological terms. The essay should include a short introduction and if possible a short concluding paragraph to tie things together or reiterate particularly important points.

There are a number of possible approaches to answering this question, each has its merits and drawbacks and the choice will depend very much on the candidate's own preference and strengths. Unfortunately some candidates overestimate their own abilities to write good essays on these broad topics. They write the essay without adequate planning by noting down points as they occur to them and without regard to providing a balanced range of examples. The most logical approach to avoid this is to consider the phrase 'essential for life' and approach the answer by looking at the seven essential characteristics of life: reproduction, growth and repair, nutrition, respiration, excretion, movement and sensitivity. If the importance of water to each of these processes is considered in turn and a note made of plant and animal examples for each, then a plan can be easily drawn up. By reviewing the examples these can be altered to give a good range and an equal balance between plants and animals. This approach should produce relatively good results and a range of examples. It may, however, miss certain aspects of the importance of water which do not fit clearly into the seven categories used, e.g. use of water in temperature control.

A second approach is to consider the main functions of water, i.e. its roles in metabolism, as a solvent, as a lubricant and in support. If these are considered with the characteristics of life in mind the essay should incorporate most of the important points. A range of examples is still essential.

Whichever approach is used the essay should begin with an introduction, mentioning that life arose in water and evolved from it. When organisms moved on to land they effectively took the watery environment with them, covering it to prevent desiccation. A very brief review of the properties of water could be made, with reference to how water owes many of its special properties to hydrogen bonding.

Another useful addition, whatever method is used to answer the question, is a 'miscellaneous' group of points at the end. It is almost inevitable that in any classification of features there are some points which do not clearly fit into a category. These can be included in this final group rather than wasting time trying to justify their inclusion elsewhere. The essay should conclude with a few sentences restating the importance of water and its essential nature for the existence of life.

In the answer plan that follows the second of the approaches discussed above is used, although this should not be taken to indicate that it is the better one. Candidates should be guided by their own preferences and choose whichever format suits them best. There are many other methods, equally valid and effective. The important consideration is to include all the facts, with examples, in a clear and logical way. Some specific examples are given in the plan where the feature relates to specific organisms but not where it applies to many groups. In these cases candidates should use examples which they know.

Answer plan

Introduction

1 Life arose in water
2 Some organisms still live in water
3 All are made up of at least 50% water
4 Properties – very briefly

Metabolic role

1 All reactions occur in an aqueous medium (see 'Thermal capacity' in essential information)
2 Hydrolysis reactions, e.g. proteins to amino acids
3 Diffusion – moist surfaces, e.g. gaseous exchange and digestive absorption
4 Osmosis – importance
5 Photosynthetic substrate

Solvent

1 Transport – blood plasma/tissue fluid/lymph, e.g. excretory, respiratory and digestive products; hormones; phloem, e.g. sucrose; xylem, e.g. mineral salts
2 Removal of excretory products, e.g. ammonia, urea
3 Cytoplasm contains many dissolved chemicals
4 Secretions, e.g. digestive juices, tears, venom in snakes

Lubricant

1 Mucus – external, e.g. movement in snail, earthworm; internal, e.g. gut wall, vagina
2 Synovial fluid, pleural fluid, pericardial/periviceral fluid

Support (movement)

1 Amniotic fluid, e.g. mammals
2 Vitreous/aqueous humors of the eye
3 Hydrostatic skeleton in earthworm
4 Coelom in many invertebrates
5 Tube feet, e.g. echinoderms
6 Turgor pressure in herbaceous plants
7 Erection of penis
8 Dispersal of seeds, e.g. squirting cucumber
9 Medium in which organisms live
10 Transference of gametes, e.g. mosses and ferns

Miscellaneous

1 Temperature control, e.g. sweating and panting
2 Medium for dispersal, e.g. larval stages of terrestrial organisms
3 Endolymph/perilymph in ear – hearing and balance

NB In all cases show how the role of water is essential for life.

4 The diagram shows two chambers, A and B, separated by a membrane that only allows the passage of water across it. Tubes A and B are open to the air. The water potential of pure water is zero.

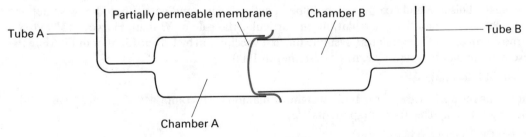

Fig. 8.1(b)

(a) In what units should water potential be measured?

(b) If the two chambers are filled with pure water to the same levels in tubes A and B, how will water molecules behave in relation to the membrane?

(c) If chamber A and tube A contain pure water and chamber B and tube B contain a salt solution, and if the levels in the tubes are equal, how will water molecules behave in relation to the membrane?

Mark allocation 4/80   Time allowed 6 minutes   *AEB 1983, Paper I, No. 4*

Water potential is a pressure and as such used to be measured in atmospheres. The unit now used is the Pascal (Pa) but this is rather small for most purposes in biology. The kilopascal (kPa), which is equal to 1 000 Pa, is a more practical unit. One atmosphere is equivalent to about 100 kPa (101.32 kPa, to be precise). The best answer is therefore kilopascals (kPa).

There is a potential pitfall in part (b). The unwary candidate may assume that as pure water is present on both sides of the membrane, and because it has a water potential of zero, then the water potential on both sides is equal at zero, and no exchange takes place. In fact water will move from chamber A to chamber B and vice versa. There will be no change in the levels in the tubes A and B, because the volume of water moving from A to B is equal to that moving in the reverse direction. The answer is therefore that water molecules cross the membrane in both directions in equal quantities.

In (c), chamber B contains salt which lowers the water potential in that chamber. As the water potential in chamber A is now greater than that in chamber B, more water molecules move from A to B than in the reverse direction. It would be worthwhile making clear in the answer the fact that some water molecules still move from B to A; but that more move from A to B.

8.2   BLOOD AND CIRCULATION

Assumed previous knowledge   Histology of the blood (Section 2.2 Plant and animal tissues)
Definition and types of immunity.
Natural and artificial immunity.

Underlying principles

As animals grew in size and complexity, tissues and organs with specific functions developed, each organ dependent on the others for some essential process or chemical. The need arose for a system to transport materials, especially food, oxygen, carbon dioxide and wastes, between the various organs. The system must service every cell either by flowing freely over them (open system) or having capillaries within diffusing distance of them (closed system). The fluid must be circulated by cilia, body movements or a specialized pump, the heart. The heart takes many shapes and forms depending on the organism and its environment. It may be tubular (insects), two chambered (fish), three chambered (amphibians) or four chambered (birds and mammals). Both the beating of the heart and the distribution of the blood must be carefully controlled in order to meet the varying demands placed upon it at different times. The blood itself must be capable of carrying large volumes of many different substances, particularly oxygen. Pigments for carrying oxygen have developed and are either suspended in the plasma or, if there is a possibility of removal during excretion, in special cells. In addition the blood contains the body's defence and immune system because it is ideally situated to fight infection in all parts of the body. Finally the liquid nature of the blood, necessary for transportation, creates problems if the body is damaged and leakage occurs. The blood has the capacity to clot in such circumstances.

Points of perspective

It would be an advantage for candidates to have seen by dissection or demonstration the heart and major blood vessels of a fish and a mammal. It would likewise be an advantage to have dissected or have had demonstrated the heart of a mammal such as a pig or a sheep. A thorough understanding of the antigen–antibody immune system and its role in blood groups, immunization against disease and taxonomy would be beneficial. The candidate should know something of the history of blood and circulation, in particular the role of William Harvey (1578–1657) in the discovery of the circulatory system, the first vaccination by Edward Jenner in 1796 and the discovery of blood groups by Karl Landsteiner in 1900.

Essential information

Blood makes up 7–10% of the body weight of mammals. It comprises 45% corpuscles (cells) and 55% plasma. The main components are:

1 Erythrocytes (red blood cells)
2 Leucocytes (white blood cells)
3 Platelets
4 Plasma

Erythrocytes   The main function of red blood cells is the transport of respiratory gases. Because the solubility of oxygen in water is low (0.58 cm^3/100 cm^3 at 25°C), active organisms have developed substances – respiratory pigments – with a high affinity for oxygen.

Table 8.1 Respiratory pigments

Pigment	Location of pigment	Colour	Group of animals
Chlorocruorin (contains iron)	Plasma	Green	In some polychaetes
Haemocyanin (contains copper)	Plasma	Blue	Molluscs, some crustacea, arachnids
Haemoerythrin (contains iron)	Corpuscles	Red	Some nematodes and polychaetes
Haemoglobin (contains iron)	Corpuscles	Red	Vertebrates
	Plasma		Some annelids and molluscs

Respiratory pigments need to carry sufficient oxygen to supply actively respiring tissues but at the same time they must be able to release oxygen readily to those tissues.

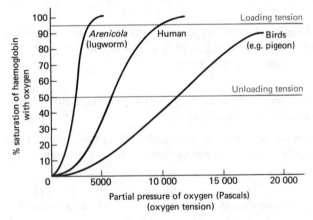

Fig. 8.2 (a) Oxygen dissociation curves for haemoglobin in three organisms

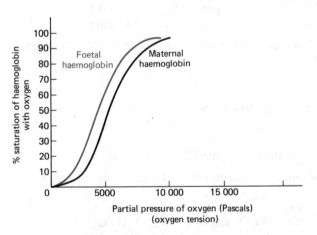

Fig. 8.2 (b) Comparison of oxygen dissociation curves for foetal and maternal haemoglobin of man

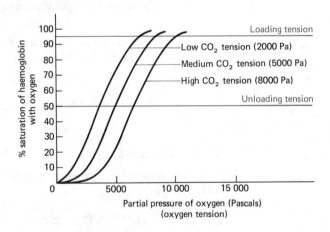

Fig. 8.2 (c) Oxygen dissociation curves at different carbon dioxide concentrations – the Bohr effect

Notes on Figure 8.2a

1 When a respiratory pigment is exposed to a gradual increase in oxygen tension it absorbs oxygen rapidly at first and then progressively more slowly.

2 Because 100% saturation of haemoglobin does not occur at normal environmental partial pressures of oxygen the **maximum loading tension** is usually taken at 95% saturation.

3 The **unloading tension** is where the pigment is 50% saturated.

4 The partial pressure of oxygen in the lungs is usually less than in the atmosphere because air in the lungs includes the residual volume air from which some oxygen has already been removed.

5 Because there is less oxygen in water than in air the dissociation curves for aquatic organisms

are usually to the left of those for terrestrial ones. This is accentuated in *Arenicola* which lives in muddy waterlogged burrows where circulation of the water is limited.

6 The dissociation curve for birds is to the right of that of man because the oxygen is given up to the respiring tissues by the pigment more readily, i.e. at a high partial pressure of oxygen. This is necessary if birds are to obtain oxygen fast enough to allow the high metabolic rate required by flight.

Notes on Figure 8.2b
The curve for foetal haemoglobin is to the left of the mother's because if the foetus is to obtain oxygen through the placenta its haemoglobin must have a greater affinity for oxygen than the mother's haemoglobin.

Notes on Figure 8.2c
1 The partial pressure of carbon dioxide in the blood alters the oxygen dissociation curve – the Bohr effect.

2 Where there is a higher concentration of carbon dioxide, i.e. in respiring tissues, the haemoglobin will release its oxygen more readily. Therefore a higher partial pressure of carbon dioxide in the blood shifts the oxygen dissociation curve to the right.

3 This is the main factor responsible for oxygen being taken up in the lungs and released to respiring tissues.

4 Animals living in regions of low oxygen tensions are less sensitive to carbon dioxide effects as any shift to the right would mean they could not absorb oxygen from the low partial pressure of oxygen in their environment.

Carbon dioxide transport
Blood takes up carbon dioxide at increasing rates as the carbon dioxide concentration increases. The carbon dioxide is carried in three ways:

Formation of bicarbonate compounds Respiratory pigments do not carry just oxygen but also some carbon dioxide. Because haemoglobin can form a potassium salt, carbon dioxide can combine in the following way:

$$\underset{\text{carbon dioxide}}{CO_2} + \underset{\text{water}}{H_2O} \rightleftharpoons \underset{\text{carbonic acid}}{H_2CO_3}$$

$$\underset{\text{carbonic acid}}{H_2CO_3} + \underset{\substack{\text{potassium salt}\\\text{of haemoglobin}}}{KHb} \rightleftharpoons \underset{\substack{\text{potassium}\\\text{bicarbonate}}}{KHCO_3} + \underset{\substack{\text{reduced}\\\text{haemoglobin}}}{HHb}$$

Direct combinations with carbon dioxide Carbon dioxide may combine with amino groups in the protein part of the haemoglobin molecule.

$$Hb-N\overset{\displaystyle H}{\underset{\displaystyle H}{\Big\backslash}} + CO_2 \rightleftharpoons Hb-N\overset{\displaystyle H}{\underset{\displaystyle COO^-}{\Big\backslash}} + H^+$$

These carbamino compounds carry 2–10% of the carbon dioxide in the blood.

Carbonic anhydrase This is an enzyme found in the corpuscles and it increases the rate of the following reaction:

$$H_2O + CO_2 \underset{\text{carbonic anhydrase}}{\rightleftharpoons} H_2CO_3 \rightleftharpoons H^+ + HCO_3^-$$

The dissociation of the H_2CO_3 would lead to a rise in hydrogen ion concentration, especially since the HCO_3^- readily diffuses out of the corpuscle into the plasma.

However, the balance is restored by the movement of chloride ions from the plasma into the corpuscle. This is called the chloride shift.

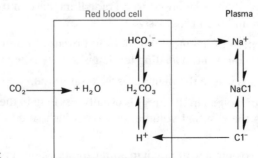

Fig. 8.3 The chloride shift

Leucocytes These are able to migrate out of the blood system and their main function is defence of the body, which they achieve in two main ways:

1 Phagocytosis: the first leucocytes to arrive at the site of an injury are usually neutrophils and they engulf foreign particles by phagocytosis.
2 Production of antibodies: an antigen is a substance which causes the body to manufacture specific proteins which agglutinate, precipitate, neutralize or dissolve foreign organisms and materials. The specific proteins so manufactured are called antibodies. Antigens are usually proteins but may be other large molecules such as polysaccharides. An antigen stimulates a lymphocyte to produce antibodies which act against foreign particles in one of a number of ways:
   (a) make them more vulnerable to phagocytic attack
   (b) interfere with the harmful activities of the foreign particles
   (c) break them down.

Platelets The function of the platelets is in clotting and an outline of the process is given below. Prothrombin and thrombin are plasma proteins.

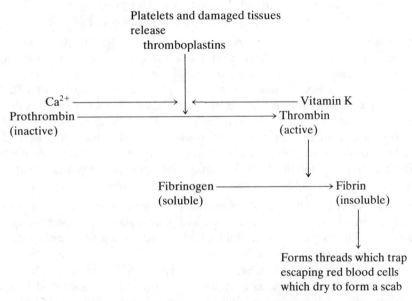

Plasma This straw coloured liquid is an important transport medium within the body. Most of the carbon dioxide combined as HCO_3^- in the red blood cells is carried in the plasma. It also carries:

1 the end products of digestion, especially glucose and amino acids
2 waste products, especially urea and uric acid
3 hormones
4 heat from very actively respiring tissues such as the liver and muscles to cooler parts of the body.

Blood groups The chemical composition of the cell membrane of erythrocytes is variable. If erythrocytes of one type are introduced into the blood stream of another type they act as antigens. There are 256 known blood groups but two main grouping systems – the ABO system and the rhesus system.

ABO system
Erythrocytes may contain either of two types of antigen A or B, both, or neither. Antibodies, a and b, are present in the plasma.

Table 8.2 Blood group antigens and antibodies

Blood group	Antigen (in corpuscle)	Antibody (in plasma)
A	A	b
B	B	a
AB	A and B	None
O	None	a and b

Incompatibility between donor's corpuscles and recipient's plasma causes agglutination and must be avoided. There seems to be little effect if the recipient's corpuscles are incompatible with the donor's plasma.

Table 8.3 British blood group frequencies

Blood group	% population of Britain	Groups from which blood can safely be transfused
A	42	A, O
B	9	B, O
AB	3	A, B, AB, O
O	46	O

Group O is known as the universal donor and group AB as the universal recipient.

The proportions of populations with a particular blood group vary in different parts of the world, e.g. in Thailand 35% of the population are group B and in some North American Indian tribes group AB is unknown.

Rhesus system

A substance present in the blood of rhesus monkeys is also present in some humans.

Present (85% of population) = Rhesus positive (Rh$^+$)
Absent (15% of population) = Rhesus negative (Rh$^-$)

Transfusion of Rh$^+$ blood to an Rh$^-$ recipient is dangerous but Rh$^-$ to Rh$^+$ is safe.

During pregnancy there is some leakage of foetal blood into the mother's blood stream. If there is A, B, O incompatibility the red blood cells are quickly broken down, together with any Rh factor they may carry. If however the ABO group is compatible but a Rh$^-$ mother is carrying a Rh$^+$ foetus the mother builds up Rh$^+$ antibodies. This will not affect this pregnancy but the Rh$^+$ antibodies remain in the mother's blood stream after the birth. If there is a subsequent Rh$^+$ foetus in the same mother, these antibodies pass into the foetus causing agglutination and destruction of the foetal erythrocytes. Immediately after the birth of the first child the mother may be injected with Rh antibodies which destroy the foetal erythrocytes before they cause a natural build-up of anti Rh factors in her blood.

Lymph Owing to the relatively high blood pressure in the capillaries a watery solution of low protein content, some salts, nutritive materials and phagocytic cells leaves the capillaries and bathes the tissues. This is called tissue fluid. Having had the nutritive materials and oxygen removed and wastes added by the cells, most of the tissue fluid re-enters the capillaries by osmosis. About 1–2% is returned in separate vessels called lymphatics. This is the lymph. Lymph moves along these vessels as a result of hydrostatic pressure and respiratory and muscular movements which squeeze the lymph vessels. The direction of flow is controlled by valves in the lymph nodes, especially in the armpits and groin. These nodes are also the site of lymphocyte production. The lymphatic vessels finally drain into the vena cava near its entrance to the heart via two major vessels, the right lymphatic duct which drains the upper right side of the body and the thoracic duct which drains the remainder of the body.

Blood systems These may be either open or closed.

Open blood system

In an open blood system, which occurs in arthropods and most molluscs, there are no capillaries connecting the arteries with the veins. Arterial blood passes into sinuses so that major tissues are bathed in circulating fluid. These fluids slowly work their way back to the open ends of veins or to the ostia of the heart. The organs lie directly in the blood-filled haemocoel.

Closed blood system

In a closed blood system, as found in vertebrates and annelids, a continuous network of minute capillaries unites the smaller arteries with the veins. Nutrients are transferred to cells by fluids filtering through the walls of the capillaries into the tissue spaces.

Single circulatory system (fish)

Gills   Heart   Body

Blood passes through the heart only once per circuit of the body. Being forced through the gill capillaries lowers the blood pressure considerably and it still has to pass across another capillary system before returning to the heart.

Double circulatory system (birds and mammals)

Lungs   Pulmonary   Heart   Systemic   Body
          circulation   circulation

Blood passes through the heart twice per circuit of the body. This enables blood pressure to be maintained.

The heart The simplest form of heart is a muscular contracting region of a blood vessel, e.g. in an earthworm. In fish this contractile region is S-shaped and divided into a receiving region, the atrium, and a pumping region, the ventricle. In amphibians there are two atria and a single ventricle, whereas in birds and mammals there are two atria and two ventricles arranged so that the left atrium and ventricle carry oxygenated blood and are completely separated from the right atrium and ventricle which carry deoxygenated blood. The left atrium and ventricle are separated by the bicuspid or mitral valve and the right atrium and ventricle by the tricuspid valve. The aorta and pulmonary arteries have semilunar valves to prevent backflow of blood into the ventricles. The left ventricle pumps oxygenated blood to the whole body except the lungs and has a thicker muscular wall than the right ventricle which pumps deoxygenated blood to the lungs only. The heartbeat consists of a contraction phase (systole) and a relaxation phase (diastole). In a resting adult human the heart pumps about 70 times per minute, pumping about 60 cm³ of blood per beat from each ventricle. The total volume of blood pumped each minute is called the **cardiac output** and varies according to the amount of physical exercise undertaken and the emotional state of the individual.

Control of heartbeat (cardiac rhythm) A vertebrate heart continues to beat in a co-ordinated way even when removed from the body. The heartbeat must, therefore, be initiated from within the heart muscle itself (myogenic heart) rather than by separate nervous initiation (neurogenic heart). In myogenic hearts the beat is initiated by a 'pacemaker' called the **sinoatrial node** (SA node) – a group of specialized cardiac muscle cells, 2 cm by 2 mm, in the right atrium, near the point where the venae cavae enter. The wave of excitation from the SA node spreads outwards causing the atria to contract, emptying their blood into the ventricles. The wave is picked up by a mass of similar tissue in the right atrium near the interatrial septum. This second group of cells is called the **atrioventricular node** (AV node). To ensure that the ventricles contract from the apex upwards and so allow blood to be forced out through the arteries, the wave of excitation is conducted to the apex of the ventricle by the **Purkinje fibres** (bundle of His).

Factors modifying heartbeat (cardiac rhythm)

Chemical control
To maintain its rhythm for any length of time the heart requires the correct balance between the ions of calcium, sodium and potassium. The chemicals adrenalin and noradrenalin and drugs such as digitalis increase cardiac output. An increase in the level of carbon dioxide in blood (effectively a fall in pH) also causes an increase in cardiac output. Acetylcholine slows myogenic hearts while accelerating neurogenic ones.

Nervous control
Receptors in the aorta and carotid arteries respond to an increase in blood pressure by sending nervous impulses to the cardio-regulatory centre in the brain, which in turn sends impulses along

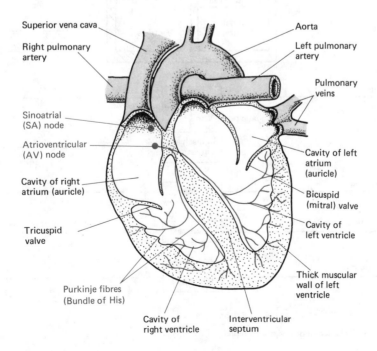

Fig. 8.4 Diagram of a vertical section through the human heart (ventral view)

the efferent vagus nerve to decrease cardiac output. An increase of pressure in the right atrium causes receptors to send impulses along the afferent vagus nerve to the cardio-regulatory centre which then sends impulses to increase cardiac output. The heartbeat is therefore speeded by the sympathetic nervous system and slowed by the parasympathetic nervous system.

The maintenance and control of blood pressure and circulation

In a healthy human at rest the arterial systolic blood pressure is 120 mm Hg and the diastolic pressure is 80 mm Hg. The pressure is created initially by the contraction of the ventricles of the heart. As blood is forced into the arteries, the elastic walls expand causing distension. The recoil of the elastic walls pushes blood away from the heart, creating distension of the artery at a point further away from the heart. (The blood is prevented from returning to the heart by the semilunar valves.) The 'pulses' continue throughout the arterial system and help to maintain blood pressure. The hormone vasopressin increases blood pressure in vertebrates, though higher levels are needed to produce this effect in humans. The return of blood to the heart in the veins is maintained in a number of ways:

1 The residual heart pressure – usually 10 mm Hg or less.
2 The contraction of muscles squeezes veins and forces blood towards the heart; flow in the reverse direction is prevented by pocket valves.
3 Inspiratory movements – when breathing in, the low thoracic pressure helps to draw blood along the major veins towards the heart.
4 Gravity will help return blood from those regions above the heart.

The 5 l of blood in a human is inadequate to supply the maximum needs of all regions of the body at the same time. Different organs make different demands on the blood both in quality and quantity.

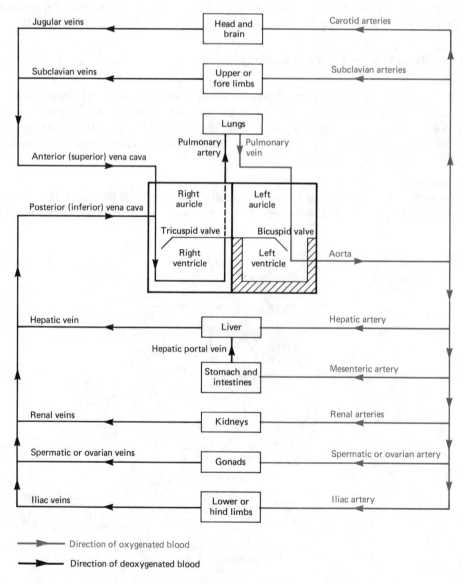

Fig. 8.5 Outline diagram of mammalian circulation

Table 8.4 Blood flow to various organs

Organ	Blood flow in cm³ to the organ (when body is at rest) per 100 g of organ
Heart	80
Liver	90
Brain	55
Muscle	3
Skin	10
Kidney	400
Remainder	2.5

Clearly, during exercise muscles require considerably more blood than they do when at rest. Control of blood flow is achieved by contraction and relaxation of muscles in the smaller artery (arteriole) walls. Control of these muscles is by the autonomic nervous system, the sympathetic system contracting the vessels and decreasing blood flow, the parasympathetic system dilating them and increasing the flow. Apart from changes in blood flow to meet physiological needs, certain psychological events control flow in humans, e.g. fear, embarrassment and erection of the penis or clitoris due to erotic stimulation.

Table 8.5 Similarities and differences between blood vessels

Artery	Vein	Capillary
Collagen fibres — Lumen; Elastic fibres (also collagen fibres and smooth muscle) — Endothelium (1-cell thick)	Collagen fibres — Lumen; Elastic fibres (few) — Endothelium (1-cell thick)	Endothelium (1-cell thick) — Lumen
Similarities		
Tubular	Tubular	Tubular
Endothelium present	Endothelium present	Endothelium present
Transports blood	Transports blood	Transports blood
Differences		
Thick wall (muscle present)	Thinner wall (muscle present)	Thinnest wall (no muscle present)
More elastic tissue	Less elastic tissue	No elastic tissue
Smaller lumen relative to diameter	Larger lumen relative to diameter	Largest lumen relative to diameter
No valves (except in aorta and pulmonary artery)	Pocket valves throughout	No valves
Can constrict	Cannot constrict	Cannot constrict
Not permeable	Not permeable	Permeable
Carries blood from heart	Carries blood to heart	Carries blood to and from heart
Carries oxygenated blood (except pulmonary artery)	Carries deoxygenated blood (except pulmonary vein)	Carries oxygenated and deoxygenated blood
High pressure (80–120 mm Hg)	Low pressure (less than 10 mm Hg)	Pressure lower than arteries but higher than veins (32–12 mm Hg)
Blood moves in pulses	No pulses	No pulses

Link topics

Section 2.2 Plant and animal tissues (for blood histology)
Section 4.2 Heredity and genetics (for genetics of blood groups)
Section 7.2 Gaseous exchange
Section 10.1 Hormones and homeostasis

Suggested further reading

Chapman, C., *The Body Fluids and their Functions*, Studies in Biology No. 8, 2nd ed. (Arnold 1980)
Inchley, C. J., *Immunobiology*, Studies in Biology No. 128 (Arnold 1981)
Neil, E., *The Human Circulation*, Oxford/Carolina Biology Reader No. 82, 2nd ed. (Packard 1979)
Vassaller, M., *The Human Heart*, Oxford/Carolina Biology Reader No. 8, 2nd ed. (Packard 1979)
Wood, D. W., *Principles of Animal Physiology*, 3rd ed. (Arnold 1983) (Chapter 4)

QUESTION ANALYSIS

5 The order of blood vessels in a portal system is best represented by:
   A Heart, arteries, capillaries, veins, heart
   B Heart, arteries, capillaries, veins, capillaries, veins, heart
   C Heart, arteries, capillaries, veins, capillaries, arteries, heart
   D Heart, arteries, capillaries, veins, heart, arteries, capillaries
   Mark allocation 1/40   Time allowed 1½ minutes   *In the style of the Cambridge Board*

This is a multiple choice question testing understanding of the term 'portal system'. The naming of blood vessels often confuses candidates, especially since there are 'exceptions' to many of the rules. One firm rule is that arteries carry blood away from the heart ('a' for artery: 'a' for away) and veins carry blood to the heart. In a correct response to this question, where a word precedes 'heart' it must be 'vein' and where it follows 'heart' it must be 'artery'. This allows option C to be rejected since the second time 'heart' appears it is preceded by 'artery'. All the remainder satisfy the rules. In a 'portal system' blood passes through two sets of capillaries before returning to the heart. This excludes option A which is a normal circulatory sequence. Option D may also be rejected since the blood returns to the heart before passing through the second set of capillaries. D is in fact the sequence of blood vessels in a double circulatory system. Option B is therefore the correct response.

6 (a) State three main functions of the lymphatic system.
    (b) State two factors which assist in producing a directional flow of fluid in lymph vessels.
   Mark allocation 5/100   Time allowed 6 minutes
   Associated Examining Board November 1980, Paper I, No. 14

This is a short structured question testing recall of factual information on the lymphatic system. In part (a) the candidate should attempt to find three rather different functions. To state 'defence against...' and list three infectious diseases will probably bring only a single mark. The main function is as a centre for the production of lymphocytes which ingest foreign material, especially in the lymph nodes. In this sense it defends the body against infection from disease-causing bacteria and other pathogens. A second function is as a 'filter'. Lymph nodes filter out not only bacteria, but also other particles such as soot or tobacco smoke, before the lymph is returned to the main circulation. Finally it is the means by which the tissue fluid which does not re-enter capillaries returns to the main circulation. Without the lymphatic system, therefore, major swelling of the tissues would occur.

In part (b) the factors looked for are firstly muscular movements of the body squeezing the lymph vessels. These vessels have valves to prevent backflow and so lymph is returned to the vena cava. For the second factor the candidate could give either the residual hydrostatic pressure that remains from the tissue fluid being forced out of the capillaries or inspiratory movements which reduce thoracic pressure helping to draw the lymph towards the vena cava and hence the heart.

7 With reference to a mammal, describe the means by which blood circulation is maintained and controlled. How is the blood maintained at constant temperature? Review the part played by the blood in conferring immunity from infectious disease.
   Mark allocation 25/100   Time allowed 45 minutes
   Southern Universities Joint Board June 1980, Paper II, No. 5

This is an essay question testing knowledge of the physiology and defence functions of mammalian circulation and thermoregulation. In the first part the key words are 'maintained and controlled'. Although the two processes are often connected it may be easier, for planning purposes at least, to deal with them separately. Remember to limit the answer to 'a mammal'. The maintenance of circulation is achieved mainly by the pumping action of the heart. Details of the process are given in the essential information under 'Control of heartbeat'. The essential details are the action of the SA node, AV node and Purkinje fibres. Any details of the heart structure should be brief and limited to aspects which relate to its function. A short account of features of cardiac muscle such as its moderately powerful action and the fact that it is not fatigued would be better than an elaborate diagram showing the heart and its major blood vessels. The remaining factors maintaining circulation are dealt with under 'the maintenance and control of blood pressure and circulation'. The essential points are recoil of elastic walls of arteries, and the veins' residual heart pressure, muscular contraction (explain this fully), inspiratory movements and gravity (in some situations).

The 'control' of circulation can be conveniently divided into control of heartbeat and control of peripheral flow (blood vessels). In 'factors modifying heartbeat' in the essential information, chemical and nervous controls are dealt with briefly. The control of peripheral flow involves the autonomic nervous system controlling vasoconstriction and vasodilation of blood vessels.

The second part of the question deals with thermoregulation. Although this is not totally concerned with the blood system, the blood plays an essential part as the distributor of heat. The maintenance of constant blood temperature is achieved via mechanisms for losing heat when blood temperature is above normal and conserving heat when it is below it. The amount of information required by the question as a whole does not allow, in the time given, a full account of each mechanism. The major mechanisms are listed in the answer plan and a small amount of detail should be given to show how each affects blood temperature. For instance, to describe shivering the candidate should state that the involuntary muscular contractions of the body generate heat which is then transferred to the blood. The range of mechanisms should be broad. Many candidates will satisfy themselves with two or three main factors and omit behavioural mechanisms. All factors should relate to the particular mammal chosen. Hair erection, for example, has little if any effect in conserving heat in humans or whales. A classic error, even at A level, is to refer to blood capillaries moving nearer to or further from, the skin surface when discussing vasodilation or vasoconstriction. The widening of the superficial blood vessels (vasodilation) increases the flow of blood near to the body surface and allows more heat to be lost to the environment. The blood flows either in vessels close to the surface or in ones deeper into the skin. The vessels themselves do not move. Finally it is important to mention control of these processes, in particular the role of the hypothalamus in detecting changes in blood temperature and the thermoregulatory centres of the brain.

In the last part of the question the key word is 'review'. The question as a whole involves many processes and details and one danger is that candidates will over-run the time allowance and so penalize themselves by having too little time to complete the paper. The examiners, clearly conscious of this, only require a 'review' and not a long detailed account. It is therefore important to isolate the essential points. The other important word is 'immunity'. Many candidates would give a complete account of the defence mechanisms of the body, when only those processes that 'confer immunity' are required. The answer should include a review of antigen–antibody action (see second point under 'leucocytes' in the essential information). It should then include points such as the continued production of antibodies even after the initial infection has been overcome and how these immediately destroy the same infective agents when they enter the body on a subsequent occasion. More specific knowledge of the exact role and mechanism of the antigen–antibody reaction would be useful, as indicated in the points of perspective.

Answer plan

Maintenance of circulation

Arteries	Veins
Mechanism of heartbeat (SA node, AV node, Purkinje fibres, cardiac muscle features)	Residual heart pressure
	Muscular action and role of pocket valves
Recoil action of arterial wall	Inspiratory movements
	Gravity

Control of circulation

Control of heartbeat	Control of blood flow
Chemical including hormones	Vasoconstriction
Nervous	Vasodilation
	Role of autonomic nervous system

Regulation of blood temperature

1 Role of hypothalamus in detecting blood temperature changes

Mechanisms to lower blood temperature	Mechanisms to raise blood temperature
Sweating	Increased metabolic activity
Vasodilation	Shivering
Lowering of hairs (some mammals)	Vasoconstriction
Behavioural mechanisms, e.g. being nocturnal,	Hair erection (some mammals)
avoiding direct sunlight	Behavioural mechanisms, e.g. lying in sunlight
2 Control by thermoregulatory centres	
of the brain	

Immunity

1 Antibody production by lymphocytes in response to foreign antigen
2 Production continues for some time and any subsequent infective agent is immediately destroyed.

8 Graph A shows the dissociation curves for human oxyhaemoglobin at three partial pressures of carbon dioxide.

* a pascal (Pa) is a unit of pressure. A pressure of 100 000 pascals is approximately equal to atmospheric pressure (760 mm Hg).

(a) What effect does an increase in carbon dioxide pressure have on the oxygen carrying capacity of haemoglobin?   (2)

(b) State where in the mammalian body the partial pressure of carbon dioxide is likely to be

(i) low

(ii) high   (2)

(c) What effect will variation in the partial pressure of carbon dioxide in different parts of the mammalian body have on the transport of oxygen?   (2)

(d) Graph B shows the dissociation curve for myoglobin (the respiratory pigment in muscles) compared with that of haemoglobin in the same animal.

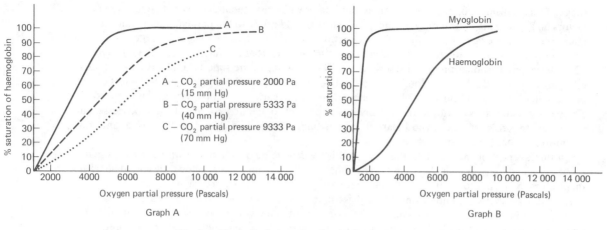

Fig. 8.6 Dissociation curves for (a) haemoglobin and (b) myoglobin

State the physiological significance of the relationship between these two pigments.   (4)
Mark allocation 10/175   Time allowed 7 minutes   *London Board June 1980, Paper I, No. 10*

This is a structured question testing the ability to interpret graphical information, draw conclusions from it and apply them to the transport of oxygen in mammals. The candidate should study Graph A carefully. A similar graph may well have been learned by the candidate but the units and the partial pressures of carbon dioxide may differ from those learned. It is a common failing in this type of question to write answers using learned information rather than that provided in the question.

In (a) it is important to appreciate that the carbon dioxide concentration increases from A to B to C. There are two parts to the answer, firstly the capacity for haemoglobin to take up oxygen is reduced and secondly the oxygen is released more easily for a given partial pressure of oxygen. Effectively these two answers amount to the same point, but vary according to whether the haemoglobin is at the respiratory surface or at the tissues.

For (b) the partial pressure of carbon dioxide in a mammal is low where the carbon dioxide is removed from the body, i.e. the lungs. It is high where the carbon dioxide is produced, i.e. actively respiring tissues such as kidney, liver or working muscles. It is necessary to be specific here for full marks; 'tissues' is too vague an answer at this level.

In (c) the candidate is required to correlate the answers to (a) and (b). Where the partial pressure of carbon dioxide is low (i.e. the lungs) haemoglobin takes up oxygen more efficiently to form oxyhaemoglobin. Where the partial pressure of carbon dioxide is high (i.e. the tissues) the haemoglobin releases its oxygen readily to the respiring tissues that need it.

To answer (d) the candidate should again study the graph provided and answer the question using it rather than one that has been memorized for the examination. The 'physiological significance' is that myoglobin – a respiratory pigment found in muscles – is able to readily accept oxygen from haemoglobin, because its dissociation curve is to the left of the one for haemoglobin. The myoglobin is saturated with oxygen at very low partial pressures and can act as a store of oxygen for immediate use when the muscles need suddenly to become active. This reduces the oxygen debt they might otherwise incur.

8.3   UPTAKE, TRANSPORT AND LOSS IN PLANTS

Assumed previous knowledge   The importance and properties of water (Section 8.1)
Structure of xylem and phloem (Section 2.2)
Structure of the leaf (Section 6.1)

Underlying principles

Most organisms comprise a variety of different cells grouped into tissues and organs. These tissues and organs have become specialized to perform particular functions. The product of one tissue may be required by another and consequently a transport system between the two is required. In addition there are certain advantages for plants in being large, for instance, they can compete more readily for light. Some have become extremely tall (over 100 m) and have thereby obtained competitive advantage. This has meant that the organs collecting the water, the roots, are some considerable distance from the leaves that require it for photosynthesis. Again this necessitates a transport system between the two structures. The sugars manufactured in the leaves must be transported in the opposite direction to sustain respiration in the roots. Unlike most animals, plants do not possess contractile cells such as muscle cells. They are therefore dependent to a large degree on passive rather than active mechanisms for transporting material. Evaporation of water from leaves creates an osmotic gradient along the leaf mesophyll cells that draws water in from the xylem. As water is drawn from the xylem, the cohesive properties cause it to be pulled up in a continuous column. An osmotic gradient in the roots brings water from the soil to the xylem. Only its entry into the xylem involves the expenditure of energy produced by the plant itself. The downward transport of sugars is less clearly understood. Certain theories, e.g. mass flow theory, favour a totally passive process, while others, e.g. transcellular strand theory, involve energy expenditure.

Points of perspective

Candidates with practical experience of using a potometer will be at an advantage. It would be beneficial to have read widely on the theories of transport in phloem and to be able to evaluate each one critically. A broad knowledge of xeromorphic adaptations, described with reference to specific examples would be useful. Knowledge of the problems of hydrophytes and halophytes and the methods they use to overcome them might gain one or two additional marks.

Essential information

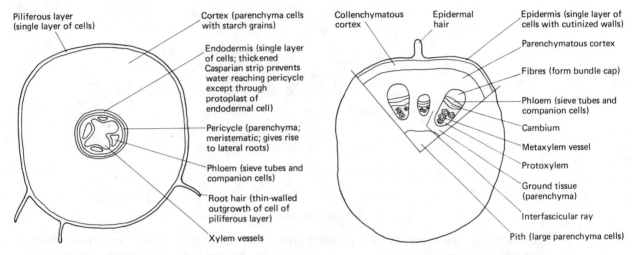

Fig. 8.7 (a) TS root of *Ranunculus*   **Fig. 8.7 (b)** TS stem of *Helianthus*

Water uptake by roots Available soil water is absorbed from the water spaces between the soil particles and the water film surrounding them. Absorption is primarily by the root hairs arising from the piliferous layer of the root – a thin layer with hairs increasing the surface area. Near the apex of the root there are no hairs but the absence of a cuticle makes direct absorption possible. The cells of the piliferous layer and root hairs contain cell sap with a more negative water potential (Ψ) than the available soil water thereby allowing water to move into them by osmosis.

Movement across the cortex The cortex comprises closely packed parenchyma cells. There are three main theories to explain water movement across this region:

1 Osmosis – as water is drawn from a parenchyma cell into the vascular stele the water potential (Ψ) of that cell decreases and water flows into it from an adjacent cell by osmosis.
2 Diffusion through the cytoplasm – plasmodesmata are strands of cytoplasm linking adjacent cells; if one cell loses water it may be replaced by the diffusion of water from an adjacent cell through the plasmodesmata (symplastic pathway).
3 Diffusion along the cell wall – diffusion may occur through adjacent cell walls and small intercellular spaces to replace water lost from a parenchyma cell to the vascular stele (apoplastic pathway).

It is possible that all three mechanisms operate to some extent and there may also be some active transport of water within the parenchymatous cortex.

Movement into the xylem The vascular cylinder is separated from the cortex by a highly specialized cylinder of cells, the endodermis. Each endodermal cell, as well as having a cellulose cell wall, has a Casparian strip of suberin which prevents water passing across its radial and horizontal walls. This means that all water passing from the cortex to the xylem must pass through the cytoplasm of an endodermal cell. As the cell ages heavy thickening covers the Casparian strip but some cells remain in the primary condition and are known as passage cells. When a stem is cut just above soil level the cut end exudes water for some time. This is evidence for water being 'pushed' into the stem from the roots by root pressure. Such movement of water from the cortex into the xylem is thought to be an active (energy requiring) process since the application of a metabolic poison prevents it. The energy for the process may come from the oxidation of starch in the endodermis.

Movement within the xylem The structure of xylem vessels and tracheids is given in Section 2.2 ('Plant and animal tissues'). Xylem vessels and tracheids lack cell contents and so their hollow lumina permit an unimpeded flow of water through the plant. Lateral movement within the xylem is possible through the pitted sidewalls of both vessels and tracheids. The forces of adhesion and cohesion (Section 8.1) maintain a long, unbroken column of water within the xylem.

Movement into the leaf The theories for the movement of water into the leaf are the same as those for movement within the cortex, i.e. osmosis and diffusion through the cytoplasm or along the cell wall. In all cases the necessary gradient is produced as water is lost from the leaf of the plant through the stomata.

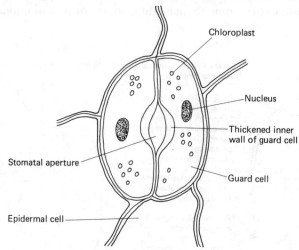

Fig. 8.8 Surface view of a stoma

Stomata These are found in leaves and herbaceous stems. Each stoma comprises a pair of specialized epidermal cells called guard cells and an elliptical pore, the stomatal aperture.

The frequency of stomata varies with environment and species. They are generally more numerous on the abaxial (under) surface of leaves than on the adaxial (upper) side. Stomata usually open in the light and close in the dark. Although it is not without criticism the main theory to explain this is as follows:

1 In the light carbon dioxide is used in photosynthesis and so the concentration of carbon dioxide in the tissues falls. This causes an increase in the pH.
2 The rise in pH in the guard cells causes:

$$\text{starch} + H_3PO_4 \xrightarrow{\text{starch phosphorylase}} \text{Glucose-1-phosphate}$$

3 The glucose-1-phosphate is hydrolysed

$$\text{Glucose-1-phosphate} + H_2O \longrightarrow \text{glucose} + H_3PO_4$$

which causes the osmotic potential to rise.
4 The water potential (Ψ) of the guard cells decreases and water moves into them by osmosis from the surrounding cells. Turgor increases and the stomatal pore opens.

The reverse process occurs in darkness, i.e. levels of carbon dioxide rise, pH falls, glucose-1-phosphate is converted to starch which is osmotically inactive, the guard cells lose their turgidity and the pore closes.

A second theory suggests that the necessary decrease in Ψ of the guard cells in order to cause opening of the stomata is due to the active transport of ions, in particular potassium ions, from the surrounding cells.

Transpiration While stomata facilitate gaseous exchange they are also the main site of water loss by evaporation from a plant. This loss of water from plants is called transpiration. It also occurs through the cuticle and lenticels. The upward movement of water from the roots is partly dependent on the pull exerted by transpiration, the transpiration pull.

Factors affecting the rate of transpiration

The rate of transpiration is dependent on the size of stomatal aperture and the diffusion gradient between the leaf and the atmosphere. The latter is affected by:

1 Humidity: the lower the humidity outside the plant the steeper the diffusion gradient between the leaf and the atmosphere. Transpiration is faster the lower the relative humidity of the atmosphere.

Table 8.6 Adaptations in xerophytes

Mechanism	Adaptation	Examples	Notes
Reduces transpiration	Thick cuticle	Most evergreens	Reduces cuticular transpiration; often shiny so reflects sun causing leaf temperature and therefore transpiration to fall
	Depression of stomata	*Ilex* (holly), *Pinus*	Lengthens diffusion path and therefore reduces diffusion pressure gradient; may trap still, moist air
	Rolled leaves	*Ammophila* (marram grass), *Calluna*	Folding lengthens diffusion path and traps moist air
	Protective hairs	*Ammophila*	Often accompanies rolling of leaf; traps moist air
	Leaves small or absent	*Pinus*	Small and circular in cross section to give low surface area/ volume ratio and structural support to prevent wilting
		Opuntia	No leaves, flattened stem photosynthesizes
	Variations in leaf positions	*Lactuca* sp. (European compass plant)	Leaf positions adjusted so that sun strikes them obliquely; lowers transpiration because lowers temperature
	High osmotic potential of cell sap	Many xerophytes	Thought to reduce evaporation from cell walls
Succulence	Succulent stem and possibly leaves	Cacti	Stores water
	Diurnal closing of stomata	Many xerophytes	Reduces transpiration; requires metabolic modifications
	Shallow, wide root system	Many xerophytes	Gains maximum benefit from light rain
	Vegetative propagation well developed	Many xerophytes	Seed germination requires water
Extensive root systems		Most xerophytes	Especially developed in the most arid conditions
Resistance to desiccation	Increased lignification (correspondingly reduced leaves) allows resistance to wilting	*Ruscus*	Flattened stem (cladode) takes over photosynthesis
		Acacia	Lamina lost; petiole flattened to form phyllode
		Cacti	Leaves reduced to spines
	Reduced cell size	Many xerophytes	Less likely to wilt than fewer large ones

2 Wind speed: an increase in wind speed normally increases the rate of transpiration since saturated air is blown away from the stomatal pore and a diffusion gradient is maintained. This is partly counteracted by the cooling effect of wind.

3 Temperature: if the temperature of the air and leaf increase simultaneously, the consequent increase in vapour pressure of the leaf increases the diffusion gradient and transpiration increases.

Xerophytes These are plants adapted to survive conditions of unfavourable water balance and they may show the adaptations shown in Table 8.6.

Uptake and transport of mineral salts Like water, mineral salts are absorbed through the root hairs and other young parts of the root. The application of metabolic poisons prevents the

Table 8.7 Mineral elements necessary for plant growth

Macro/micro nutrient	Mineral	Form in which element is absorbed	Function	Effect of deficiency
Macro-nutrient	Nitrogen	NO_3^-	Component of amino acids, proteins, nucleotides, chlorophyll, some plant hormones	Chlorosis (yellowing of leaves); stunted growth
	Phosphorus	$H_2PO_4^-$	Component of proteins; needed for ATP, nucleic acids and phosphorylation of sugar	Stunted growth; dull, dark green leaves
	Potassium	K^+	Component of enzymes and amino acids; needed for protein synthesis and in cell membranes	Yellow-edged leaves; premature death
	Calcium	Ca^{2+}	Calcium pectate in mid-lamella of cell walls; aids translocation of carbohydrates and amino acids; affects permeability of cell membranes	Death of growing points, therefore stunted roots and shoots
	Magnesium	Mg^{2+}	Constituent of chlorophyll; activator of some enzymes	Chlorosis
	Sulphur	SO_4^{2-}	Constituent of some proteins	Chlorosis; poor root development
	Iron	Fe^{2+}	Constituent of cytochromes; needed for synthesis of chlorophyll	Chlorosis
Micro-nutrient	Boron	BO_3^{3+} or B_4O^{2+}	Aids germination of pollen grains and uptake of Ca^{2+} by roots	Diseases such as heart rot of celery, internal cork of apples
	Zinc	Zn^{2+}	Activator of some enzymes; essential for leaf formation; needed for synthesis of IAA	Malformation of leaves
	Copper	Cu^{2+}	Constituent of some enzyme systems	Growth abnormalities, e.g. dieback of shoots
	Molybdenum	Mo^{4+} or Mo^{5+}	Affects nitrogen reduction through enzyme system	Reduced yield of crop
	Chlorine	Cl^-	In osmosis and important in anion-cation balance of cells	Difficult to demonstrate
	Manganese	Mn^{2+}	Enzyme activator	Type of chlorosis; leaves mottled with grey patches

uptake which indicates that the process is active. There is also evidence that salts are taken up selectively (proportions of minerals inside and outside the plant are different) and against the concentration gradient. The mechanism and route of movement across the cortex to the xylem are not certain. There are a number of theories relating to the mechanism of transport across membranes but most postulate the existence of a carrier molecule, possibly a protein.

Once inside the xylem the ions travel upwards in the transpiration stream. Although xylem is principally responsible, mineral transport also occurs in the phloem and lateral movement from xylem to phloem is possible. When the mineral ions reach the leaves and meristematic areas of the plant they diffuse out of the xylem vessels and are absorbed by the cells for various metabolic functions, e.g. building up proteins and amino acids.

Translocation of organic materials

Evidence for movement in the phloem

1 There are diurnal variations of sucrose concentration in the leaf and, after a time lag, these are reflected in the phloem sieve tubes.

2 If a stem is 'ringed' so that phloem is removed, sugars accumulate above the ring. Dyes however move normally.

3 If a plant is given $^{14}CO_2$ and ringed to remove the phloem, the sucrose accumulating above the cut contains ^{14}C.

4 Aphids use needle-like mouthparts to obtain sugars from phloem sieve tubes. Removal of the aphid, leaving its mouthparts inserted like a pipette in the phloem, allows analysis of the sieve tube contents. Sieve tubes are found to contain a solution of amino acids and sucrose, the latter showing the expected diurnal variations in concentration (see 1 above).

5 The transport in the phloem is normally downwards. However, by darkening or removing the upper leaves, the sugars can be shown, by ringing, to move upwards in the phloem.

Contents of sieve tubes
It has been found using the aphid technique that sieve tubes contain a solution containing
1 up to 30% by weight sucrose
2 up to 1% by weight amino acids
3 traces of
   (a) sugar alcohols
   (b) ionic phosphate and potassium
   (c) hormones
   (d) viruses

Theories of phloem transport
Diffusion is too slow to account for the observed rates of transport; therefore, a number of theories have been proposed:

Mass flow hypothesis (Munch, 1930) Material travels from a region of high concentration, e.g. photosynthesizing leaf chloroplast, to a region of low concentration, e.g. storage plastids in a root. In the region of high concentration water is taken up by osmosis and there is a difference in hydrostatic pressure between this region and the storage area. This causes mass flow within the sieve tube lumina. The theory ignores the membrane barrier between the sieve tube and the plastid and assumes an empty sieve tube lumen and fully open sieve plate pores.

Transcellular strands (Thaine) Thaine proposes that protein fibrils surrounding endoplasmic reticulum tubules pass from one end of the sieve tube to the other and has suggested that solutes pass along these fibrils due to the peristaltic action of the protein sheath, in a manner resembling cytoplasmic streaming. This is an active process. Although the existence of these transcellular strands is not above doubt it is the only theory which accounts for transport of solutes in both directions in one sieve tube.

Electro-osmosis (Spanner) Spanner suggests that the flow of nutrients is produced or maintained by electro-osmotic forces set up across the sieve plates (possibly a potassium ion gradient). This is theoretically possible but there is no direct evidence for it.

Link topics

Section 2.2 Plant and animal tissues (for structure of xylem and phloem)
Section 6.1 Autotrophic nutrition (photosynthesis) (for structure of leaf)
Section 8.1 Properties and importance of water

Suggested further reading

Heath, O. V. S., *Stomata*, Oxford/Carolina Biology Reader No. 37, 2nd ed. (Packard 1981)
Knight, R. O., *The Plant in Relation to Water*, 2nd ed. (Heinemann 1967)
Martin, E. S., Doukin, M. E. and Stevens, R. A., *Stomata*, Studies in Biology No. 155 (Arnold 1983)
Richardson, M., *Translocation in Plants*, Studies in Biology No. 10, 2nd ed. (Arnold 1975)
Rutter, A. J., *Transpiration*, Oxford/Carolina Biology Reader No. 24 (Packard 1972)
Sutcliffe, J., *Plants and Water*, Studies in Biology No 14, 2nd ed. (Arnold 1979)
Sutcliffe, J. and Baker, D. A., *Plants and Mineral Salts*, Studies in Biology No 48, 2nd ed. (Arnold 1981)
Wooding, F. B. P., *Phloem*, Oxford/Carolina Biology Reader No. 15, 2nd ed. (Packard 1978).

QUESTION ANALYSIS

9 Three beakers, 1, 2 and 3, each containing a different culture solution, had a small aquatic plant added to them. After four weeks the plant in Beaker 1 had flourished, the leaves having increased in size and become darker. The plant in Beaker 2 had the same number of leaves as before but they were all now small and yellow. In Beaker 3 the size and number of leaves had increased but they were very pale. Which of the following is the most likely:

A 1 lacked phosphorus   2 lacked nitrogen   3 lacked magnesium
B 1 lacked phosphorus   2 lacked magnesium   3 lacked nitrogen

C 1 lacked nitrogen   2 lacked phosphorus   3 lacked magnesium
D 1 lacked magnesium   2 lacked nitrogen   3 lacked phosphorus
Mark allocation 1/40   Time allowed 1½ minutes   *In the style of the Cambridge Board*

This is a multiple choice question testing knowledge of the effects on plants of nitrogen, magnesium and phosphorus deficiency. Candidates should try to extract the essential information from the account, possibly by drawing up a rough table as follows:

Beaker	Growth of leaf	Number of leaves	Colour of leaves
1	More	More	Darker
2	Less	Same	Yellow
3	More	More	Pale

Although the question does not actually state it, it is reasonable to assume that the leaves of the plants were originally green and not yellow. Having drawn up the table candidates should apply their knowledge of the roles of nitrogen, phosphorus and magnesium in the plant (see 'Summary of essential mineral requirements of plants' in the essential information).

Nitrogen deficiency causes

1 leaves to become pale (N is a component of chlorophyll)
2 growth reduction (N is a component of proteins)

Beaker 2 with its reduced growth and yellow leaves clearly lacked nitrogen. Magnesium is a constituent of chlorophyll, but does not directly affect growth. Beaker 3 with some growth and pale leaves must therefore lack magnesium. One notable feature of phosphorus deficiency in plants is the production of dull, dark green leaves. Beaker 1 is therefore deficient in this mineral. To summarize, Beaker 1 lacks phosphorus, Beaker 2 lacks nitrogen and Beaker 3 lacks magnesium. The correct response is A.

10 (a) Explain how a flowering plant obtains (i) water and (ii) ions.   (10)
    (b) Describe the pathways and mechanisms of water transport in the plant.   (10)
Mark allocation 20/100   Time allowed 35 minutes   *London Board June 1981, Paper II, No. 1*

This is a structured essay question mostly testing recall of factual information. In part (a) (i) the candidate should begin the explanation by mentioning the sources of water available to the plant. These are the available water, i.e. the water between the particles and to some extent the film of water over the soil particles. One source likely to be forgotten is metabolic water originating as a by-product of biochemical processes, such as respiration, within the plant. Many candidates will interpret the word 'obtains' as meaning 'give an account of water uptake'. While this is part of the answer there are other processes involved before the plant can actually absorb the water. One of these is hydrotropism, a growth movement of plant roots towards the wettest area of the soil. Another point to mention before discussing the actual absorption of water is the large surface area provided by the finely divided lateral roots and the root hairs. In addition the main roots branch widely to absorb water over a large region of the soil. The root hairs are the cells which absorb the water and the candidate should make references to how they are adapted for this function, e.g. they have a large surface area, are thin walled and no cuticle is present. The description of the mechanism of absorption should include reference to the higher DPD (lower water potential ψ) in the root hair cell than in the soil solution. This higher DPD (lower ψ) is due to the salts and sugars in the sap of the root hair cell. Mention should also be made of absorption into and through cell walls.

In (a) (ii) the candidate should again be aware of all the processes involved. Very often a candidate gives only one correct answer, which typically comprises an account of diffusion with no mention of the equally important active transport. Discussion of the latter should include the need for a source of energy, the fact that absorption occurs against a concentration gradient and the use of a carrier. The perhaps obvious point that the ions are obtained from solution in the soil water and are absorbed through the root hairs should not be forgotten. For part (b) the DPD gradient across the cortex and water movement via the endodermis and into the xylem should be included followed by the transpiration pull created in the leaves by the evaporation of water through the stomata. A full account of these processes is given under the headings 'Movement across the cortex', 'Movement into the xylem', 'Movement within the xylem' and 'Movement into the leaf' in the essential information. The important words to mention are diffusion pressure deficit, cohesion tension, adhesion, capillarity, evaporation and transpiration pull. The processes to explain are the osmotic gradient, movement through cell walls and plasmodesmata, active transport by the endodermis and root pressure. In all, part (b) is extensive and care should be taken to give detail and yet be concise. Annotated diagrams may help, but large detailed drawings of leaf, stem and root structure will only use precious time and gain few additional marks. The time could be more profitably spent qualifying words such as adhesion, cohesion, capillarity and transpiration pull. Candidates often use the words with no explanation. Examinations at A level set out to test candidates' understanding rather than their ability to simply learn words and phrases. Successful candidates will be those who satisfy the examiner that they understand the terms.

Answer plan

(a) (i) Water

1 Sources – available soil water, metabolic by-product
2 Hydrotropism
3 Root hair adaptations – large surface area, thin walled, no cuticle
4 Uptake – root hair cell sap has a higher DPD (lower ψ) than soil solution; osmosis carries water in; movement through cell walls alone.

(a) (ii) Ions

1 Taken in by root hairs from soil solution
2 Diffusion – along concentration gradient
3 Active transport – against concentration gradient using energy and carrier.

(b)

1 Osmotic gradient across cortex (DPD/ψ)
2 Root pressure
3 Into xylem via endodermis using active transport
4 Along xylem
   (a) continuous from root to leaves
   (b) no end walls in xylem to restrict flow
   (c) cohesion, adhesion, capillarity
5 Across leaf – osmotic gradient
6 Into atmosphere – through stomata by evaporation (transpiration pull)
7 Movement across cortex and leaf may also be through cell walls alone and plasmodesmata.

11 The rate of transpiration in green plants is affected by environmental factors such as light intensity, humidity, wind and temperature. The number of stomata and the size of each stomatal aperture also affect the rate. Show concisely how each of these factors changes the rate of transpiration in green plants.

Mark allocation 20/100   Time allowed 40 minutes   *In the style of the Joint Matriculation Board*

This is a highly structured essay which tests the candidate's ability to explain the effects of external and internal factors on transpiration rate. The six factors are listed and should be taken in turn, preferably in the order given.

In dealing with light intensity, candidates must take care not to complicate this by reference to temperature, which should be given later. The effect of high light intensity is to cause stomata to open and so to increase the rate of transpiration, the reverse being true of low light intensity or darkness. The precise mechanism causing this change is dealt with under 'Stomata' in the essential information and should be included briefly. Remember always to refer to 'high' or 'low' light intensity. Candidates all too often make meaningless statements such as 'light intensity causes stomata to open'. Humidity, air movements (wind speed) and temperature are all dealt with under 'Factors affecting the rate of transpiration' in the essential information. In all cases it is important to deal with the effects of both an increase and a decrease for each factor. Once again refer to 'high' or 'low' levels of each and do not write 'temperature increases transpiration'. Explanations should be thorough and refer to 'diffusion or concentration gradients'. Remember that the rate of diffusion is inversely proportional to the square of the distance across which it takes place. Any factor that even slightly increases the distances over which diffusion takes place will therefore greatly reduce the rate. Make reference to the kinetic theory of gases in explaining how temperature affects transpiration. At high temperatures, the molecules move faster and so escape more readily through stomata. The rate of transpiration therefore increases. The reverse is true of lower temperatures.

The internal factors involve the density of stomata per unit area of leaf surface and the size of the stomatal apertures. The water molecules leaving the edge of a stoma are less impeded by other water molecules than those leaving through the centre of it. These molecules therefore have a short diffusion path to lower humidities and escape more readily. Diffusion is more rapid near the edges than at the centre. In effect this means that the rate of diffusion is much greater through many small holes than through a few large ones of equal surface area. This is true provided the holes are at least ten times their own diameters apart. The rate of transpiration is therefore greater if there are many small stomata placed at least ten times their own aperture diameters apart. Do not make the common mistake of stating 'the transpiration rate is faster if the stomatal aperture is made smaller'. If the aperture is reduced then the total area through which transpiration takes place is reduced. The points made previously apply only when the total area of apertures remains the same. If the area is reduced then so is the rate of transpiration. Even when dealing with these factors it is essential to refer to the ways in which they affect diffusion rates. Throughout the essay therefore the point being considered should always be correlated to how it affects one or more of the following factors which affect diffusion:

1 The surface area over which it takes place
2 The concentration difference between which it occurs
3 The distance through which it takes place.

Answer plan

Light intensity: transpiration rate increases at high light intensity and decreases at low light intensity. It affects stomatal opening and closing (give details from essential information).

Condition	Low level causes transpiration to	High level causes transpiration to	Mechanism by which change occurs
Light intensity	Decrease	Increase	Affects stomatal opening and closing (details in essential information)
Humidity	Increase	Decrease	Affects diffusion gradient
Air movement	Decrease	Increase	Affects diffusion gradient. Removes saturated air from stomata. May also cool leaf
Temperature	Decrease	Increase	Affects kinetic movement of water molecules. May affect rate of photosynthesis and therefore have an indirect effect on opening of stomata
Number of stomata	Decrease	Increase	Affects total area over which diffusion occurs. Transpiration increases if stomata are conveniently spaced
Aperture size	Decrease	Increase	Affects total area over which diffusion occurs

12 An experiment to determine which tissue conducts sugar through the stem was set up using the apparatus below.

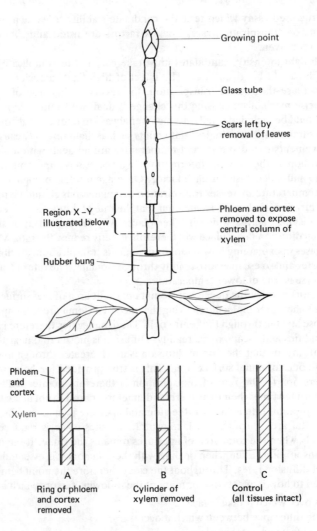

Fig. 8.9 Experiment to determine sugar-conducting tissue

The leaves were removed in the upper region of the plant, as shown. Three sets of apparatus were set up as follows:

A A ring of all tissues outside the xylem was removed at the base of the defoliated region.

B A cylinder of xylem was removed from the stem at the base of the defoliated region. The other tissues were left intact except for a small cut through which the xylem was removed.

C All tissues were left intact (control).

The cylinders were filled with fresh distilled water each day. The results obtained after a week are shown in the table below:

	A Phloem and cortex removed	B Xylem removed	C Control
Increase in stem length in mm	5.3	51.6	65.9
Total sugar content of stem above region X–Y in mg	0.06	6.43	3.36

(a) Why were the following procedures carried out?
 (i) The leaves removed from the upper part of the stem.
(ii) The stems enclosed in a glass tube containing distilled water.
(b) Explain the difference in the results between
 (i) A and B
(ii) B and C
(c) Why is an increase in the length of the stem not a particularly good measure of translocation?
Mark allocation 8/100   Time allowed 15 minutes   *In the style of the Joint Matriculation Board*

This is a structured question testing understanding of an experimental procedure and ability to interpret results and evaluate their accuracy. Candidates should study all the diagrams and results carefully. Before attempting an answer, they should fully understand the reasons why each set of apparatus was used and what each result indicated.

In (a) (i) the leaves are not removed to allow the stem to be fitted into the glass tube. Such simple responses cannot reasonably be expected to bring marks at A level. The experiment is designed to indicate which tissue transports sugars. To do this the sugars must be made to move in one direction, in this case upwards, over the experimental region X–Y. If the leaves are removed any sugars present above this region could only have been translocated from below. If the leaves are not removed it would be impossible to determine whether sugars above X–Y were translocated from below or manufactured in the leaves above the region.

In part (b) answer (i) and (ii) separately rather than confusing yourself, and the examiner, by comparing all three sets of results together. The two measurements taken are a means of measuring the relative amounts of sugar translocation that has taken place. In A both the extent of the growth and the amount of sugar above region X–Y are considerably smaller than in B, which indicates that the phloem and cortex carry much more sugar than does the xylem; the latter transports very little sugar, if any. In comparing B and C the growth is better in the control, which shows that the xylem must carry some material necessary for growth. However the sugar content is less in the control (C) than in B. It is unlikely that there is reduced translocation in C because the phloem and cortex are intact. The reduced level of sugar must therefore be due to it being used for the extra growth that occurs in C.

Part (c) requires the candidate to evaluate the accuracy of one of the methods for measuring sugar translocation. Although sugars are essential for growth in plants, so are many other substances. A lack of growth when the phloem supply is cut only indicates that some factor is not being supplied. Whether or not this missing factor is sugar can only be deduced from consideration of the second set of data obtained.

8.4 OSMOREGULATION AND EXCRETION

Assumed previous knowledge Properties and importance of water (Section 8.1)
Gross structure of the kidney.

Underlying principles

Primitive organisms probably arose in the sea. The tissue fluids of animals are dilute saline solutions with a composition similar to that of sea water. This is not mere coincidence: animals effectively bathe their tissues in sea water. Some organisms have an internal osmotic pressure that is isotonic with sea water. Even this can create osmotic problems because as a consequence of carrying out metabolic activity the internal concentration of the animal will fluctuate from that of the medium in which it lives. Some marine animals moved up estuaries into fresh water. As the external medium was then hypotonic to the animal cells, water flooded in by osmosis. A number of mechanisms to overcome these problems evolved:

1 The permeability of cell membranes to water and salt became altered.
2 The cells became enveloped in a rigid wall that prevented expansion.
3 The dilution was tolerated.
4 The water was pumped out by some means.

One or more of these methods was employed by organisms in fresh water; in all cases the problem was reduced by allowing their own internal concentrations to fall, thereby reducing the tendency of water to enter by osmosis. Some of these organisms then returned to salt water. They then had an external medium that was hypertonic to their cells and water was lost by osmosis. Two methods have evolved to overcome this. In marine teleost fish salt water is drunk and the salts are eliminated by special glands and the kidney. In marine elasmobranchs high urea and amino acid concentrations of the blood are tolerated so that the internal concentration of these fish is slightly hypertonic to sea water. When organisms moved on to land water retention became essential. This was achieved by:

1 Anatomical adaptations, e.g. waterproof body coverings.
2 Physiological adaptations, e.g. toleration of dehydration, mechanisms for controlling water balance.
3 Behavioural mechanisms, e.g. avoidance of dry areas and by aestivation during dry seasons.

During metabolic activities wastes are produced which are toxic and if allowed to accumulate would poison the organism. Many are removed by diffusion as part of some other process, e.g. carbon dioxide diffuses into the lungs during breathing. In animals, however, nitrogenous wastes resulting from the breakdown of excess amino acids pose particular problems. The ammonia produced is especially toxic. If water is readily available it can be diluted sufficiently and removed. Where water needs to be conserved, e.g. in terrestrial organisms and marine vertebrates, nitrogenous wastes need to be more concentrated before being removed. They must be made less toxic before conversion to urea. Where water is particularly scarce or flight makes the storage of watery urine impractical, the ammonia is converted to uric acid which requires almost no water for its removal. Birds and insects excrete uric acid. By virtue of their autotrophic mode of nutrition plants take in the materials they need and no more. They therefore have almost no complex excretory products – only simple diffusible ones. Their high surface area/volume ratio is suited to the removal of these simple substances, e.g. carbon dioxide (dark), oxygen (light). Any complex excretory material can be removed with the leaves when they are discarded.

Points of perspective

It would be an advantage to have dissected or seen a dissection of the urinary system of a small mammal. Experimental observations of osmoregulation in protozoa under different conditions would also be of benefit. A good candidate will know the major differences in the composition of urine and the glomerular filtrate in a mammal, and if possible be able to quote figures. A range of anatomical, physiological and behavioural adaptations to preventing water loss in terrestrial organisms would be helpful.

Essential information

Excretion is the separation and elimination of metabolic wastes usually in aqueous solution.

Excretory products in plants
1 Carbon dioxide – when the rate of respiration exceeds the rate of photosynthesis.
2 Oxygen – when the rate of photosynthesis exceeds the rate of respiration.

In plants there is no regular excretion of nitrogenous or other complex materials. Wastes such as silicates and tannins are moved to and stored in parts of the plant which will be removed, e.g. leaves and fruits.

Excretory products in animals
1 Carbon dioxide – from respiration
2 Excess water and mineral salts
3 Bile pigments
4 Nitrogenous substances, such as:

(a) **Ammonia** is highly toxic and never allowed to accumulate in living cells. It is soluble, readily diffusible and excreted in dilute form by freshwater animals and marine invertebrates. Animals excreting ammonia are said to be ammoniotelic.

(b) **Urea** is less toxic than ammonia but still needs to be diluted for elimination from the body. It is excreted by some terrestrial organisms and marine ones whose body fluids are hypotonic to sea water. These animals are ureotelic. Urea is formed from ammonia by the ornithine cycle.

(c) **Uric acid** is non-toxic. As it is highly insoluble, little water need be lost in its elimination from the body. It is a common excretory product of animals living under arid conditions. Insects and birds excrete uric acid and are said to be uricotelic.

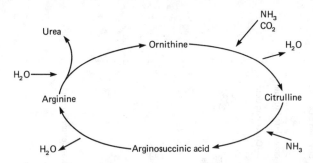

Fig. 8.10 Ornithine (urea) cycle

Ornithine (urea) cycle Excess amino acids cannot be stored. By deamination the amine group is removed from them resulting in the formation of ammonia. Ammonia is also toxic and so it is converted through a series of reactions, the ornithine cycle, to urea. The enzyme arginase catalyses the formation of urea from arginine. The purpose of the cycle is to regenerate arginine using excretory ammonia. In mammals this cycle occurs in the liver cells.

Osmoregulation in freshwater protozoa Freshwater protozoa have cell contents which are hypertonic to the medium in which they live. Since the cell membrane is semi-permeable water enters by osmosis. To prevent the animal bursting the excess water must be removed. Freshwater protozoa such as *Amoeba* and *Paramecium* have contractile vacuoles which constantly fill and empty at the surface. The water which enters the cytoplasm must move into the contractile vacuole against the concentration gradient. As the process requires energy, there are often aggregations of mitochondria near contractile vacuoles. The contractile vacuoles of freshwater protozoa moved to a medium with a higher salt concentration fill and empty less frequently than in fresh water.

Osmoregulation in freshwater teleosts Although much of the surface of a teleost fish is impermeable to water there is still a tendency for water to enter across the gills and lining of the buccal cavity and pharynx. The organ of osmoregulation in the teleost is the kidney, which reacts to the influx of water in two main ways:

1 Production of a large volume of urine – aided by the many large glomeruli in the kidney.
2 Production of very dilute urine – achieved by extensive reabsorption of salts from the renal fluid into the blood stream.

However, a freshwater teleost still suffers some loss of salts and so this is offset by the active uptake of salts from the environment by special chloride secreting cells in the gills.

Osmoregulation in marine teleosts Marine teleosts, which evolved in fresh water and have body fluids slightly hypotonic to sea water, have a tendency to lose water to the environment by osmosis. As in the freshwater teleost the areas of the body permeable to water are the gills and the lining of the buccal cavity and pharynx. To combat dehydration the kidney:

1 Produces a small volume of urine – there are relatively few and small glomeruli.
2 Produces the nitrogenous excretory product trimethylamine oxide. Freshwater teleosts excrete ammonia since they have plenty of water available for its dilution. Trimethylamine oxide is non-toxic and therefore a more suitable excretory product for marine teleosts since little water is needed for its expulsion.

Marine teleosts also drink sea water; the excess salts ingested are actively removed from the body by the chloride-secreting cells of the gills.

Osmoregulation in a marine elasmobranch Marine elasmobranchs retain urea so that the body fluids are slightly hypertonic to sea water. There is consequently a slight influx of water but the excess is easily removed by the kidney. Elasmobranch tissues are unusual in being tolerant of high levels of urea. Normally a high concentration breaks the hydrogen bonds in protein molecules, thus altering their shape and disrupting enzymatic properties. The ability to retain urea dispenses with the need to drink sea water and actively eliminate salts.

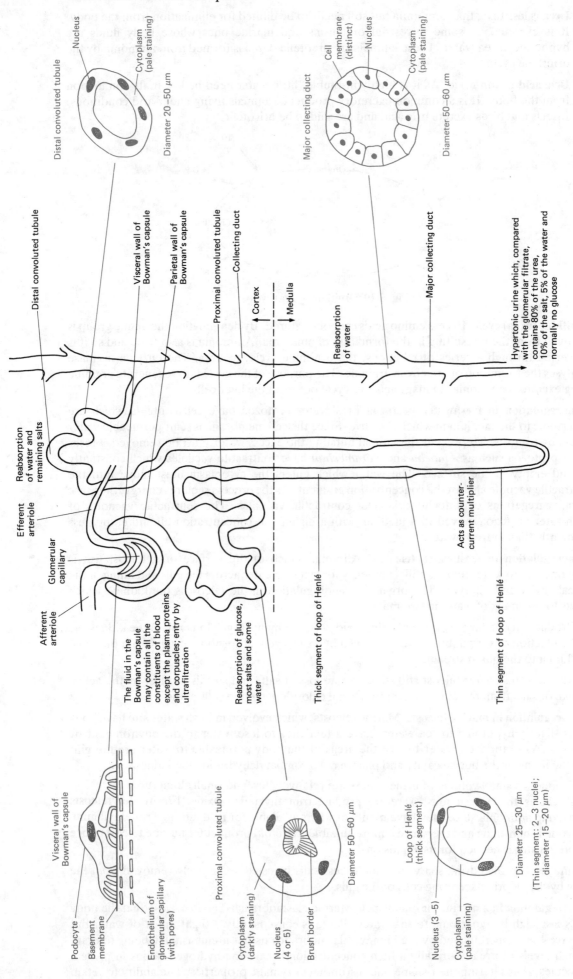

Distal convoluted tubule

Nucleus

Cytoplasm (pale staining)

Diameter 20—50 μm

Major collecting duct

Cell membrane (distinct)

Nucleus

Cytoplasm (pale staining)

Diameter 50—60 μm

Distal convoluted tubule

Visceral wall of Bowman's capsule

Parietal wall of Bowman's capsule

Proximal convoluted tubule

Collecting duct

Cortex

Medulla

Reabsorption of water

Hypertonic urine which, compared with the glomerular filtrate, contains 80% of the urea, 10% of the salt, 5% of the water and normally no glucose

Reabsorption of water and remaining salts

Efferent arteriole

Glomerular capillary

Afferent arteriole

The fluid in the Bowman's capsule may contain all the constituents of blood except the plasma proteins and corpuscles; entry by ultrafiltration

Reabsorption of glucose, most salts and some water

Thick segment of loop of Henlé

Thin segment of loop of Henlé

Acts as counter-current multiplier

Major collecting duct

Visceral wall of Bowman's capsule

Podocyte

Basement membrane

Endothelium of glomerular capillary (with pores)

Proximal convoluted tubule

Cytoplasm (darkly staining)

Nucleus (4 or 5)

Brush border

Diameter 50—60 μm

Loop of Henlé (thick segment)

Nucleus (3—5)

Cytoplasm (pale staining)

Diameter 25—30 μm

(Thin segment: 2—3 nuclei; diameter 15—20 μm)

Fig. 8.11 Diagram of a mammalian nephron

Osmoregulation in a terrestrial insect All terrestrial organisms are liable to water loss through evaporation. In an insect the loss is reduced in three main ways:

1 Having an impermeable surface – the cuticle is coated with wax. Retention of water can never be complete and there is bound to be some loss of water from respiratory surfaces.
2 Production of uric acid – because this nitrogenous waste product is non-toxic and insoluble, it can be eliminated from the body as a semi-solid without the loss of water.
3 Reabsorption of water by the Malpighian tubules and rectal gland. The Malpighian tubules are a bunch of blind-ending tubules at the junction of the mid-gut and rectum. Potassium urate is produced by insect tissues and this enters the Malpighian tubules with water and carbon dioxide. In the tubules they form uric acid and potassium bicarbonate. The potassium bicarbonate is reabsorbed into the haemolymph. Reabsorption of water from the uric acid occurs in the Malpighian tubules and the rectal gland so that the urine eliminated is in a very concentrated form.

Mammalian kidney The basic unit of nitrogenous excretion in a mammal is the kidney tubule, or nephron. Each human kidney contains over one million nephrons and filters about $120 \, cm^3/min$. Over the length of the nephron reabsorption from the glomerular filtrate takes place and results in the formation of urine which is hypertonic to the blood.

Functions of the mammalian kidney
1 Excretion of metabolic wastes, especially urea
2 Osmoregulation
3 Maintaining the acid-base balance of the body. This is mainly brought about by the excretion or retention of H^+ and HCO_3^-; the pH of the urine can vary between 4.5 and 8.0
4 Maintaining the balance of ions, e.g. Na^+, K^+, Ca^{2+}, Mg^{2+}, H^+, Cl^-, HCO_3^-

Formation of hypertonic urine

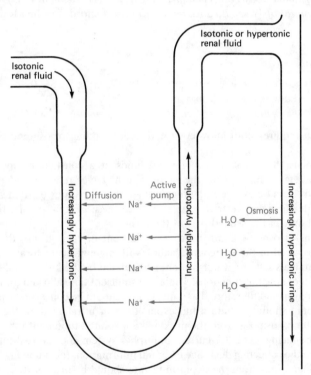

Fig. 8.12 Counter-current multiplier

The loop of Henlé, found in birds and mammals, constitutes a counter-current multiplier, the active mechanism being the transport of sodium ions from the ascending to the descending limb. This movement of sodium ions results in a very high concentration developing in and around the apex of the loop of Henlé, deep in the medulla. In this region the thin-walled collecting ducts open into the renal pelvis. The high concentration of sodium in the tissues causes water to be drawn out of the collecting ducts by osmosis. Consequently the renal fluid which has been isotonic or hypotonic to the blood through the length of the nephron becomes hypertonic in the collecting ducts. It has been estimated that 99% of the fluid filtered by the Bowman's capsule is reabsorbed by the nephron. The length of the loop of Henlé is proportional to the concentration of urine, e.g. desert mammals, such as the kangaroo rat, which produce very hypertonic urine have very long loops of Henlé.

Control of water and salt balance in mammals

Antidiuretic hormone (ADH) The permeability of the collecting ducts to water is affected by the hormone ADH, secreted by the posterior lobe of the pituitary. The presence of ADH increases the permeability of the membranes and results in the production of more hypertonic urine. If no ADH is present the membrane permeability is lower and more dilute urine is produced. The production of ADH is controlled by the hypothalamus whose osmoreceptors monitor the solute concentration of the blood. If the solute concentration of the blood is high ADH secretion is stimulated, the permeability of the membranes of the collecting ducts increases, water is reabsorbed into the blood and very hypertonic urine is produced. ADH secretion may also be triggered, via the hypothalamus, by blood volume receptors in the walls of the heart, the aorta and the carotid arteries.

Aldosterone This hormone produced by the adrenal cortex stimulates the reabsorption of sodium from the kidney. When the sodium concentration in the kidney tubule decreases the kidney produces the enzyme renin which catalyses the conversion of the blood protein angiotensinogen into angiotensin. Angiotensin stimulates the adrenal cortex to produce more aldosterone.

Link topics

Section 7.2 Gaseous exchange
Section 8.1 Properties and importance of water
Section 10.1 Hormones and homeostasis

Suggested further reading

Hardy, R. N., *Homeostasis,* Studies in Biology No. 63, 2nd ed. (Arnold 1983)
Lockwood, A. P. M., *Animal Body Fluids and their Regulation* (Heinemann 1963)
Moffat, D. B., *The Control of Water Balance by the Kidney,* Oxford/Carolina Reader No. 14, 2nd ed. (Packard 1978)

QUESTION ANALYSIS

13 What is the importance of osmotic control in animals? Describe the methods by which the salt and water content of the body is regulated in mammals and fish.
   Mark allocation 20/100   Time allowed 45 minutes   *London Board June 1985, Paper 2, No. 8*

This essay question requires both knowledge and understanding of osmotic control in a range of animals.

Osmotic control is the maintenance of body fluids at a constant composition in order that metabolic activities take place and the animal can have relative independence from the environment. Control can be exercised by regulating either the water or solute content of the body fluids or both. In this first part restrict your answer to these general points, the detailed methods should be included only in the second part of the answer.

As the problems of osmotic control are very different in each group, the second part of the question can be answered in three sections dealing with mammals, freshwater fish and marine fish, respectively. In mammals salt and water content is regulated by the kidney and therefore the answer should include detail of how this functions. There is insufficient time allowed for the detailed anatomy of a kidney or a nephron and therefore diagrams, if any, should be concerned with illustrating kidney function. The account should include a full explanation of ultrafiltration, the reabsorption of the majority of the water by osmosis from the proximal convoluted tubule into the blood. Detail of the loop of Henle in producing a counter-current multiplier system and the reabsorption of water from the distal tubule and the collecting duct. See essential information, including Fig. 8.12 for details. The reabsorption of mineral ions from the nephron filtrate should be discussed.

The key word in the second part of the question is 'regulated'. Full details must therefore be provided on the hormonal control involved. The role of the hypothalamus in monitoring blood osmotic potential, the posterior pituitary gland in producing anti-diuretic hormone (vasopressin) and its effect on the permeability of the cells lining the distal convoluted tubule and the collecting duct, should all be considered. Equally the role of the adrenal cortex in producing the mineralocorticoid hormone, aldosterone and its effects on sodium chloride and potassium ion levels must be included. The involvement of renin and angiotensin in this process are also worthy of mention. In discussing freshwater fish a good starting point would be that the solute concentration of the body is much higher than that of the surrounding water. Coupled with the highly permeable gills, this leads to a constant osmotic influx of water. Control is achieved by having highly glomerular kidneys which have a high filtration rate resulting in the production of large quantities of dilute urine. The potential loss of salt in this urine is prevented by extensive reabsorption into the blood from the nephron. Any salt lost is replaced by active uptake of mineral ions against the concentration gradient by chloride secretory glands on the gills.

In marine teleosts, the osmotic potential of the body fluids is lower than that of the surrounding seawater. Water is therefore lost by osmosis, mainly from the gills. To replace this the fish swallow seawater and eliminate the salts taken in, through the chloride secretory glands on the gills, by active transport. The use of trimethylamine oxide as an excretory substance conserves water as it is non-toxic and therefore need not be diluted before removal. The kidneys have few glomeruli and a low filtration rate.

Marine elasmobranchs, e.g. sharks, also have a body fluid osmotic potential lower than the surrounding water. They employ, however, a different method of control. They retain urea in the blood in order to raise its osmotic potential almost to that of the seawater. The remaining small influx of water is easily eliminated by the kidneys. The body cells have a high tolerance to urea.

As an additional point it would be worth mentioning the osmotic problems which are experienced by fish migrating between fresh and sea water. Examples like the eel or salmon could be quoted.

Answer Plan

Importance of osmotic control:
  Maintenance of constant body fluid composition for metabolic activities
  Regulation of water and solute content
  Gives increased environmental independence
Regulation in mammals:
  Kidney functioning, including ultrafiltration, reabsorption of water by osmosis, sodium pump, counter-current multiplier, reabsorption of sodium and chloride ions.
  Control: role of hypothalamus, pituitary, ADH in water balance and adrenal cortex and aldosterone in salt balance.
Regulation in freshwater fish:
  Body fluid OP > surrounding water; osmotic entry of water through permeable gills, highly glomerular kidney producing copious dilute urine, extensive salt reabsorption by kidney, any salt loss compensated by active absorption of chloride secretory glands on gills.
Regulation in marine fish:
  Body fluid OP < surrounding water; osmotic loss of water through permeable gills. Teleosts swallow seawater and eliminate salts via chloride secretory glands on gills. Kidneys have few glomeruli, urine is concentrated and non-toxic trimethylamine oxide is excreted. Elasmobranchs retain urea, to which cells are tolerant, raising body OP, kidney eliminates the small amount of water which enters.

14 (a) How does the structure of a nephron suit it to its function? (7)
  (b) List the ways in which a mammal obtains and loses water. (4)
  (c) A shipwrecked sailor in a boat on tropical seas has had nothing to eat or drink for four days. He drinks a small quantity of whisky he has with him, but is still desperately thirsty. As a last resort he drinks large quantities of sea water. Next day he reaches land and quenches his thirst by drinking much fresh water. Describe the processes concerned with water and salt balance that have taken place in the sailor during the events described. (9)
  Mark allocation 20/100   Time allowed 40 minutes   *In the style of the Joint Matriculation Board*

This is a highly structured essay-type question which tests the application of the principles of excretion and osmoregulation with some factual recall of knowledge.

In relating nephron structure to its function the candidate must avoid simply describing the nephron and following this with an account of how it works. A logical approach would be to make a sketch of the nephron as the basis of a plan. Starting with the branches of the renal artery, work through each structure in turn and relate these structures to their function. For instance, the efferent branch of the renal artery leaving the glomerulus has a narrower bore than the afferent branch. This creates an increase in pressure which is responsible for removal of substances of relatively high molecular weight from the blood, i.e. it is the basis of the ultrafiltration process in the kidney. The glomerulus has a large surface area to increase the rate of filtration and the Bowman's capsule has special podocyte cells to reduce resistance to the flow of filtrate into the kidney tubule. The outline of the remaining structures and how they relate to their functions is given in the answer plan at the end. In the time allocated it is doubtful whether a diagram of the whole nephron could be included, but sketches of parts, e.g. podocytes, may well be the clearest and quickest way to relate structure and function, especially if they are well annotated.

The key word in part (b) is 'list'. No more than a few words are required for each. Often candidates waste time in giving accounts of the origins and mechanisms of uptake and removal of water. Obtaining water by drinking and in food are normally quoted but obtaining it as a metabolic by-product of respiration and other reactions is frequently forgotten. Depending how one interprets the word 'obtains' it could be argued that metabolic water comes ultimately from food and is not a separate source. It is probably better to play safe and include it separately. Loss of water in urine, sweat and evaporation from lungs is usually remembered. Loss of water in the faeces and external secretions is usually omitted.

Part (c) requires not only some knowledge of the control of urine production but also the effect of chemicals and external conditions on the homeostatic control of water balance. Do not forget the obvious. The sailor is in 'tropical seas' and he will be warm and sweating profusely during the day and therefore losing much water. He has no water source so the osmotic pressure of his blood will rise. The events that follow this rise in osmotic pressure must be described in detail for good marks. The role of the hypothalamus, pituitary gland, anti-diuretic hormone and the loop of Henlé must all be included. See essential information under 'Control of water and salt balance in mammals' for details. The whisky has a number of effects. It contains ethyl alcohol which inhibits the production of ADH and so allows more water to be lost in the urine (i.e. it is a diuretic). In addition the alcohol is toxic and its removal (some is metabolized) involves its dilution in water. It also causes dilation of blood vessels including the superficial vessels of the skin. This may further increase sweating. In all, the whisky only increases the sailor's dehydration. The effect of salt water on the body is often misunderstood by candidates. The salts taken in with the sea water must be removed if they are not to further concentrate the blood and create osmotic problems. Sea water is approximately a 3.5% solution of salt. At most the body can concentrate salt in the urine to 2.2%. This means that the removal of the salts takes more water than was taken in with the sea water, i.e. dehydration is yet more severe; the drinking of sea water makes the situation worse not better. An account of salt removal and its hormonal control should be made. Avoid repeating the water balance process, simply make reference to it adding 'as explained earlier'. The uptake of fresh water lowers blood osmotic pressure, reversing the processes that led to increased ADH production, which is then reduced.

Answer plan

(a) Nephron structure and function

Part of nephron	Relation of structure to function
Renal artery	Efferent branch wider than afferent therefore pressure of ultrafiltration is produced
Glomerulus	Large surface area for filtration
Bowman's capsule	Funnels filtrate into tubule
Podocytes (diagram?)	Reduce resistance to filtrate
Proximal convoluted tubule	Large surface area for absorption of minerals, glucose and amino acids
Loop of Henlé	Counter-current multiplier to increase water absorption because of hairpin shape. Sodium pump occurs
Distal convoluted tubule	Large surface area for absorption
Collecting duct	Makes counter-current multiplier possible because urine in it runs opposite way to ascending limb of loop of Henlé
Epithelial cells in tubules	Many mitochondria for energy for active transport, e.g. sodium pump. Microvilli to give large surface area for absorption

(b)

Water obtained through	**Water lost through**
Drinks	Urine
Food	Faeces
Metabolic by-product, e.g.	Sweating
respiratory product	Evaporation from lungs
	External secretions, e.g. tears

(c) No water source but water lost in all ways stated in (b). Blood OP rises, hypothalamus detects this, message to pituitary to produce ADH which causes more water reabsorption in loop of Henlé. Urine becomes concentrated. Whisky inhibits ADH production and increases sweating. Dehydration becomes more severe. Sea water causes loss of more water than it provides. Salts are removed in the urine.

Drinking fresh water lowers blood OP and reverses the previous processes. Dilute urine is again produced.

15 (a) Amino acid metabolism in animals leads to the formation of nitrogenous waste products. Explain briefly why nitrogenous waste does not normally occur in plants. (2)

(b) Analysis of the glomerular filtrate and the urine of a mammal yielded the following mean daily values:

	Glomerular filtrate	Urine
Urea	60 g	35 g
Water	180 dm	1.5 dm

(i) 150 dm of water is reabsorbed by the proximal tubules. Calculate the percentage of water from the filtrate that is reabsorbed elsewhere.

(ii) Name two other regions of the tubules where this further reabsorption of water takes place. (3)

(c) In mammalian kidneys, the relative length of the loops of Henle shows considerable variation from one species to another. Suggest, with reasons, the type of habitat in which you would expect to find species with extremely long loops of Henle. (3)

(d) Nitrogenous waste in animals may occur as ammonia, urea, or uric acid. Ammonia is very soluble and highly toxic; urea is soluble and mildly toxic; uric acid is insoluble and non-toxic. The table shows the percentage of these three compounds in the urine of four different animals.

	Ammonia	*Urea*	*Uric acid*
Freshwater fish	56	6	0
Sea water fish	7	81	0
Lizard	0	0	91
Bird	3	4	72

(i) Offer an explanation for the difference in the main excretory compound in fresh water and sea water fish. (5)

(ii) Both lizards and birds are terrestrial, egg-laying animals. How do these characteristics relate to the nature of their main excretory products? (2)

Mark allocation 15/130   Time allowed 15 minutes

Welsh Joint Education Committee June 1985, Paper A1, No. 10

This structured question tests a range of abilities, including recall of information, data analysis, the understanding of biological principles and basic mathematical skills.

Part (a) requires candidates to appreciate that nitrogenous excretory products result from the breakdown of excess amino acids consumed by animals. As plants can make all their amino acids from basic raw materials (CO_2, H_2O and NO_3) they only manufacture them as and when the need arises. Consequently they have no excess amino acids to breakdown and hence no nitrogenous waste.

Care is necessary in part (b) as it is all too easy to use the wrong values in the calculation. The key word in (b) (i) is 'reabsorbed' and it is therefore important to realise that 1.5 dm of water is present in the urine, i.e. not reabsorbed. The calculation is hence:

Water reabsorbed by proximal tubules			150 dm
Water in urine (i.e. not reabsorbed)			1.5 dm
		Total	151.5 dm
Amount reabsorbed elsewhere	=		180 − 151.5
	=		28.5 dm

$$\text{As a percentage} = \frac{28.5}{180} \times 100$$

$$= 15.83\%$$

This water is reabsorbed by the distal tubule and the collecting duct. To help answer part (c) read the account of the mammalian kidney in the essential information which shows that the longer the loops of Henle the greater the proportion of water that is reabsorbed. Long loops of Henle are hence common in animals which need to conserve water, e.g. mammals living in hot deserts.

The information to answer part (d) is given under the osmoregulation section of the essential information. While ammonia is the simplest excretory product it is very toxic. However, freshwater fish are able to dilute this to harmless concentrations using the large volume of water which passes into them osmotically and has to be removed by the kidneys. As seawater fish tend to lose water osmotically, they cannot afford to dilute the ammonia adequately. They therefore excrete urea which does not require as much water to dilute it to a non-toxic level. As eggs are enclosed by shells, excretory products cannot escape. If these products were soluble they would diffuse throughout the egg and poison the embryo. The insoluble uric acid is therefore the only safe form in which excretory products can be stored.

9 Locomotion

9.1 LOCOMOTION AND SUPPORT IN PLANTS AND ANIMALS

Assumed previous knowledge Structure of cilia and flagella (Section 2.1)
Structure of cartilage, bone, skeletal muscle, parenchyma, sclerenchyma and xylem (Section 2.2)
Properties and importance of water (Section 8.1)
Distribution of plant tissues in a stem and root (Section 8.3)
Action of antagonistic pairs of muscles around one named joint, e.g. biceps, triceps and brachialis producing movement at the elbow.

Underlying principles

Organisms initially evolved in water which gave them support. Even so skeletons were useful as a means of protection and a rigid framework for the attachment of muscles. When organisms became terrestrial it became necessary to use the skeleton for support also. For a plant the support needs to be substantial because competition for light has led to the evolution of trees over 100 m in height. Woody plants are supported by their xylem. As xylem is a dead tissue it makes no energy demands on the organism and may therefore be massive without any disadvantage to the plant. Herbaceous plants and the herbaceous parts of woody ones rely on the hydrostatic pressure created by water entering cells by osmosis. This pressure is termed turgor. Hydrostratic pressure is also utilized by animals. Earthworms for instance take advantage of the fact that the liquid in their coelom, like all liquids, is virtually incompressible and can act as a skeleton. Other animals use specialized tissues which, unlike those of plants, are living and require energy. For this reason, and because most animals move from place to place, animals and their skeletons are much smaller. Where very large size has been attained the organisms have returned to water in order to gain support for their bodies, e.g. blue whale.

Locomotion is a feature of animals and some algae. The mode of nutrition in plants makes movement unnecessary because the essentials of water, carbon dioxide and light can all be obtained while remaining in one place. Indeed to move from place to place would involve being relatively small and this could put them at a disadvantage in the competition for light. Being sessile however creates certain problems especially in the transfer of gametes during sexual reproduction. In animals locomotion is necessary in order to obtain food, although some aquatic organisms can filter food from the ambient water. Although cilia and flagella can be used to achieve locomotion in unicellular organisms, the remainder are dependent on some musculo-skeletal system. The actual arrangement of this system is dictated by the mode of locomotion, e.g. burrowing, walking, crawling, jumping, climbing, gliding, flying or swimming.

Points of perspective

It would be an advantage to have studied and drawn all parts of the mammalian skeleton and to know the parts of different bones and their function. To have carried out microscopic examination of cartilage, bone, xylem, parenchyma, collenchyma and sclerenchyma would also be useful. Observation of different methods of locomotion at first hand would be beneficial. Candidates with an understanding of the mechanics of support will be at a considerable advantage. It would be beneficial to have a detailed knowledge of the theories of amoeboid movement and the exact nature of muscle contraction.

Essential information

Support in flowering plants This is achieved in two main ways:

1 Mechanical strengthening
2 Turgor pressure

Mechanical strengthening (see Sections 2.2 and 8.3)
Some cell walls are strengthened with lignin as well as cellulose. These include collenchyma and the heavily thickened sclerenchyma. Most support is provided by the xylem sclerenchyma which constitutes the 'wood' of woody plants. In dicotyledonous stems the mechanical tissue forms a cylinder close to the perimeter. A hollow cylinder of mechanical tissue provides the

best resistance to horizontal force (e.g. wind). In roots the central core of strengthened tissue is ideally placed to resist the vertical forces met as the root pushes through the soil.

Support in animals As an animal increases in size the soft tissues of the body require support. On the whole aquatic animals need less extensive skeletal support than terrestrial ones. There are three basic types of skeleton:

1 Exoskeleton   **2** Endoskeleton   **3** Hydrostatic skeleton

Exoskeleton (external skeleton), e.g. arthropod cuticle

An exoskeleton provides a more or less complete protection for the internal organs and a large area for the attachment of muscles. In arthropods the exoskelton is composed of a three-layered cuticle secreted by the underlying epidermal cells. The outer layer is a thin, waxy, waterproof layer (epicuticle). Immediately above the epidermis is a flexible layer made of chitin which is a polymer of glucosamide and acetic acid (endocuticle). Between the two is a rigid layer of chitin impregnated with tanned proteins (exocuticle). The proteins together with lipids help to make the chitin impermeable to water. Crustaceans also have calcium carbonate deposited in this layer to give additional strength.

The inflexible plates of the exoskeleton are separated by flexible regions where the rigid exocuticle is absent; this permits movement. Sensory hairs project through the exoskeleton which is also interrupted by the openings of various glands and the digestive, respiratory and reproductive systems. The most serious limitation imposed by an exoskeleton is on growth. Chitinous plates do not grow but must be periodically shed by a process known as ecdysis (moulting).

Endoskeleton (internal skeleton)

The endoskeletons found in vertebrates are cellular although the bulk of the substance is a non-cellular matrix secreted by the cells. Cartilage forms the entire skeleton of elasmobranch fishes. It also forms the first skeleton of mammalian embryos but later most of it is converted to bone. Cartilage combines rigidity with a degree of flexibility. Most vertebrates are supported by a bony endoskeleton, bone providing a strong, rigid, framework requiring the presence of joints if movement is to take place.

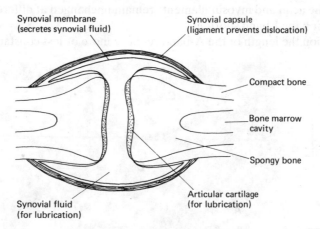

Fig. 9.1 Generalized synovial joint

Hydrostatic skeleton

This is found in certain invertebrates, e.g. earthworm, where the coelomic fluid forms an incompressible but flexible material to aid locomotion. The muscles are arranged segmentally and act on the coelomic fluid. Contraction of the circular muscles makes the body longer and thinner while contraction of the longitudinal muscle makes the body shorter and thicker. Chaetae on each segment anchor the appropriate part of the body so that these alternate contractions bring about movement. Not all the circular or longitudinal muscles contract at one time; waves of contraction pass back along the body. This movement is controlled by intersegmental reflexes.

Types of locomotion

Amoeboid movement

This is exhibited by amoeboid protozoans and cells of vertebrates such as lymphocytes. Amoeboid cells have an outer plasmalemma which adheres to the surface. Beneath the plasmalemma is a clear fluid layer which thickens into the hyaline cap at the anterior end. The bulk of the cytoplasm is divided into a granular endoplasm surrounded by a less granular ectoplasm. For a pseudopodium to move forwards gelated proteins at the posterior end are converted to the sol state and flow forwards. They accumulate and gel near the hyaline cap.

The sol to gel theories are rather simplistic and more modern ideas employ active contraction of protein filaments to explain the changes of state. The position of the motive force differs in the three main theories:

1 The posterior end contracts and pushes the endoplasm forwards.
2 The anterior gel contracts and pulls the core of endoplasm forwards.
3 Sliding filament ratchets on the edge of the gel push the endoplasm forwards.

There is no evidence for a centre to determine the flow of the sol.

Movement by cilia and flagella

Cilia and flagella have the same basic 9+2 fibre arrangement (see Section 2.1). They are remarkably similar in the ciliate and flagellate protozoa and in the cells of multicellular plants (only flagella) and animals.

The outer membrane of the cilium or flagellum is continuous with the plasma membrane of the cell bearing it and the fibre system extends into the cortex of the cell, ending in a basal granule or kinetosome. Flagella are up to fifty times the length of cilia and exhibit an effective stroke beginning at the base and travelling to the tip which exerts a pushing force on the external medium. Cilia beat as stiff slightly curved rods in the effective stroke and have a slower recovery stroke in which they are flexible and very curved. Groups of cilia contract in a co-ordinated way known as metachronal rhythm.

It is not known how movement in cilia and flagella is brought about, but the absence of the two central fibres in cilia which only have a sensory function has suggested that these are the contractile elements. No movement appears to be possible without the basal granule therefore it is possible that this provides the source of energy required.

Muscular movements

Sliding filament theory of muscular contraction Basically this supposes that the muscle filaments do not shorten when the muscle contracts but they slide between one another. The strongest evidence for this theory comes from electron microscope studies of muscles fixed at different degrees of tension. These show:

1 The lengths of the actin and myosin filaments remain unchanged at different tensions
2 During contraction the I-band shortens
3 During contraction the length of the A-band remains more or less constant.

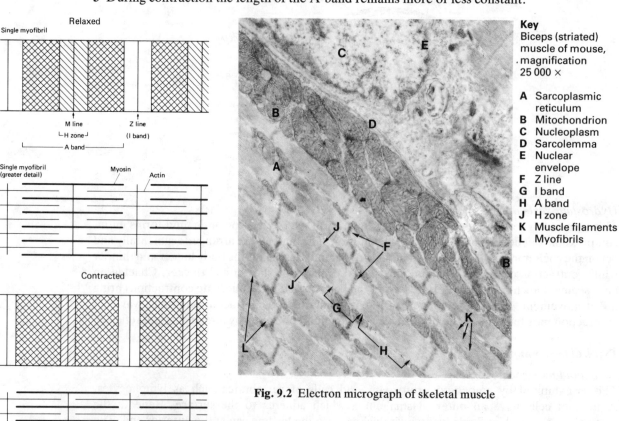

Fig. 9.2 Electron micrograph of skeletal muscle

Key
Biceps (striated) muscle of mouse, magnification 25 000 ×

A Sarcoplasmic reticulum
B Mitochondrion
C Nucleoplasm
D Sarcolemma
E Nuclear envelope
F Z line
G I band
H A band
J H zone
K Muscle filaments
L Myofibrils

Fig. 9.3 Relaxation and contraction in skeletal muscle

There are two main views to explain the sliding filament theory:

1 There are cross bridges linking the actin and myosin filaments and sliding may be brought about by the rapid joining and breaking of these bridges in succession. A cross bridge joins an actin filament, contracts and then detaches; it rejoins the actin farther along the filament. The cumulative effect of this ratchet-like action is to bring about contraction. At rest the cross bridges would need to be angled in order to bring about contraction. The attachment, contraction and detachment of each bridge is thought to require one molecule of ATP. There are therefore numerous mitochondria present to supply the considerable energy demands of a contracting muscle.

2 Myosin filaments are negatively charged and actin filaments have equal numbers of positive and negative charges and are therefore electrically neutral. During activity calcium ions are released from the sarcoplasmic reticulum and these neutralize some of the negative charges, thus giving the actin a net positive charge. The negative charges on the myosin and the positive ones on the actin attract one another causing sliding of the actin filaments. This theory does not involve the cross bridges which have been observed.

Variety of muscular locomotion

Swimming In fish the vertebral column prevents the body shortening when the segmentally arranged muscles contract. Each vertebra is joined to the next in such a way that only movement in one plane is permitted so muscular contraction brings about a movement of the body from side to side. The head movement is very small so the head acts as a fulcrum for the lever-like action of the tail.

Speed is proportional to the length of the body and the frequency of tail beats but since the frequency of the tail beat decreases with length a maximum is soon reached. The efficiency of locomotion is improved by the streamlined shape of fish. Buoyancy is provided by the swim-bladder in teleosts and by the action of the paired fins, especially the pectorals, and by the caudal fin in elasmobranchs.

Stability is provided by the various fins:

1 Yawing is prevented by vertical fins and in teleosts by the lateral flattening of the body.
2 Rolling is prevented by vertical and horizontal fins.
3 Pitching is prevented by horizontal fins.

Insect flight The figure-of-eight wing movement of insects is brought about indirectly. There are two sets of muscles, occupying most of the volume of the thorax, whose contractions alter the shape of the thorax and thereby the wings. The very rapid wing movement of some insects cannot be explained by the thoracic muscles twitching on arrival of each nerve impulse. Instead an oscillatory contraction is set up as the walls of the thorax click in and out, like a dented tin, moving the wings as they do so. Each click deactivates the myofibrils and tension falls. The deformation of the thorax is then restored by the antagonistic flight muscles.

The most elaborate flight is possible in the two-winged flies (Diptera) which have the hind wings modified to form halteres. These act as a gyroscope to orientate the insect.

Walking and running These forms of locomotion are particularly developed in terrestrial animals. It is important that such animals are able to balance, support the body and move.

In insects the exoskeleton supports the body and the three pairs of legs are under the centre of gravity. The head and abdomen act as counterweights. The insect moves forward by supporting its weight on a tripod formed by the anterior and posterior legs of one side and the middle one of the other while moving the other three legs forwards.

Most land vertebrates run on four legs. The weight of the body is slung from the vertebral column (often compared to a cantilever bridge). The four legs are moved in a definite pattern so that when one leg is lifted the centre of gravity lies over a triangle formed by the other three. Therefore the order of limb movement is: left fore; right hind; right fore; left hind. This pattern can only be changed if movement is fast. Speed is determined by the length of stride × the rate of stride. Large animals have a large stride but slow rate because large muscles contract more slowly than small ones. Limbs need to support the animal as well as propel it. Large animals may need such large legs to support their mass that their movement is slow, e.g. elephant. Fast running animals such as horses have the following adaptations to increase their speed:

1 High lever ratio of muscle to bone in leg, i.e. long leg and short muscle, attached to bone close to shoulder and hip joint.
2 As many joints as possible are moved in the same direction.

3 The bulk of the limb is reduced by:
   (a) eliminating unnecessary digits
   (b) reducing other digits
   (c) confining muscle to where movement is least.

Bipedal locomotion requires a high degree of balance and is compatible with high speed since it represents reduction of the limbs to a minimum.

Link topics

Section 2.1 The cell
Section 2.2 Plant and animal tissues
Section 8.1 Properties and importance of water
Section 8.3 Uptake, transport and loss in plants

Suggested further reading

Neville, A. C., *The Arthropod Cuticle,* Oxford/Carolina Biology Reader No. 103 (Packard 1978)
Pringle, J. W. S., *Insect Flight,* Oxford/Carolina Biology Reader No. 52, 2nd ed. (Packard 1983)
Wilkie, D. R., *Muscle,* Studies in Biology No. 11, 2nd ed. (Arnold 1976)

QUESTION ANALYSIS

1 Discuss the reasons why animals have to move from place to place. What problems are associated with large size in organisms? How have these been overcome?
Mark allocation 20/100   Time allowed 35 minutes   *In the style of the London Board*

This is an essay question testing the candidate's knowledge of locomotion and the ability to isolate the problems of large size and their solutions. It requires a wide range of knowledge to answer well and candidates should be sure they have a large number of points to make before answering a question of this type. To give two or three reasons for animals moving from place to place is inadequate – ten are listed in the answer plan. While some are more important than others it is nevertheless the candidate who can give a range of seven or eight reasons who will do well. It is difficult for candidates to know whether they have a good enough range of reasons and many may answer the question believing that the three they have chosen represent the sum total. This problem can be overcome by reading widely during your A-level course. Only in this way can you hope to achieve a broad and deep enough knowledge of biology to attempt confidently a question like this, knowing that you have covered all the relevant points. Give more than the list shown in the answer plan. Each reason could be qualified by a short explanatory statement explaining exactly why movement is necessary and its possible advantage to the animal.

   The problems of large size and the solutions to them should be dealt with together. Having made a brief statement of the problem, show why it creates difficulties. For each difficulty it causes give at least one way in which it has been overcome. If possible include more than one way and always give a specific example. To take one problem, that of support, begin by indicating that in terrestrial organisms air gives little support. Any increase in size will therefore make it difficult for the organisms to remain upright and rigid, and in animals this may make locomotion impossible. To overcome this 'skeletons' have developed. Then briefly review the different skeletal systems with examples, e.g. bony endo-skeleton in mammals and birds, with additional support from cartilage, chitinous exoskeletons in insects, hydrostatic skeleton in earthworms and xylem in plants. An alternative solution is to return to water to live, e.g. blue whale. A similar approach should be adopted for the other problems listed in the answer plan. One important point which escapes the notice of many candidates is that the first part of the question says 'animals', while the second part says 'organisms'. Take care to include a number of plant examples in the second part. The advice again must be to read the question carefully before attempting the answer and also while answering it.

Answer plan

Reasons for locomotion

   1 **To obtain food:** the food requirements of most animals cannot be supplied in their immediate vicinity
   2 **To capture food:** apart from the food supply being inadequate for the animals' needs carnivores must often chase and capture their prey; without locomotion this type of animal cannot survive
   3 **Escape from predators:** essential to survival
   4 **To find a mate:** essential to survival of the species
   5 **Distribution of individuals:** each individual has a different genotype. Movement to new areas allows this variation to be exploited and realizes its evolutionary potential
   6 **Reduction of competition:** prevents overcrowding and intraspecific competition

7 **To find shelter:** from both biotic and abiotic factors
8 **To maintain position:** paradoxically sharks must swim to stay still (this involves movement from place to place because the shark moves horizontally to maintain a vertical position)
9 **Reduced vulnerability to disease:** a scattered population is less likely to suffer epidemics of disease
10 **Escape from waste products:** these are toxic and may carry disease.

Problems of large size

Problem	Solution
More support needed	Bone, cartilage, chitin, xylem, become aquatic
Greater food requirements	Make centre hollow or fill it with dead tissue, e.g. xylem
Movement is difficult	Become sessile or slow moving, return to water, e.g. whale
Surface area/volume ratio is reduced	Develop internal or external surfaces to increase surface area, e.g. gills, lungs, long gut
Transport between surface and centre	Develop blood system, circulate fluids, develop transport mechanisms, e.g. respiratory pigments. Phloem and xylem in plants
More waste to dispose of	Develop excretory organs
Difficult to shelter, e.g. from predators	Large size may itself deter predators although camouflage and protective mechanisms will also help

2 (a) Give an illustrated account of the structure of a skeletal muscle fibre as shown by:
(i) high-power magnification with a light microscope (4)
(ii) an electron microscope (6)
(b) Describe the sequence of events involved in stimulation and contraction of a skeletal muscle fibre. (Start your answer at the point where the impulse reaches the end of the motor fibre.)
Mark allocation 20/100   Time allowed 30 minutes   *London Board 1984, Paper II, No. 1*

This structured essay question demonstrates the paramount importance of reading questions carefully. In (a) the key words are 'illustrated account'. 'Illustrated' means that diagrams are essential in both parts (i) and (ii). As marks are probably available for the quality of the drawing as well as its content, a little time should be spent in producing clear, neat, well-annotated diagrams, a suitable diagram being Fig. 2.6(a) (iii) in Section 2.2. Note that the question says 'a skeletal muscle fibre', and so diagrams of whole muscle sections (e.g. Fig. 2.6(a) (i)), or of a single myofibril, are unsuitable. A large number of candidates think labelled diagrams will suffice; the word 'account' should indicate that more is required. It is true that this additional information could be added to the diagrams as annotations, but these would need to be comprehensive. The alternative is a separate written account. Some of the required points are given in the answer plan.

In (a) (ii) much the same points apply. Suitable diagrams would be the first two in the series in Fig. 9.3 in Section 9.1. Both diagrams are needed and should be clearly labelled. Again some of the additional information required is given in the answer plan.

In part (b) the words in brackets at the end are intended to help the candidate avoid giving information on the transmission of the nerve impulse for which no marks are available. Heed the advice. In answering this part, two important points about answering questions at this level should be borne in mind. Firstly, marks are not given for single words or even groups of words, but for 'packets' of information which reveal that the candidate understands an idea, concept, sequence of events, etc. For example, the first 'packet' might be '. . . upon reaching the neuro-muscular junction the impulses cause vesicles to fuse with the pre-synaptic membrane and release a transmitter substance called acetyl choline'. Secondly, detail is essential. Remember that O-level detail will not bring A-level success. Give specific terms rather than general ones. Specific examples in this answer might be: 'cholinesterase' rather than 'enzyme' breaking down acetyl choline; 'acetyl choline' rather than 'transmitter substance'; 'actin and myosin' rather than 'muscle proteins'. An outline of the necessary sequence is given in the answer plan. The candidate need not worry where one 'packet' ends and the next begins. If the sequence is complete and accurate, maximum marks should be attained.

Answer plan

(a) (i) Diagram – Fig. 2.6(a) (iii)
Appearance shows: many cross striations
            many nuclei in specialized cytoplasm called sarcoplasm
            thread-like myofibrils run parallel along length of fibre

(ii) Diagram – Fig 9.3 (top pair only)

Appearance shows: Alternate dark (anisotropic) and light (isotropic) bands

       Dark bands comprise actin and myosin, light bands actin only.

       H zone occurs in dark bands and comprises myosin only

       H zone possesses M line

       Light band possesses Z line to which actin filaments are attached

       Distance between adjacent Z lines is called a sarcomere

       Numerous mitochondria present in sarcoplasm

(b) Nerve impulse arrives at nerve-muscle junction – vesicles fuse with pre-synaptic membrane – acetyl choline released (later broken down by cholinesterase) – depolarizes muscle end-plate – with critical threshold value exceeded – then action potential created in fibre – mitochondria produce ATP – helps release of calcium ions – activate troponin – displaces tropomyosin which has prevented actinomyosin bridges being formed – bridges now form between actin and myosin filaments – these contract – pull actin filaments along stationary myosin ones – bridges break and reattach further along (ratchet mechanism) – H zone and I bands shorten – sarcomeres shorten – muscle fibre contracts.

3 (a) The figure below is a diagrammatic representation of a longitudinal section through the head of a human femur.

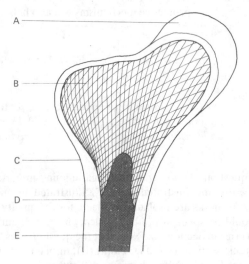

Fig. 9.4 LS head of femur

State the names of the five parts labelled A, B, C, D, E.   (5)

(b) Explain how the bone structure represented in the diagram assists in:

 (i) reducing the overall weight of the skeleton without reducing its strength.

 (ii) allowing smooth friction-free movement at the joints.

 (iii) distributing the weight of the animal from bone to bone when the limb bones are in different

   relative positions.   (8)

(c) (i) State three ways in which the structure of a long bone such as a femur and the stem of a herbaceous plant are similar.   (3)

 (ii) Name two different types of stress which operate both on long bones and on plant stems. (2)

(d) The tensile strength of tendon is much greater than that of muscle. What is the significance of this difference in tensile strengths?   (2)

Mark allocation 20/140 Time allowed 25 minutes

        Associated Examining Board June 1980, Paper II, No. 5 Alternative 1

This long structured question tests knowledge of support in general and requires the ability to relate structure to function and compare support in plants and animals.

  The labels for part (a) are articular cartilage (A), spongy or cancellous bone (B), periosteum (C), compact bone (D) and marrow cavity (E). These parts must be used in part (b) which states 'structure represented in the diagram'. Care is needed to limit comment to the diagram and not to include additional information. To answer (b) (i) it is necessary to remember that the bone is a 'human femur'. As humans are bipedal the femur supports much of the weight of the body and is under compression. Forces of compression are resisted most efficiently by vertical, hollow cylindrical supports. The reasons are complex, but basically a structure that is likely to experience compression, bending and torsion forces (as the femur is) is subject to the greatest stress around the outside. The stress and strain forces in the central region of a cylinder are very slight so there is little loss of support if this area is hollow. There is, however, the considerable advantage of a reduction in weight. The femur has compact bone around its circumference but a hollow marrow cavity. The ends of the bones have forces acting in a number of directions and the stresses are more widely distributed. No one place bears exceptional stress, but all parts bear some. The end regions therefore comprise spongy bone which has some ability to withstand stress. Its diffuse nature with spaces between bony material provides adequate support but again reduces weight.

For (b) (ii) the candidate needs to refer to the smooth rounded shape of the head of the femur and the articular cartilage. Do not mention synovial fluid. Although this is relevant it is not shown on the diagram. In (b) (iii) it is the overall shape of the head of the femur that is important. Firstly the head forms the ball of a ball-and-socket joint permitting a wide degree of freedom in movement at this point. This allows the limb to be swung forward, back and to some degree sideways, thereby moving the centre of gravity of the human and transferring the weight onto the other limb. The shape of the femur causes the limbs to stand out slightly to the side of the body, which means the weight is distributed over a wider base and the structure is more stable.

For part (c) (i) the candidate should simply state the three similarities; no explanations are needed except perhaps to indicate how they are similar, e.g. both have their major supporting tissue arranged cylindrically around the outside. In bone the compact bone is peripheral as are the vascular bundles which provide support in a herbaceous stem. The other similarities are a relatively light non-supporting central tissue and even the overall shape, i.e. long, thin cylinders. For the purposes of this question it is necessary to keep to similarities regarding support. To state that both comprise cells or that both have an outer covering may be correct but is clearly not the answer the examiner is seeking. The key words are 'structure' and 'herbaceous'. Typical mistakes by candidates include giving functional differences and comparing woody stems.

The main stresses referred to in (c) (ii) are vertical ones due to the weight of the organism they support and horizontal ones due to wind in a herbaceous plant and the action of certain muscles on the long bones.

In (d) the candidate must understand what is meant by 'tensile strength'. It is a measure of the ability to resist a pulling or a stretching force. Once a muscle has contracted it must be stretched again before being contracted a second time. It therefore has a low tensile strength. Tendons however transmit the muscular contractions to bones and so effect movement. These muscular forces can be very powerful. If tendons did not have a high tensile strength, the contractions of muscles would result in the tendons snapping or simply being stretched. In either case the bone would not be moved.

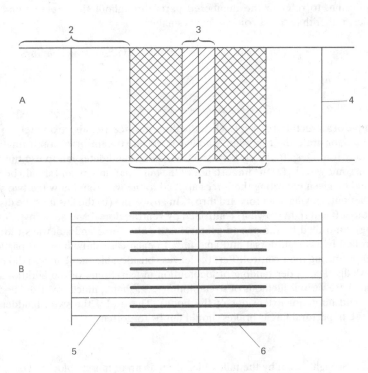

Fig. 9.5 Striated muscle fibre in a relaxed state

4 The above diagrams represent (A) a sarcomere of a single myofibril from a striated (striped) muscle fibre in the relaxed state and (B) the comparable region of a myofibril in the same state as seen in greater detail.
   (a) Name the regions numbered 1–4 in diagram A. (4)
   (b) Name the filaments labelled 5 and 6 in diagram B. (2)
   (c) Explain the appearance of the myofibril in A in terms of the detailed filament structure shown in B. (3)
   (d) (i) Draw a sketch of the sarcomere and of its constituent filaments as they appear when the myofibril is in the contracted state. (2)

(ii) Indicate briefly any significant differences from the appearance of these structures in the relaxed state shown in A and B.   (3)

Mark allocation 14/200   Time allowed 10 minutes   *London Board June 1981, Paper I, No. 10*

This is a structured question testing the candidate's knowledge of the structure of striated muscle and an understanding of the mechanism of its action.

The regions 1–4 required in (a) are stated in the essential information. Candidates may remember the terms H zone, Z line, anisotropic (A) band and isotropic (I) band well enough but not be able to decide which is which on the diagram. One useful method is to remember that the A and I bands are dark and light when viewed under a microscope. The word 'dark' has the vowel 'a' and this is A band. The word 'light' has the vowel 'i' and this is the I band. Each band is made of either actin or myosin and has either an H zone or a Z line in it. To remember which is which use the words HAM and ZIA.

H – H zone	Z – Z line
A – Anisotropic band	I – Isotropic band
M – Myosin	A – Actin

In the diagram Region 1 is darker and must be the anisotropic band, which means that 3 must be the H zone. Region 2 is light in colour (isotropic band) and has in the centre 4 which must be the Z line.

In (b) the same scheme can be used. Candidates should realize that Diagram B is a comparable region of A. This means 6 is the protein filament of the dark (anisotropic) band labelled 1 in Diagram A. Using HAM and ZIA the candidate can deduce that 6 must be the myosin filament and 5 the actin filament.

In (c) the answer needs to relate the relative darkness of the bands in Diagram A to the density of filaments in Diagram B. For instance, the darkest region in A corresponds to where the thick myosin (labelled 6) and the thin actin filaments (labelled 5) overlap. The next darkest region (labelled 3) is where there are only the thick myosin filaments. The lightest region (labelled 2) is where there are only the thin actin filaments.

For (d) (i) a diagram which illustrates the appearance of the sarcomere when the myofibril is contracted is given in the essential information (Figure 9.3).

The differences referred to in (d) (ii) are in the contracted state region: Regions 3 (H zone) and 2 (I band) are smaller, and the lines 4 (Z lines), though themselves the same size, have drawn closer together. In case the identification of the numbered parts is incorrect, it is always better, and on the whole less confusing, to refer to the numbered parts throughout the explanations, i.e. Region 3 becomes smaller, rather than the H zone becomes smaller.

5 Describe locomotion in bony fish. Write your answer under the following headings:
   (a) propulsion,
   (b) directional control,
   (c) additional features.

Mark allocation 17/125   Time allowed 20 minutes   *In the style of the Scottish Higher II*

This is an example of a structured essay. Each of the three parts carries approximately equal marks. The answer should demonstrate the candidate's ability to recall straightforward information concerning swimming in bony fish. Despite the clear instructions some candidates fail to use the headings given or adapt them in some way. In (a) the importance of the tail is paramount and should be fully discussed. Diagrams should be given indicating the forces applied to the water and how the two sideways forces cancel each other out, leaving a net forward thrust. In answering (b) the fins must be discussed in turn, i.e. movement to left and right (yawing) controlled by ventral, dorsal and anal fins, movement up and down (pitching) controlled by the paired pectoral and pelvic fins, and sideways tilting of the body (rolling) controlled by ventral, dorsal and anal fins. In each case show how the particular fins used are effective in producing their control. Part (c) leaves considerable scope for candidates to display a thorough knowledge and understanding of locomotion in fish. Some of the additional features that could be included are various methods of streamlining (e.g. shape, mucus over the body, backwardly directed scales and no major extensions of the body). The use of the swim bladder in controlling depth and the use of pectoral fins as brakes should not be forgotten.

Answer plan

(a) Propulsion is brought about by the tail, which is made up of muscle blocks. This organ is large in relation to other parts of the body, is laterally flexible and is further aided by the flattened caudal fin, which gives a large increase in surface area. The tail and caudal fin together give forward thrust, alternating sideways waves cancelling the sideways thrust.

(b) Directional control is brought about by fins. Paired pectoral and pelvic fins control pitch while the unpaired dorsal, ventral and anal fins prevent rolling and yawing.

(c) Drag is reduced by streamlining, mucus and by the backwardly directed scales. The swimbladder is important for changing overall density of the fish, so aiding depth control. Pectoral fins may serve as brakes.

10 Co-ordination

10.1 HORMONES AND HOMEOSTASIS

Assumed previous knowledge Position of endocrine glands in body.
Structure of mammalian skin.
For information on both these topics, see 'Suggested further reading'.
Metabolic rate.

Underlying principles

An organism needs methods of communication to co-ordinate its organ systems and behaviour. In an animal, information about both the environment and the body's internal state is integrated by the endocrine and nervous systems. Appropriate responses to changes in the environment are essential for the animal's survival.

Efficiency of organ systems in all animals is improved by the maintenance of a constant internal environment – **homeostasis** – so that metabolic processes can be regulated more easily. All animals and plants have some ability to control their osmotic pressure, chemical constitution and body temperature, but homeostatic mechanisms are best developed in the birds and mammals.

Points of perspective

Even if the syllabus does not mention it specifically, it would be useful for the candidate to be able to recognize microscope slides of the thyroid, pancreas and liver.

The presumed mechanism of hormonal action in the cell, via cyclic AMP, should be considered.

A knowledge of the effects, symptoms and treatment of excess and deficiency of all the hormones studied would be helpful background information, together with some idea of the experimental work which has led to the present understanding of endocrine function.

Information about the biochemical processes taking place in the liver, especially glycolysis and deamination, would also be useful.

Essential information

Hormones and the endocrine system Hormones are organic substances produced in one part of an organism and transported to another, where they exert their effects. In animals they are usually formed in the endocrine or ductless glands and secreted directly into the bloodstream, which carries them to their target organs.

The glands are integrated into the endocrine system, which interacts with the nervous system to provide the communication, co-ordination and control within the body.

Comparisons between the endocrine and nervous systems At some point in both these systems a stimulus triggers the chemical transmission of a message; this produces a response. They each carry out these basic functions rather differently (see Table 10.1).

Table 10.1 Comparison of the endocrine and nervous systems

	Hormonal communication	*Nervous communication*
Origin of stimulus	Gland	Sense receptor
Nature of stimulus	Hormone	Nervous impulse
Means of transmission	Bloodstream	Nerve fibre
Destination of stimulus	All over body	To a specific point
Receptor	Target organ	Effector (muscle or gland)
Speed of transmission	Usually slow	Rapid
Effects	May be widespread	Localized
Duration	Usually long-lasting	Usually brief

Co-ordination

Hormones and glands The pituitary gland at the base of the brain co-ordinates the activities of most of the other endocrine glands, and is itself a unit functionally integrated with the hypothalamus. The hypothalamus constantly monitors the composition of the blood and in response to any changes produces factors which affect hormonal release from the pituitary.

Table 10.2 Some hormones and their effects

Gland	Hormone	Effects
Pituitary Anterior lobe	Somatotrophin (Growth hormone, GH)	Increases growth rate in young animal and maintains size of body parts in adult (See Section 5.5)
	Thyroid stimulating hormone (TSH)	Acts on thyroid gland, thereby controlling metabolic rate
	Adrenocorticotrophic hormone (ACTH)	Controls activity of adrenal cortex
	Melanocyte stimulating hormone (MSH)	Darkens skin in frogs; function in humans unknown
	Follicle stimulating hormone (FSH)	Together known as the gonadotrophins, control development, maintenance and hormonal functions of the gonads. (See Section 5.2)
	Luteinizing hormone (LH), or interstitial cell stimulating hormone (ICSH)	
	Prolactin, or luteotrophic hormone (LTH)	Mammary gland development and milk production; maternal behaviour in birds
Posterior lobe	Anti-diuretic hormone (ADH)	Stimulates water reabsorption from the kidney tubules
	Oxytocin	Causes contraction of smooth muscles in uterus of a pregnant female and the release of milk during suckling
Thyroid	Thyroxine	Increases metabolic rate; induces metamorphosis in frogs
	Calcitonin	Lowers plasma calcium level
Parathyroids	Parathormone	Raises plasma calcium level
Pancreas	Insulin	Lowers blood glucose level
	Glucagon	Raises blood glucose level
Adrenals Cortex	Aldosterone	Stimulates sodium reabsorption from the kidney tubules
	Cortisol	Helps body resist stress, partly by raising blood glucose level
Medulla	Adrenaline	Prepare body for activity
	Noradrenaline	
Ovaries Testes	See Section 5.2	

Specialized cells in the gut produce the digestive hormones gastrin, secretin and cholecystokinin-pancreozymin, which stimulate the release of digestive enzymes (see Table 6.3).

Feedback mechanisms These are control mechanisms in which the response to a stimulus itself affects that stimulus. In *negative feedback,* the response reduces the strength of the original stimulus, while in *positive feedback* the opposite occurs and the response increases the stimulus.

Negative feedback

This type of feedback is very important in the maintenance of homeostasis. Its self-regulating characteristics make it an especially useful method for controlling release of hormones.

(a) Negative feedback

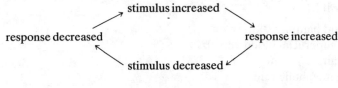

(b) Insulin release

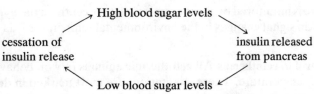

Lack of insulin therefore leads to permanent high blood sugar levels. As insulin reduces blood sugar partly by stimulating the active transport mechanism of glucose entry into the cell, the cells are deprived of glucose and diabetes results.

(c) Anti-diuretic hormone release

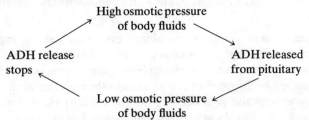

Osmoreceptor cells in the hypothalamus monitor water content of the body fluids and therefore control ADH release.

Positive feedback

As a control method, this is less often found, as it intensifies both stimulus and response. However it is put to good use as the smooth muscle of the uterus contracts during labour. Oxytocin stimulates contraction, and contraction, in turn, stimulates the release of more oxytocin. Consequently, contractions become more and more violent.

The liver Many homeostatic functions are performed by the liver. The most important ones include:

1 **Maintenance** of steady blood sugar levels by the conversion of glucose to glycogen and *vice versa,* under the influence of insulin and glucagon
2 **Breakdown** or removal of excess or used materials such as
   (a) old red blood cells by phagocytosis
   (b) excess lipids
   (c) excess cholesterol
   (d) sex hormones and adrenaline
   (e) excess amino acids by deamination
   (f) poisons by detoxification
3 **Manufacture** of
   (a) plasma proteins such as fibrinogen
   (b) cholesterol
   (c) bile
4 **Storage** of iron, copper, potassium and vitamins A, D and B_{12}
5 **Production** of heat.

To be able to carry out these activities, the liver has a very rich blood supply. As well as oxygenated blood from the aorta it receives blood containing digested food in the hepatic portal vein, coming directly from the small intestine (see Figure 6.7f Section 6.2).

Temperature control in endotherms Birds and mammals are capable of maintaining a relatively constant body temperature so that their metabolic rate need not vary with environmental temperature changes. This has been assisted by the evolution of feathers and fur for insulation. Mammals **raise** their body temperature by:

1 shivering
2 constriction of superficial blood vessels
3 raising their hairs to trap an insulating layer of air next to the skin surface
4 increasing metabolic rate

and they **lower** it by:

1 sweating and panting
2 dilation of superficial blood vessels
3 lowering hairs
4 decreasing metabolic rate

Blood temperature is monitored by the thermo-regulatory centre in the hypothalamus. Thermo-receptors in the skin signal changes in the environmental temperature via the sensory nerves to the brain.

Temperature control in ectotherms All ectothermic animals rely on behavioural adaptations to avoid extremes of temperature. These adaptations are most marked in desert reptiles who are active in the early morning and evening, spend the hottest parts of the day in the shade and at night hide in crevices warmed by their own metabolic heat.

Temperature control in plants Plants are more tolerant of temperature change than animals, but even so there is always an optimum temperature at which each species is most likely to thrive. As their metabolic rate is so low they are unable to raise their temperature much above that of the environment in cold weather, but in warm conditions transpiration may exert some cooling effect.

Adaptations to seasonal change Animals and plants, especially those living in temperate parts of the world, must be able to adapt to seasonal changes if they are to survive. When conditions are good, usually in spring and summer, they can feed, grow, reproduce and disperse. However in autumn a strategy must be adopted to see organisms through the winter when food may be unavailable. A common response is to become dormant. Many plants lose their leaves, while annuals overwinter as seeds. Insects may survive in the egg or pupa stage. Small mammals, on account of their large surface area relative to volume, lose so much heat that it is difficult to find enough food to supply their needs; they are then obliged to hibernate, and their body temperature drops. Birds often migrate to areas where they can find food.

Link topics

Section 5.2 Reproduction in mammals
Section 5.4 Growth and development
Section 5.5 Control of growth

Section 7.2 Gaseous exchange
Section 8.4 Osmoregulation and excretion
Section 10.2 Nervous system and behaviour

Suggested further reading

Buckle, J. W., *Animal Hormones* Studies in Biology No. 158 (Arnold 1983)
Clegg, A. G. and Clegg, P. C., *Hormones, Cells and Organisms,* (Heinemann 1970)
Hardy, R. N., *Homeostasis,* Studies in Biology No. 63, 2nd ed. (Arnold 1983)
Randle, P. J., and Denton, R. M., *Hormones and Cell Metabolism,* Oxford/Carolina Biology Reader No. 79, 2nd ed. (Packard 1983)
Young, J. Z., *The Life of Mammals,* (Oxford University Press 1970)

QUESTION ANALYSIS

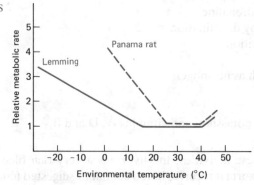

Fig. 10.1(a) Relationship between environmental temperature and metabolic rate

1 The graph above shows changes in the metabolic rate of the Panama rat (tropical rodent) and Lemming (arctic rodent) with different environmental temperatures. With reference to the graph:
(a) (i) Describe the sequence of physiological events which brings about an increase in the metabolic rate of these mammals when the environmental temperature falls below a certain minimum.

(5)

(ii) Explain why no change is needed in the metabolic rate of either animal when the environmental temperature falls from 35°C to 25°C. (3)

(iii) Explain why the metabolic rate increases when the environmental temperature rises above 40°C. (3)

(iv) Account for the difference between the two graph lines at temperatures below 20°C. (2)

(b) Species of the wood rat *Neotoma* which live in cold environments are larger than *Neotoma* species living in warmer environments. Eskimos are of stocky stature with short limbs, whilst the Masai tribesmen from Kenya are tall and thin with long limbs. Comment on these observations. (7)

(c) Water has a high thermal conductivity and so rapidly removes heat from any object of a higher temperature which is placed in it. Nearly all fish are 'cold blooded' ectotherms yet dolphins living in the same environment are 'warm blooded' homoiotherms.

Account for the capacity of dolphins to maintain a higher temperature than their surroundings. (5)

Mark allocation 25/140   Time allowed 30 minutes

Associated Examining Board November 1980, Paper II, No. 2 (Alternative ii)

This is a longer type structured question about thermoregulation which tests the candidate's ability to understand and interpret the information provided by the graph and to give explanations based on their knowledge of the biological principles of control.

The key words are 'with reference to the graph' which appear before the questions. While candidates must draw on their own knowledge the answer should, wherever possible, make reference to the graph.

In (a) (i) the certain minimum temperature referred to is called the low critical temperature. When the environmental temperature drops an endothermic animal uses physical and behavioural means to maintain body temperature, e.g. shivering, vasoconstriction, sheltering from the cold. However at the low critical temperature these mechanisms alone are inadequate. The body temperature falls and the blood passing the hypothalamus has a temperature below normal. The thermoregulatory centre located in the hypothalamus initiates a series of complex changes resulting in an increase in the metabolic rate of many organs, especially the liver. The purpose is to generate additional heat adequate to maintain normal body temperature.

The environmental temperature range of 35–25°C referred to in (a) (ii) where the metabolic rate remains constant, is termed the 'efficiency zone' and within this range endotherms can control their body temperatures by physical and behavioural means. No change is required in their metabolic rates. In answer to (a) (iii) candidates should refer to the high critical temperature, explaining that the physical and behavioural processes are inadequate to keep the body temperature as low as normal. The body temperature rises as do all the metabolic processes within the cells. This is a direct consequence of the rise in temperature. Furthermore the rise in metabolic rate produces yet more heat and so initiates a further rise in temperature (positive feedback).

The candidate should note that for environmental temperatures below 20°C, the line for the lemming does not rise as immediately or as steeply as that for the Panama rat. This is because the lemming in its arctic habitat would be unable to survive if its metabolic rate increased as quickly as that of the Panama rat. It would reach a lethal level at temperatures frequently encountered in the arctic.

The answer to (b) involves surface to volume ratios. In hot climates heat needs to be lost easily and so tall, thin individuals with their greater surface to volume ratios are at an advantage. The reverse occurs in cold regions. Furthermore in cold regions food is more scarce and the need to metabolize it to produce heat is greater. The need to store food is therefore of more importance in cold than in hot regions. The rats in cold environments are larger.

For (c) candidates should make reference to the dolphin's low surface to volume ratio, their thick layer of insulating blubber beneath the skin, their normal body temperature (which is slightly lower than in terrestrial endotherms) and the use of a **rete**. In a rete the blood in transit to the extremities transfers its heat to blood returning from them. As a result of this arrangement the blood in the extremities is cooler than blood at the core of the body and so less heat is lost to the water.

2 The diagram shows the methods of heat transfer in a small terrestrial mammal.

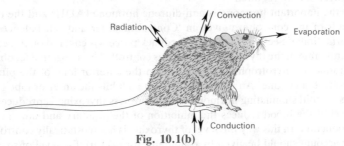

Fig. 10.1(b)

(a) Describe the various ways in which this mammal reduces the amount of heat lost in a cold environment. (5)

(b) (i) To what extent does the need to lose heat in a hot dry environment involve reversal of the mechanisms described in (a)? (3)

(ii) Explain ONE other mechanism that a mammal may use to increase heat loss. (2)

(c) Homeostatic processes involve receptors, control mechanisms and effectors. Where are the

receptors and control mechanisms located and how do they act in temperature regulation?   (4)
(d) Summarize the relation between body size and heat regulation.   (3)
(e) The metabolic activities of fish, like those of mammals, generate heat. Explain how in a cold environment the temperature difference between ectothermic (poikilothermic) fish and endothermic (homoiothermic) mammals arises.   (3)
Mark allocation 20/120   Time allowed 30 minutes   AEB June 1983, Paper II, No. 3 (Alternative II)

The diagram, resembles a rodent such as a mouse or a rat, and it would be reasonable to use this as a working basis for your answer, without actually making a specific reference to any one animal.

The answer to (a) should include not only short-term processes such as constriction of superficial blood vessels (vasoconstriction), and raising the hair to trap a thicker layer of insulating air next to the skin, but also such long-term measures as having a thick coat of fur and a thick insulating layer of subcutaneous fat. Shivering and an overall increase in metabolic rate should NOT be included. The question specifically refers to reducing heat loss, whereas these two processes are mechanisms for generating additional heat. Certain behavioural mechanisms may be included, such as sheltering in nests, especially at night. It may also hibernate in winter.

(b) (i) requires the candidate to look at each answer in (a) and consider whether a reversal of each process is possible, and if so, effective in increasing heat loss. Vasodilation would definitely be effective, and lowering of the hair would probably be a small help. Having a thin coat of fur and little subcutaneous fat would certainly assist heat loss. The mammal may become nocturnal, sheltering during the heat of the day, and undergoing summer hibernation, called aestivation. The best mechanism for explanation in (b) (ii) would be sweating, the consequent evaporation helping to reduce body heat.

The receptors referred to in (c) are located within the hypothalamus and the skin. Those in the hypothalamus monitor change in the temperature of the blood, whereas the thousands of receptors throughout the skin monitor environmental temperature changes. The control mechanisms lie in the hypothalamus, although the cerebral cortex plays an auxiliary role. The skin receptors are of two types; warm receptors which are stimulated by a rise in temperature, and cold receptors which respond to a fall in temperature. Both send nervous messages to the hypothalamus and the cerebral cortex. The hypothalamus has a heat-gain centre which activates mechanisms for retaining heat, and a heat-loss centre which activates mechanisms for losing heat. In (d) the important point to remember is that while heat is generated and stored within the volume of the body, it is lost through the surface. The greater the volume for a given surface area, the more heat is conserved; the greater the surface area for a given volume the easier it is lost. What matters, then, is the surface area to volume ratio.

The answer to (e) requires an appreciation of how the heat-generating metabolic activities are controlled by enzymes. Enzymes work more efficiently at higher temperatures, their rate doubling for each 10°C rise in temperature up to an optimum. The mammal can maintain its temperature, and hence the rate at which heat is generated. The fish, however, is less able to control heat loss. Its body temperature falls and the enzymes involved in generating heat work less efficiently. Less heat is generated and the temperature falls further. The process continues until the body temperature of the fish is at, or close to, that of its surroundings.

3 (a) How are each of the following feedback processes controlled in humans
  (i) the content of water in the blood   (5)
  (ii) the secretory rate of the thyroid gland   (5)
  (iii) the level of sugar in the blood?   (5)
  (b) Explain briefly the feedback processes inside living cells.   (5)
Mark allocation 20/80   Time allowed 35 minutes   In the style of the Cambridge Board

This is a structured essay question largely testing recall of factual information. In part (a) the candidate should restrict the answer to mechanisms of feedback control. The actual functions of water, the thyroid and sugar are not required.

In (a) (i) the important hormone is anti-diuretic hormone (ADH) and the organ involved is the kidney. Details of the control are given in 'Control of water and salt balance in mammals' in the essential information of Section 8.4 and the feedback process is covered on page 218 of this section.

The secretory rate of the thyroid gland (a) (ii) is controlled by a negative feedback process. Thyroid stimulating hormone (thyrotrophic hormone) from the anterior lobe of the pituitary stimulates the thyroid to produce thyroxine. An excess of thyroxine inhibits the anterior lobe of the pituitary which produces less thyroid stimulating hormone and in turn less thyroxine is produced. A lowering of the thyroxine level of the blood reduces the inhibition of the pituitary and ultimately leads to a rise in thyroxine production. In this way the level of thyroxine is homeostatically controlled.

A similar account should be given by candidates for (a) (iii) 'the level of sugar in the blood'. For details candidates should refer to page 218 in the essential information and also to Question 2 in this question analysis section. Reference must be made to both insulin and glucagon.

In part (b) a key word is 'briefly'. Although it would be possible to write at length on this topic, candidates must confine themselves to making the essential points—in particular the fact that the end product of a metabolic pathway often inhibits an enzyme at the start of the pathway. When the product is in excess it prevents its own production, until its level falls so low that it ceases to inhibit the enzyme and its production commences again. In the same way a metabolic pathway leading to the production of a substance may be inhibited by the presence of that very substance. In *Escherichia coli* the amino acid tryptophane is produced by a particular metabolic pathway. When *E. coli* is grown on a medium containing this amino acid, such production ceases within seconds, but recommences rapidly upon removal of the tryptophane.

Answer plan

(a) (i) ADH and the kidney (see Section 8.4)
   (ii) Anterior lobe of pituitary produces thyroid stimulating hormone (TSH); thyroid produces thyroxine which inhibits TSH production by pituitary.
   (iii) Pancreas: insulin and glucagon
(b)   End product may inhibit enzyme within pathway (negative feedback). Production of essential material may be inhibited by the presence of that material in the medium or diet (e.g. *E. coli* and tryptophane).

10.2   NERVOUS SYSTEM AND BEHAVIOUR

Assumed previous knowledge   The structure of the neurone (Section 2.2)
Characteristics of nervous communication (Section 10.1)

Underlying principles

The nervous system permits rapid passage of information from one part of the body to another, so that suitable responses to stimuli can be made at once. In its simplest form it comprises a network of nerves connecting each part to every other part. With an increase in the number of parts to be connected this system proved inadequate and a central nervous system (CNS) developed. Each part was connected to this central 'switchboard' which then connected incoming messages to the appropriate effector organs. In bilaterally symmetrical animals it was appropriate to have the CNS running along the length of the body with paired nerves branching from it to each segment. As animals developed a definite head region which encountered stimuli before other parts, it obviously became the main location for sense organs. To deal with this increased volume of nervous information at the anterior end, the CNS in this region swelled. The swelling or cerebral ganglion later developed into a brain whose function evolved beyond simple co-ordination to include storage of learned information and development of intelligent behaviour.

Points of perspective

It would be an advantage to be aware of the experimental techniques which have been used in the study of brain function.

It is essential to be able to quote specific examples of reflex actions, and also to know the functions of the lobes of the cerebral hemispheres.

The differences between innate and learned behaviour patterns should be appreciated.

Essential information

The nervous impulse   In invertebrates, the speed at which the impulse travels is dependent on the diameter of the axon. Neurones innervating muscles involved in escape or defence reactions therefore tend to have very wide axons. This problem of size has been overcome in vertebrates with the development of the myelin sheath, which greatly increases rates of conduction. However most research on the nature of the impulse has been carried out on giant fibres of invertebrates such as earthworms.

The impulse itself follows a wave of increased permeability passing along the axon membrane. In the resting membrane sodium ions are kept outside the cell by an active transport mechanism known as the sodium pump, so that the outside of the membrane carries a positive electrical charge. As soon as the permeability increases these ions pass into the axon by diffusion so that the membrane charge changes. However the membrane quickly recovers and the sodium ions are ejected. This changing charge moving along the membrane is known as the **action potential**. (See Figs. 10.2a/b.)

The synapse   Neurones are not in direct contact with their neighbours; there are tiny gaps known as synapses between them. The action potential stops at the synapse, but its arrival causes the release of a transmitter substance from the neurone which diffuses across the gap. Arriving at the post-synaptic membrane, it causes a change in its permeability, so that a new action potential begins. The transmitter is then broken down enzymatically, to prevent it from continuously initiating action potentials.

Some stimuli do not initiate the production of enough transmitter substance to cross a synapse but enough may be produced by several knobs acting simultaneously (**spatial summation**). If several impulses arrive in rapid succession at a synaptic knob they may initiate the production of sufficient transmitter substance to generate a post-synaptic action potential

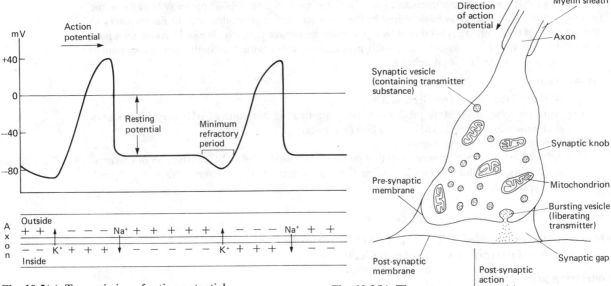

Fig. 10.2(a) Transmission of action potential

Fig. 10.2(b) The synapse

even if one impulse was not sufficient. This is known as **temporal summation** and involves a gradual accumulation of transmitter substance until there is sufficient to create an action potential, a process known as **facilitation**. Regardless of the strength of the initiating stimulus, the action potential itself is always of uniform size–**'the all or nothing law'**.

Table 10.3 Transmitter substances

Transmitter	Location	Breakdown enzyme
Acetylcholine	Nerve/muscle junction	Cholinesterase
Noradrenaline	Sympathetic nervous system	Monoamine oxidase

Several transmitter substances have been discovered, but two important ones are acetylcholine and noradrenaline. Not all transmitters excite the post synaptic membrane, some inhibit it so that it is more difficult for it to carry an impulse. Blocking of nervous pathways in this way helps the direction of impulses through the nervous system.

Invertebrate nervous systems

Coelenterates show all the elements of a nervous system necessary to perform reflex actions, i.e. receptors, neurones and effectors. They have multi-polar neurones forming a net in contact with both receptors and effectors. The arrival of one action potential at a synapse facilitates the passage of the next and so on until the whole net is stimulated. This is known as **interneural facilitation**. In some coelenterates, e.g. sea anemones, there are also through conduction pathways which transmit a single impulse the length of the body. This provides the main function of a central nervous system: rapid transmission.

Annelids and arthropods have a nervous system which is more centralized with a solid ventral

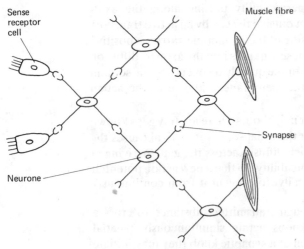

Fig. 10.3 (a) The coelenterate nerve net

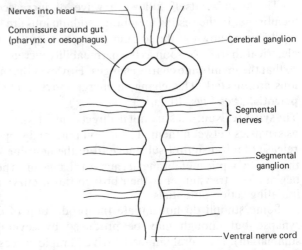

Fig. 10.3 (b) Nervous system of an annelid or arthropod

nerve cord running the length of the body. This increases the speed with which impulses arriving from the receptors can be carried to the appropriate effectors. The nerve cord is swollen at the anterior end but this primitive brain plays little part in co-ordination, it merely deals with impulses arriving from the anterior receptors and messages to the anterior effectors.

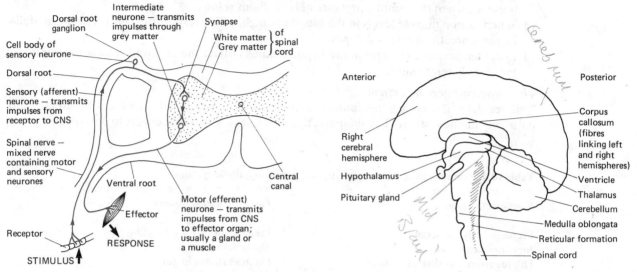

Fig. 10.4(a) Reflex arc   **Fig. 10.4 (b)** VS human brain

Vertebrate reflex arc

A reflex is an automatic response to a stimulus, via the central nervous system.

$$\text{RECEPTOR} \xrightarrow[\text{neurone}]{\text{sensory (afferent)}} \text{CNS} \xrightarrow[\text{neurone}]{\text{Motor (effector)}} \text{EFFECTOR}$$

$$\underset{\text{stimulus}}{\uparrow} \qquad\qquad\qquad\qquad\qquad\qquad \underset{\text{response}}{\downarrow}$$

In most reflex actions the sensory and motor neurones are connected via one or more intermediate neurones in the brain or spinal cord (the knee jerk reflex is an exception to this).

There are many reflex arcs in the body and these are connected to each other and to the higher centres of the brain, so that, by conscious effort, it is possible to suppress a normal reflex action.

The central nervous system in humans

In vertebrates the brain developed as a swelling at the anterior end of the dorsal nerve cord, the rest remaining as the spinal cord.

The **spinal cord** consists of tracts of ascending (sensory) and descending (motor) fibres which carry information between the brain, and sense receptors, muscles and glands.

The **brain** may be subdivided into the following main regions each with their own specific functions:

Forebrain	olfactory tracts	concerned with sense of smell; found deep in forebrain
	cerebral hemispheres	very enlarged and important area of human brain. Electrical stimulation has led to 'mapping' of the most superficial area, the cerebral cortex
'Tweenbrain	thalamus	contains ascending and descending tracts linking forebrain with spinal cord
	hypothalamus	largely controls the pituitary gland; seat of basic emotions or 'drives' such as hunger, thirst, fear, rage and sex.
Midbrain	corpora quadrigemina (optic lobes of lower vertebrates)	concerned with sense of sight
	red nucleus	helps control movement and posture
Hindbrain	medulla oblongata	controls voluntary movement and also has control centres for activities such as swallowing, salivation, heartbeat, vascular constriction and dilation, and respiration
	cerebellum	regulation and co-ordination of muscular activities.

[handwritten annotations: responsible for ½ of all bodily activities – coordination – memory also found here. / involuntary sensors – & beat. / balance & coordination]

Stimulation of brain cells Neurones in the brain are normally stimulated by transmitter substances released at the synapse by a neighbouring cell. However chemical or physical changes in the body can sometimes affect brain cells directly. For example:

1 Osmoreceptor cells in the hypothalamus will stimulate the pituitary to release anti-diuretic hormone when the osmotic pressure of body fluids is low.
2 When carbon dioxide levels in the blood are high, reflex respiratory centres in the medulla increase breathing rate and depth.
3 Local temperature changes in the hypothalamus initiate the required responses to raise or lower body temperature.

The autonomic nervous system
This regulates the body's involuntary activities, and it is divided into the sympathetic and parasympathetic systems. The sympathetic system has comparable effects to adrenaline.

Table 10.4 Comparison of the sympathetic and parasympathetic systems

Sympathetic system	Parasympathetic system
Prepares body for action	Prepares body for relaxation
Increases heartbeat	Slows heartbeat
Dilates arteries in skeletal muscles	Dilates arteries in gut
Slows gut movements	Speeds gut movements
Dilates bronchioles	Constricts bronchioles
Dilates pupil	Constricts pupil
Causes sweat glands to secrete	Causes tear and salivary glands to secrete
Causes hairs to stand erect	
Bladder and anal spincters contract	Bladder and anal spincters relax

Behaviour

It is difficult to categorize behaviour patterns rigidly but many of them involve elements of reflex action, orientation and learning.

Reflex action Reflex actions have the advantage of being fast and automatic, and they are important in escape and avoidance reactions in all animals.

Orientation Orientation often involves reflex actions but they are normally developed into behaviour patterns. There are two main forms of orientation found in animals:

(a) *Kinesis* This is an increase in random movement under unfavourable conditions. Woodlice move rapidly and turn frequently in dry conditions but the animals tend to congregate under humid conditions as their rate of movement slows.

Table 10.5 Comparison of learned and instinctive behaviour

Instinctive behaviour	Learned behaviour
Inborn and not acquired during an animal's lifetime	Acquired during an animal's lifetime
Fixed and not adaptable although some minor modification over a long period may be possible	May be easily and rapidly adapted to suit changing circumstances
Similar amongst all members of a species	Varies considerably amongst different members of the same species
Unintelligent and there is no appreciation of the functions of the behaviour	May be intelligent and the animal often appreciates the function of a particular action
Often comprises a chain of actions in which the completion of one acts as the trigger for the start of the next	No fixed sequence of actions and the completion of one need not necessarily affect which action should follow
Apart from minor modifications the behaviour is permanent	Usually a temporary and short-lived form of behaviour, although it may be reinforced, thus making it more or less permanent
Although there are many forms of instinct the basic form of this behaviour is the same for all organisms	Wide range of learned behaviour ranging from simple taxes or imprinting to complex forms of intelligence and reasoning

(b) *Taxis* This is the directional movement of the whole organism in direct response to a stimulus. Such stimuli may be light (phototaxis) or chemicals (chemotaxis), e.g. movement of planarians towards food; location of the ovum by the sperm of lower animals.

Learning This is a change in behaviour based on experience and it is the most adaptable form of behaviour. Five subdivisions are often recognized:

(a) *Habituation* This is the diminishing of a response as a result of repeated stimulation. It is specific to a particular stimulus, e.g. snail, *Helix*, withdraws its tentacles in response to mechanical stimulation but ceases to do so after repeated stimulation.

(b) *Associative learning* In this the animal learns to associate a reward or punishment with a particular form of behaviour or stimulus. Such a pattern may be established either as a result of **classical conditioning** (the work of Pavlov on the salivation of dogs) or through **trial and error**.

(c) *Imprinting* Young animals tend to follow, or imprint on, their parents. Experiments have shown that they may follow the first moving thing they see.

(d) *Exploratory learning* Even without a specific reward or punishment an animal learns features of its environment which may benefit it later.

(e) *Insight learning* This involves the production of a new adaptive response as a result of 'insight'. It is a difficult subdivision to define and any definition of it results in controversy. It is thought to apply only to man.

Other factors affecting behaviour

Hormones may modify behaviour patterns; testosterone, for example, may increase the incidence of aggressive behaviour, and prolactin releases nestbuilding responses in birds.

Animals may show rhythmic changes in behaviour, according to the time of day or the season.

Characteristic behaviour patterns may be released when an animal meets another member of its own species. A male bird, meeting a female on his territory in spring, may begin a courtship display, but should he meet another male he will threaten him until the intruder withdraws, so that competition for food or a mate is removed. Intraspecific interactions are especially important to humans, who owe their success largely to their ability to live in co-operative groups.

Link topics

Section 10.1 Hormones and homeostasis
Section 10.3 Sense organs

Suggested further reading

Adrian, R. H., *The Nerve Impulse,* Oxford/Carolina Biology Reader No. 67 2nd ed. (Packard 1980)
Brockfield, A. P., *Animal Behaviour,* (Nelson 1982)
Gray, E. G., *The Synapse,* Oxford/Carolina Biology Reader No. 35 2nd ed. (Packard 1977)
Jones, D. G., *Neurons and Synapses,* Studies in Biology No. 135 (Arnold 1981)
Messenger, J. B., *Nerves, Brains and Behaviour,* Studies in Biology No. 114 (Arnold 1979)
Usherwood, P. N. R., *Nervous Systems,* Studies in Biology No. 36 (Arnold 1973)

QUESTION ANALYSIS

4 (a) (i) Name TWO hormones produced by each of the following endocrine glands.
   1 Pancreas 2 Pituitary 3 Adrenal
   (ii) For each gland, state the main function of ONE of the two named hormones. (3)
(b) Describe the effect of both the sympathetic and the parasympathetic nervous systems on each of the following structures.
   1 Heart 2 Gut muscles 3 Bladder sphincter 4 Bladder wall 5 Anal sphincter
6 Bronchioles (3)
(c) Compare and contrast hormonal and nervous communications under the following headings.
   (i) Speed
   (ii) Specificity of destination
   (iii) Number of organs responding
   (iv) Length of persistence time (8)

The graph overleaf shows the membrane potential of the post-synaptic membrane of a neurone after application of acetyl choline to the synapse, thereby producing an excitatory post-synaptic potential (the curve labelled EPSP).
   (i) What is the cause of the resting potential of −70 mV?
   (ii) Describe the molecular events which result in the changes in membrane potential shown in the graph. Your answer should explain the significance of the threshold potential.
   (iii) A different transmitter substance was applied to the synapse. The change in membrane potential of the post-synaptic membrane is illustrated by the curve labelled IPSP. What is the cause of this? (6)

(d)

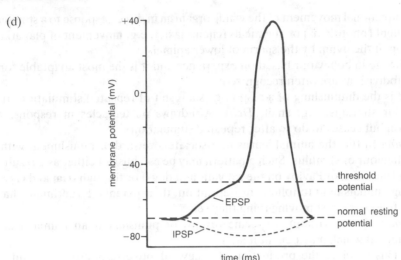

Mark allocation 20/80   Time allowed 45 minutes   *Northern Ireland 1985, Paper 2, No. 3*

This long-structured question tests knowledge of coordination, both nervous and hormonal. Parts (a) and (b) test recall of factual information and the answers can be obtained by studying Tables 10.2 and 10.4, respectively. Table 10.1 in the essential information provides the basis for the answer to part (c) although the mark distribution suggests that the response should include examples or other appropriate detail.

It is important in part (c) to again look at the mark distribution before attempting an answer. While the response could be long and detailed, with only 6 marks available the answer must be detailed but concise. The resting potential referred to in part (d) (i) is due to the uneven distribution of ions either side of the neurone membrane. In particular sodium ions are actively removed resulting in an overall external positive charge relative to the inside.

The molecular events which result in the changes shown are caused by the acetyl choline depolarizing the post-synaptic membrane. The acetyl choline combines with receptors on the membrane but has little effect until the threshold value is exceeded, above this point, however, there is a sudden influx of sodium across the post-synaptic membrane. This causes depolarization of the membrane resulting in an action potential of around +40 mV. At the same time potassium ions move outwards only much more slowly. These in time repolarize the membrane returning to its resting potential of −70 mV.

The alternative transmitter substance mentioned in (d) (iii) causes the post-synaptic membrane to become more negative inside because it makes the membrane even less permeable to sodium ions. This means they are even more likely to remain excluded from the neurone. This Inhibitory Post-Synaptic Potential as it is called, makes it even less likely that the threshold value will be exceeded and a new impulse transmitted.

5 (a) Describe the structure of the various types of neurone in a mammal.   (10)
  (b) Describe the sequence of events which takes place when impulses travel along a motor nerve fibre to a muscle causing it to contract.   (10)
Mark allocation 20/100   Time allowed 30 minutes   *London June 1981, Paper 2, No. 2*

Part (a) of this essay question needs careful interpretation. 'Various types of neurone' must be taken to mean that an account of a wide range of neurones is needed, e.g. uni-polar, bi-polar, multi-polar. The word 'structure' must be considered in its widest context, i.e. the answer should not be limited to the general shape of different neurones, but should include parts of the neurone such as dendrites, the cell body and the axon, and also internal detail such as mitochondria, neurotubules, Nissls granules etc. The emphasis should be on giving a range of structure associated with different neurones, e.g. motor end plates (present on motor neurones to muscles), detail of the synaptic knob (on most neurones) and axon terminal detail. As with most questions on structure, relevant diagrams, provided they are adequately labelled and annotated, can help considerably and earn marks. Any diagrams should provide general detail specific to neurones, e.g. a diagram showing detail of a synaptic knob, motor end plate or whole neurone. Detailed diagrams of unit membranes, mitochondria or nuclei are not sufficiently specific to neurones to be worthy of inclusion.

In part (b) a description of 'resting potential' is the obvious starting point, as an impulse is only the change in potential along a neurone, relative to the resting potential. Be detailed and give values for the resting potential, i.e. 60 millivolts. Make it clear that the inside is negative relative to the outside.

The action potential should then be described, making mention of the change in the axon membrane permeability on stimulation, the rapid influx of sodium ions and the slower outward movement of potassium ions. These events cause depolarization of the membrane. A graph to illustrate these changes (see Fig. 10.2a) could be included. Finally the restoration of resting potential by means of the sodium pump, involving ATP as an energy source, should be discussed.

The question refers to 'when impulses travel along a motor nerve fibre'. Clearly more is needed than an action potential in a neurone. The role of the myelin sheath in speeding transmission should be mentioned and the importance of the nodes of Ranvier in allowing the action potential to 'jump' from node to node should be included. Candidates may easily omit to complete the question which requires the muscle to contract at the end of the process. Detail of changes at the motor end plate are therefore needed, including a mention of transmitter substances such as noradrenalin and acetylcholine, the activation of the sarcolemma and the release of calcium ions causing the contraction of all sarcomeres. Details of the actin/myosin sliding filament theory of muscle contraction are beyond the scope of this question and should be omitted.

Answer plan

(a) Uni-, bi- and multi-polar neurones (diagrams).

Detail of single neurone (gross structure).

Cell body, axon, dendrites (well-labelled diagram).

Internal (electronmicrograph) detail.

Cell body detail, large nucleus, Nissl granules, many mitochondria, endoplasmic reticulum. neurotubules and permeable membrane with large surface area.

Synapse – detail of synaptic knob (diagram).

Motor end plate, described.

Axon terminals, described.

(b) Resting potential – 60mV -ve inside.

Action potential – membrane permeability increased, Na^+ moves in, K^+ moves out, membrane becomes depolarized. Diagram and/or graph.

Restoration of resting potential – renewal of sodium pump, ATP provides energy.

Transmission along neurone – role of myelin sheath and nodes of Ranvier.

Contraction of muscle – motor end plate detail; acetycholine/noradrenalin as transmitters; cholinesterase to breakdown acetycholine; activation of sarcolemma, contraction of all sarcomeres on release of Ca^{++}

6 (a) Distinguish clearly between:

(i) tropic and reflex responses, and (ii) instinctive and learned behaviour. (8)

(b) Describe ONE example of a tropism and ONE of a reflex response. (8)

(c) Describe the significance of the examples that you have described to the life of the organism concerned. (4)

Mark allocation 20/100   Time allowed 30 minutes   *London Board, January 1982, Paper II, No. 4*

This structured essay examines knowledge of coordination and responses in plants and animals. A reasonable mark could not be obtained on this question by a candidate with knowledge of either plant or animal coordination only.

In (a) (i) the key word is 'distinguish'. Only clear differences need be included. Tropisms are slow responses, reflexes rapid. The differences between instinctive and learned behaviour are set out clearly in Table 10.5.

In (b) the example should be specific, e.g. phototropism in a stem, rather than a general account. Choose the example carefully, bearing in mind part (c), and keep strictly to it. Table 10.5 provides a wide range of possible tropism examples to choose from. The example of a reflex could be a withdrawal reflex. Remember that if the knee-jerk reflex is used there is no intermediate neurone involved, and Fig 10.4 (a) should be modified accordingly.

The significance required in (c) will depend upon the choices made in (b). The phototropic response of a stem, for example, ensures that a plant grows towards light in order to obtain maximum illumination for photosynthesis. This allows maximum growth and hence an advantage over its competitors. The reflex involving withdrawal from an unpleasant stimulas has a protective function. By the rapid removal of a foot from a nail, further injury is prevented.

Answer plan

(a) (i)

Tropism	Reflex
Plant response	Animal response
Movement due to growth	Movement due to muscular contraction
Response is slow	Response is rapid
Always involves movement of part of a plant	May occasionally involve movement of whole organism
Direction of response is related to direction of stimulas	Direction of response can be unrelated to direction of stimulas

(a) (ii) See Table 10.5

(b) Choose examples of tropism from Table 5.3. Describe how stimulus is perceived, type of response (including role of auxins). Typical reflex exmple is withdrawal reflex. Use Fig 10.4(a) as a guide and give account of the process in seven stages: 1 stimulus, 2 receptor, 3 motor (afferent) neurone, 4 intermediate (connector) neurone, 5 sensory (efferent) neurone, 6 effector, 7 response.

(c) Answer depends on examples chosen in (b), but in the cases above expand on the following: phototropism allows plant to grow to light for greater photosynthesis, better growth and a competitive advantage. Withdrawal reflexes allow rapid escape from potentially harmful situations.

10.3 SENSE ORGANS

Assumed previous knowledge Basic structure of the eye and the ear as studied at O level (see 'Suggested further reading').

Underlying principles

Animals have evolved specialized sense receptor cells to provide them with essential information about aspects of their environment. Touch, pressure, temperature and pain receptors are necessarily spread all over the body, while those for taste and smell are only in the mouth and nose. Light, sound and gravity receptors are grouped together to form highly specialized sense organs.

Points of perspective

An appreciation of the significance of colour vision to various animal groups, e.g. in food finding and courtship, is important.

An understanding of dark and light adaptation, and the formation of after images would be useful, together with colour blindness and defects of vision in humans.

Essential information

Information about the environment is provided by receptors. Since there are many environmental stimuli the range of receptors must be varied. In man there are three main groups of receptors: chemoreceptors, photoreceptors and mechanoreceptors.

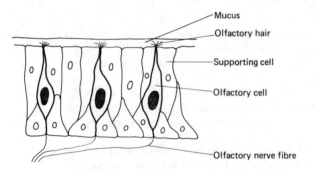

Fig. 10.5(a) Olfactory epithelium

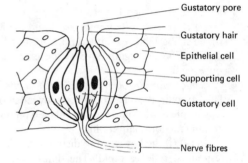

Fig. 10.5(b) Taste cells

1 Chemoreceptors

(a) *Olfactory epithelium* (see Fig. 10.5(a)) Olfactory cells are more or less evenly distributed between the supporting cells in two patches of epithelium in the upper nasal cavities. They are derived from the cell bodies of bipolar nerves. They respond to chemicals which are volatile at ordinary temperatures and which are soluble in fat solvents. The sense of smell is easily fatigued.
(b) *Taste cells* (see Fig. 10.5(b)) Taste cells are grouped into taste buds, found mainly round the base of the papillae of the tongue and also on the soft palate, epiglottis and around the opening of the oesophagus. Taste buds are sensitive only to chemicals which are soluble in water and although they appear similar each bud is sensitive to only one taste: sweet, sour, bitter or salt.

2 Photoreceptors

The eye As light enters the eye it is refracted by the curved, transparent cornea before passing through the pupil. This varies in size according to light intensity, as the muscles of the iris contract or relax:

Table 10.6 Action of the iris muscles

Light	Dim	Bright
Circular muscles of iris	Relax	Contract
Radial muscles of iris	Contract	Relax
Pupil	Dilates	Constricts

The iris tends to be darkly pigmented in diurnal animals such as dogs, sparrows and cows, and pale in nocturnal ones such as cats and owls, so that more light can reach the retina.

Behind the iris is the elastic lens. Its shape is changed by the action of the ciliary muscles, so that the image can be focused on the retina.

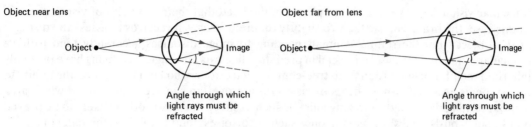

Fig. 10.6 Accommodation

Table 10.7 Accommodation of the eye

Object	Near eye	Far from eye
Light must be refracted	A lot	A little
Lens must be	Thick	Thin
Ciliary muscles	Contract	Relax
Suspensory ligaments	Slack	Tense

Reflection of the light, once it has passed through the lens, is prevented by the dark, vascular choroid layer.

The light sensitive layer of the retina consists of two types of photoreceptors, rods and cones, containing pigments which are bleached by light. The bleaching process can initiate an action potential. There are certain differences between the two types of cell.

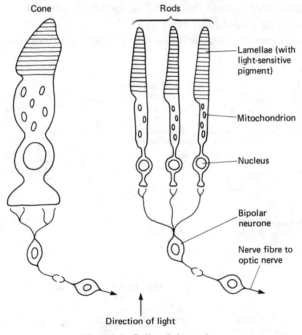

Fig. 10.7 Cells of the retina

Table 10.8 Comparison of rods and cones

Rods	Cones
More numerous	Less numerous
Usually around periphery of retina	Usually located in centre of retina
Arranged in functional units served by one bipolar neurone, therefore acuity low	Each cone served by its own bipolar neurone, therefore acuity high
Very sensitive to low levels of illumination	Only stimulated by bright light
One type of rod only, stimulated by most wavelengths of visible light except red	Three types of cone, each selectively responsive to different wavelengths, therefore allowing colour perception
Rapid regeneration of light sensitive pigment, therefore can perceive flicker well	Slower regeneration of light sensitive pigment, therefore less responsive to flicker

Nocturnal animals may have only rods in their retinas, so although they may lack colour vision they will have a heightened sensitivity to movement.

Binocular vision The eyes may be so placed in the head that their two fields of vision overlap. In this case, the brain can use the two slightly different images it receives to assess the distance of the object from the retina. Animals needing to judge distances accurately tend to have binocular vision like this, and it is usual in predators like cats and eagles. Humans have probably inherited their binocular vision from tree-climbing ancestors, who later developed the ability to throw spears. Animals whose fields of vision do not overlap can see over a much wider area without turning their heads. Grazing animals such as sheep and cows, others likely to be preyed on like small birds, and slow-moving ones such as tortoises, tend to fall into this category.

3 Mechanoreceptors

(a) *The Ear* The receptor cells in the ear are equipped with sensitive hairs which are believed to initiate action potentials when they are moved. The pressure changes of sound waves move the hairs on one type of receptor, while movements or changes in position of the head deflect them on other types.

Hearing The pinna, which in most mammals can be directed towards the source of sound, channels the sound waves towards the drum, which is sunk well into the head down a bony shaft.

The vibrations are relayed across the air-filled middle ear by the ossicles, which amplify them at the same time. Muscles attached to the ossicles prevent excessive vibration in response to a very loud noise. The eustachian tube opening into the back of the throat helps to equalize air pressure in the outer and middle ears.

The vibrations enter the fluid-filled cochlea via the oval window. The round window beneath it also vibrates, thereby neutralizing pressure changes in the cochlear fluid. The vibrating basilar membrane transfers its movements to the sensitive hair cells of the organ of Corti, which press against the tectorial membrane above them, and an impulse is generated in the auditory nerve.

It is believed that cells in different regions of the organ of Corti respond to sounds of different pitch, those nearer the oval window being more sensitive to high notes than those further away from it.

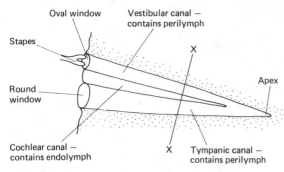

Fig. 10.8(a) Uncoiled cochlea

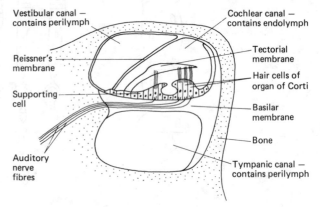

Fig. 10.8(b) TS cochlea (line X–X on Fig. 10.8(a))

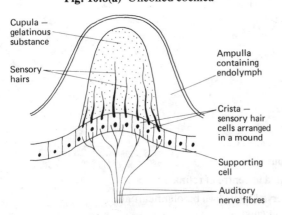

Fig. 10.8(c) Ampulla

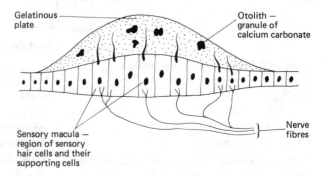

Fig. 10.8(d) Utricle

Balance Head movements are registered by the ampullae inside the semi-circular canals. The three canals are arranged at right angles to each other so that head movements in one direction will cause corresponding movements in the fluid inside the canals. The gelatinous cupula of the ampulla will be deflected by the moving fluid, and the hairs on the receptor cells will be moved.

Positions of the head are monitored by sensory hair cells in the utricule and the saccule. Otoliths of calcium carbonate balanced on these cells are pulled downwards in response to gravity when the head is tilted.

(b) *Dermal receptors*
 (i) Free nerve endings–these are pain receptors. The terminal branches of the sensory nerve cells may extend beyond the dermis into the epidermis
 (ii) Pacinian corpuscle–this is a pressure receptor found in the subcutaneous tissue. It takes the form of an axon whose non-myelinated tip is enclosed in a fluid-filled capsule.
(iii) Meissner's corpuscle–this responds to touch and is found in the dermis.

Link topics

Section 10.2 Nervous system and behaviour

Suggested further reading
Friedman, L., *The Human Ear,* Oxford/Carolina Biology Reader No. 73, 2nd ed. (Packard 1979)
Weale, R. A., *The Vertebrate Eye,* Oxford/Carolina Biology Reader No. 71, 2nd ed. (Packard 1978)

QUESTION ANALYSIS

7 The graph below shows the number of receptor cells (types A and B) in the human retina along a horizontal line from the nasal side of the eye to the outer side. Distances are expressed in arbitrary units.

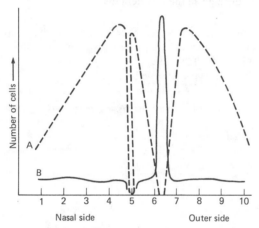

Fig. 10.9 Receptors in the human retina

(a) (i) Identify the types of receptor represented by A and B.
 (ii) Explain why there are no receptor cells at position 5.
 (iii) What is the name of the region of the retina at position 6?
 (iv) Explain why the greatest concentration of receptor cells of type B occurs at position 6.   (9)
(b) (i) When a person moves from bright surroundings into a dimly-lit room, objects in the room cannot at first be seen but they gradually become visible. Explain this occurrence.
 (ii) In the dimly-lit room, objects are only visible in black and white. Explain this.   (9)
(c) Describe the particular features of receptor cells which allow objects to be viewed in colour.   (2)
(d) The flowers of three species of cinquefoil, *Potentilla,* are similar in form and all appear to have the same uniformly coloured yellow petals. When photographed in ultraviolet light, each species shows a different pattern on its petals. Using this information, explain in detail how bees are able to distinguish between the flowers of the three species of cinquefoil while man cannot do so.   (5)
Mark allocation 25/40   Time allowed 30 minutes
Associated Examining Board November 1979, Paper II, No. 5

This longer type of structured question tests the ability of candidates to apply their biological knowledge of photoreception to the graph provided and to explain a number of observed facts. The graph should be interpreted in general terms before attempting the questions. To answer (a) (i) candidates need to know that cones are highly concentrated at the fovea centralis (yellow spot) and that this occurs in the centre of the retina, i.e. about position 6 on the graph. It is receptor cell type B that is abundant here. B therefore represents the cones and A the rods. (a) (ii) refers to position 5 where there are no receptor cells. This is the blind spot. Candidates must note the word 'explain' and not satisfy themselves with simply naming the part. The neurones from the rods and cones pass in front of the retina. Where they converge to form the optic nerve they are so densely packed that no receptor cells occur.

For (a) (iii) the term blind spot is required. The reason required in (a) (iv) for the large concentration of receptor cells of type B at position 6 is that it lies on the optical axis directly opposite the centre of the lens. It is here that the greatest concentration of light occurs when a person is looking directly at an object. Type B receptor cells (cones) are sensitive to high light intensities, hence their concentration where the light intensity is greatest (i.e. at the fovea).

In (b) it is the action of the iris diaphragm that explains the occurrence. In bright surroundings its circular muscles are contracted and the pupil is constricted. This reduces the amount of light entering the eye and reduces overstimulation of the retinal cells. On entering a dimly lit room the radial muscles of the iris contract and the pupil slowly dilates to allow the maximum amount of light to enter the eye. The process takes a little time during which so little light enters the eye that the threshold value for stimulating the rods (which are sensitive to light of low intensity) is not reached and so nothing can be seen. As the pupil dilates fully the threshold value for rods is reached and objects become visible. The rods, however, do not respond to light of various wavelengths in the same way that the three types of cone cell do. For this reason objects in the dimly-lit room are visible only in black and white (b) (ii).

In answer to part (c) candidates need to make reference to the three types of cone cell, each sensitive to light of a different wavelength. They respond to red, green and blue light, respectively. Through differential stimulation of these three types the brain can interpret the image of the retina in the whole range of the colour spectrum.

For (d) candidates should explain that the ultraviolet light comprises a range of wavelengths in much the same way that visible light does. The retinal cells of humans cannot distinguish these wavelengths and the light from the petals stimulates the red and green cones, giving the appearance of yellow. In bees the ommatidia of their compound eyes can distinguish different wavelengths of ultraviolet light. The petals must have a pattern of pigments that reflects the ultraviolet light differently. The different wavelengths reflected produce a pattern when perceived by the bees. In each species this pattern must vary, allowing the bees to distinguish the wavelengths.

8 One theory of colour vision suggests there are three different types of cone cell in the human retina, each containing a different variety of the colour-sensitive pigment iodopsin. There are three varieties of iodopsin, one sensitive to red light, one to green and one to blue. The absorption of different wavelengths of light by the three types of cone is given below:

Wavelength (nm)	Amount of light absorbed as a percentage of maximum		
	Red cones	Green cones	Blue cones
660	5	0	0
600	75	15	0
570	100	45	0
550	85	85	0
530	60	100	10
500	35	75	30
460	0	20	75
430	0	0	100
400	0	0	30

(a) From the data, explain the following:
 (i) Light of wavelength 430 nm appears blue
 (ii) Light of wavelength 550 nm appears yellow
(iii) Light of wavelength 570 nm appears orange
(b) From your knowledge of the retina explain why small objects close together can be more easily distinguished by cones than by rods.
Mark allocation 8/100   Time allowed 7 minutes   *In the style of the Joint Matriculation Board*

This is a short structured question testing the ability to interpret a table of data and the understanding of the structure of the retina of the eye. Part (a) must be explained 'from the data'. To make clear that this has been done it is advisable to quote figures from the table. In (a) (i) for instance, looking along the row opposite 430 nm, shows no light absorbed by the red and green cones whereas the blue cones absorb their maximum amount of light (100%). Consequently only the blue cones are stimulated by light of 430 nm and send a nerve impulse to the brain which gives the sensation of blue light. For (a) (ii) the situation is not so simple. Light of wavelength 550 nm is absorbed by both red and green cones to the same extent (85% of their maximum absorption). The brain therefore receives impulses of equal frequency from red and green cones but none from the blue cones. Because red and green colours mixed in equal quantity produce yellow, the brain interprets the same frequency of impulse from red and green receptors as yellow. The argument in (a) (iii) is basically similar, except that the red cones have their maximum absorption (100%) at light wavelengths of 570 nm whereas the green cones have only about half their maximum absorption (45%) at this wavelength. The blue cones again absorb no light. The brain therefore interprets the impulses received as predominantly red with some green included. Such a mixture is interpreted as orange.

In part (b) the candidate is required to know the cellular detail of the retina. Of particular importance is the way in which each cone cell has its own separate neurone communicating with the brain whereas groups of rod cells share a neurone. For full marks candidates need to relate this to the ability to distinguish small objects close together. The light from such objects will enter the eye as two parallel beams very close together. If these stimulate adjacent cone cells two separate impulses will pass to the brain, which will interpret the impulses as two separate objects. If, however, two adjacent rod cells are stimulated the chances are that they will be connected to the same neurone. The simultaneous stimulation of these two rod cells therefore results in a single neurone sending impulses to the brain, which will interpret it as a single object, i.e. it cannot distinguish the two objects.

9 What is meant by a receptor? (3)
  Describe the general features common to all receptors. (5)
  Using the mammalian ear as an example, show how a receptor organ functions. (12)
  Mark allocation 20/100   Time allowed 30 minutes   In the style of the Oxford and Cambridge Board

This structured essay tests candidates' knowledge of receptors in general and the ear in particular. The mark allocation in the first part indicates that only a brief statement of what is meant by 'a receptor' is needed. A receptor is a cell or tissue which transforms chemical, radiant, electrical or mechanical energy into electrical energy in the form of an action potential in a neurone. In doing so it informs the CNS of changes in the internal and external environment of an organism.

In describing the features common to all receptors candidates must look for more than a repetition of the answer to the first part of the question. It is true they all convert some form of energy to an action potential but considerably more than this is required to obtain five marks. Another point that could be included is that they are each specialized to receive one type of energy only. In fact, in some circumstances a receptor may respond to another stimulus, e.g. heat may stimulate some mechano-receptors, but the point is sufficiently general to be included as a 'general feature'. All receptors when stimulated produce a generator potential which is caused by localized depolarization of the membrane of the receptor cell. The size of the generator potential is proportional to the size of the stimulus. All receptors have a threshold value. If the stimulus is not adequate to produce a generator potential above a fixed level, no action potential is produced in the axon leading from the receptor. The stimulus required just to produce such an action potential is called the threshold value. All receptors show adaptation, that is they gradually cease to create an action potential if continuously stimulated. They vary, however, in the speed with which they become adapted. Receptors are sensitive to stimuli of low intensity, although again variations between types occur. It may finally be worthwhile mentioning some 'general features' concerned with structure that are common to all receptors. In effect, receptor cells are like all other cells in possessing a nucleus, cytoplasm and cell membrane. Commonly one end of the cell is drawn out into a process that receives the stimulus.

In the final part candidates should not spend time on elaborate drawings of the ear. The question is specifically concerned with the functioning of the ear and while its structure is obviously related to its function it should not become the over-riding consideration. At most a simple diagrammatic re-presentation should be used. The ear functions both as an organ of hearing and of balance. It is possible to 'show how a receptor organ functions' by reference to either although it is probably better to refer largely to hearing because it illustrates more easily the functioning of a receptor. One approach would be to give a straightforward account of how sound energy is transformed into an action potential in the cochlear nerve, but with emphasis on the 'general features' common to all receptors described in the previous section. An outline of how the ear functions is given in the essential information and is summarized in the answer plan.

Answer plan

Receptor

1 Transforms various forms of energy into action potential.
2 Informs CNS of external and internal changes.

General features

1 Transforms energy to action potential
2 Specialized in structure and position
3 Creates generator potential
4 Has a threshold value of stimulation
5 Becomes adapted
6 Sensitive to low intensity stimulation
7 Has general cellular features but one end of cell is frequently extended.

Ear as receptor

1 Sound energy
2 Vibrates tympanic membrane
3 Ear ossicles amplify vibrations at the oval window
4 Vibrations transmitted across organ of Corti
5 Different sound frequencies take a different route
6 Sensory cells between basilar and tectorial membranes are stimulated
7 Action potential occurs in a branch of cochlear nerve.

11 Ecological Relationships

11.1 ECOLOGY

Assumed previous knowledge The concept of ecology as the study of interrelationships between living organisms.

The concept of the environment of an organism as its physical surroundings combined with the effects of other living organisms.

Underlying principles

The Law of Conservation of Energy Energy may be transformed from one form into another but is neither created nor destroyed.

Biotic factors and abiotic factors are interrelated. Biotic factors which influence an environment include feeding interrelationships, i.e. food chains, food webs, and biomass pyramids. Examples are

1 Grazing
2 Predator-prey balance
3 Parasitism
4 Symbiosis.

The effects of abiotic factors will vary according to the environment. Table 11.1 lists some of these.

Table 11.1 Abiotic factors

Physical	Chemical	Edaphic
Temperature	Oxygen	Chemical and physical
Light	pH	factors related to soil type
Wave action (marine)	Minerals	and important in terrestrial
Desiccation (littoral)		environments
Substrate		

Points of perspective

A knowledge of practical field techniques will be essential for an understanding of ecology.

The ability to use keys to identify the organisms in the environment studied is a vital part of ecology.

Essential information

1 **Biosphere:** the sum total of the living organisms on Earth and their environment.
2 **Biome:** an aggregation of similar ecosystems in a particular region of Earth. It could be a habitat or zone such as grassland, desert or ocean.
3 **Ecosystem:** a localized group of communities and their physical environment.
4 **Community:** a localized group of several populations of different species.
5 **Population:** a geographically localized group of individuals of the same species.

One way of investigating biotic factors which affect any environment is to classify the various organisms according to their methods of nutrition. The basic distinction is between autotrophs and heterotrophs.

1 Autotrophs: organisms able to manufacture organic foods from inorganic nutrients.
2 Heterotrophs: organisms unable to manufacture food and dependent on autotrophs for food.

The heterotrophs can be sub-divided according to the nature of their food.

1 Herbivores: feeding on plants.
2 Omnivores: feeding on plants and animals.
3 Carnivores: feeding on animals.

In addition to the predatory carnivores, the grazing herbivores, and the omnivores that do both, there is another group of heterotrophs which obtain food from living organisms. These are the parasites, which live either in or on their hosts.

Any animal or plant may act as a host to a number of individuals of one or more parasitic species. These, in turn, may be parasitized themselves. Organisms which are parasitic on parasites are called hyperparasites. In this way, special category food chains may be set up:

$$\text{Host} \longrightarrow \text{Parasite} \longrightarrow \text{Hyperparasite}$$

Among the organisms which feed on dead organisms, there are two major types:

1 Those that take in their food as solid masses which are digested in the gut. These are the scavengers or, when the food particles are very small, detritus feeders (detritivores).
2 Those that secrete digestive enzymes onto dead organisms or their wastes and absorb the products of digestion as nutrients. Examples are bacteria and fungi. They are called saprophytes. There are also parasitic fungi and bacteria.

In addition to breaking down the organic chemicals in the dead bodies of other organisms or their wastes, the digestive activity of saprophytes also produces inorganic materials such as water and minerals, which return to the abiotic environment. This results in putrefaction or decay.

Living organisms occupy a position in food chains as herbivores, carnivores or as producers. These chains, which always have plants as their starting point, are often called grazing chains. When an organism dies, it can act as a starting point of another series of food chains containing scavengers, detritus feeders and saprophytes. These chains are decay chains.

The saprophytes play a very important role in the circulation of energy and nutrients. By releasing energy from dead organisms and returning it to the abiotic environment so that it becomes available for the synthetic processes of plants, they form a vital community of organisms in an ecosystem. They are known as decomposers.

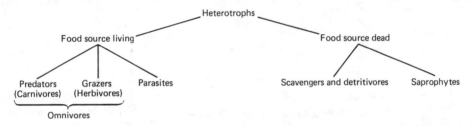

Fig. 11.1 Feeding relationships

Trophic levels Another way of categorizing organisms within an ecosystem emphasizes the role of communities. All food chains start with autotrophs, which make organic food by photo-synthesis, converting the energy from sunlight into the energy contained in food molecules. The heterotrophs are dependent on the plants for their supplies of energy-giving food molecules. For this reason, the green plants are called the producers. Next come the animals which depend directly or indirectly on the producers for their food. These are the consumers. The herbivores are primary consumers, the carnivores which feed on the herbivores, are the secondary consumers, and so on. Such groups are known as trophic levels.

Table 11.2 shows the results of a quantitative investigation of a grassland ecosystem in which the total numbers of organisms occupying each trophic level were estimated.

Table 11.2 Numbers in the trophic levels of a grassland ecosystem

Trophic level	Numbers of individuals/m^2
Producers	3100
Primary consumers	400
Secondary consumers	150
Tertiary consumers	0.02

The data can be represented in the form of a pyramid of numbers. This illustrates that the number of organisms decreases from one trophic level to another along a food chain.

When the producers are very large, relative to the primary consumers, the pyramid is inverted. One oak tree can provide food for hundreds of thousands of insects.

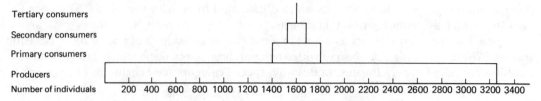

Fig. 11.2 Ecological pyramid of numbers

Energy relationships of trophic levels

Energy enters the trophic levels by photosynthesis. It is used for the synthesis of new plant material in growth, repair and reproduction. Ecologists say that it is used to increase the biomass of the trophic level. **Biomass** is the weight of living material per unit volume or area. Energy flows out of the trophic level as follows:

1 Heat loss
2 In structures synthesized by the plant but not contributing to an increase in biomass. For example:
   (a) Shedding leaves, bark, flowers, fruits.
   (b) Production of seeds for dispersal.
3 Death. When a plant dies, its remains contain some energy.
4 The plant may be eaten by a herbivore.

The remains of dead plants and the structures mentioned in 2 above, represent a loss of energy from the trophic level to the environment. However, this energy does not remain locked up in these structures. They may be broken down by the decomposer chain, so that the energy is used by saprophytes. Figure 11.3a shows energy flow through the plant community.

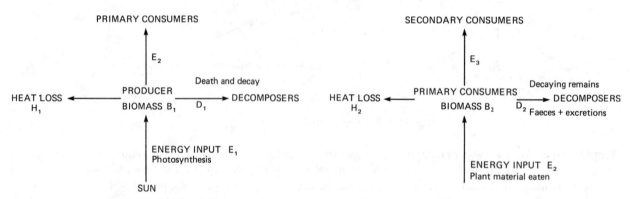

$E_1 = B_1 + D_1 + H_1 + E_2$ because of the Law of Conservation of Energy

Fig. 11.3 (a) Energy flow through plant community

Fig. 11.3 (b) Energy flow through primary consumer trophic level

It follows from this that the amount of energy available to the primary consumers (E_2) must be less than the amount originally fixed by the producers (E_1) because some is lost as heat (H_1), some is present in the producer biomass (B_1) and some occurs as non-living structures produced by the plants (D_1).

As a result, as there is less energy available for synthesis, the total biomass of the primary consumers must be smaller than that of the producers.

Energy flow through the consumer trophic levels can be considered in the same way. Energy flows into the primary consumer trophic level in the form of plant material eaten (E_2). Some of the energy contributes to the primary consumer biomass (B_2), some is lost to the environment as heat (H_2) and some passes to the decomposers (D_2) in the form of dead bodies, undigested material in faeces and excretory products. Finally, some passes on to the secondary consumer level as herbivores eaten by carnivores. Figure 11.3b summarizes the energy flow through the primary consumer trophic level.

In this way, a picture of energy flow through each trophic level and ultimately the whole ecosystem may be built up.

Ecological succession After a patch of ground has been cleared of vegetation, certain plants begin to establish a simple community. Gradually other species spread into the community where they often displace some of the original colonizers. It will be noticed that the changes take place rapidly at first, slowing gradually until dynamic equilibrium is reached and the community is stabilized. The sequence of change is called ecological succession and is characteristic of all ecosystems.

The first to become established are the **pioneers**: the equilibrium community at the end of the succession is the **climax**. During the succession, many environmental factors will change. Some change as a result of the influence of the community on the abiotic factors of the ecosystem; others may be the result of import or export of materials to or from the ecosystem. As the abiotic factors change, the community changes in response and succession proceeds under feedback control as shown in Figure 11.4a.

Progressive changes in communities during ecological succession can be illustrated with the model shown in Figure 11.4b.

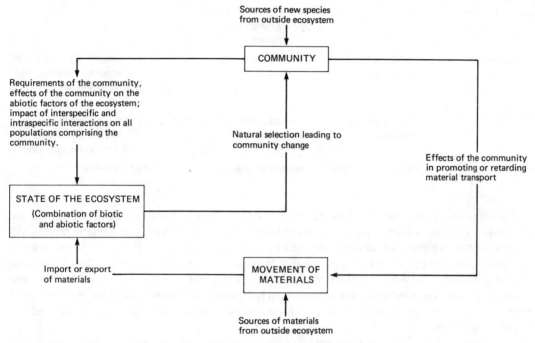

Fig. 11.4 (a) Feedback control of ecological succession

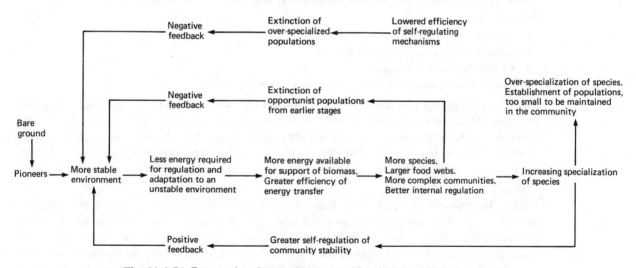

Fig. 11.4 (b) Progressive changes in communities during ecological succession

Communities Individual populations interact with one another as food supplies or as population regulators, e.g. a prey and its predators. However, populations also interact with other members of their communities in much more subtle ways which may have little direct effect in the regulations of the specific groups involved, but which are fundamental to the maintenance of the community as a whole, and hence to the existence of any species in the community. For example, a rabbit and nitrogen-fixing bacteria in a grassland ecosystem belong to different food chains and are regulated by different components of the abiotic and biotic environment; yet without nitrogen-fixing bacteria, nitrogen could not be cycled and rabbits could not exist. Interdependence among species goes far beyond the obvious interactions between one population and another. Populations exist, in fact, because they are parts of communities.

The effects of climatic, edaphic and biotic factors on distribution All organisms must maintain themselves in an external environment which could harm them because it differs from their internal environment. However, when the external environment is variable, then the problem is greater than it would be if the environment remained stable. Table 11.3 shows how the marine environment is least variable and the terrestrial environment most variable.

Table 11.3 Variability in environmental factors

Factor	*Marine*	*Freshwater*	*Terrestrial*
Light intensity	Remains constant	Remains constant	Remains constant
Oxygen and carbon dioxide	Remain constant	Remain almost constant at a concentration which is determined by local conditions	Remain constant
Temperature	Slight seasonal variations	Sizeable seasonal variations	Very variable. In some areas, the summer and winter temperatures differ by over 38°C
Mineral salts	Slight seasonal variations	Remain constant but there may be deficiencies	Remain constant for any one place, but the soil locally may be deficient in certain ions
Water	Remains constant	Remains constant	Some variation in the amount available

Distribution of organisms within the environments listed in Table 11.3 will depend on the physical and chemical factors indicated in the table. The edaphic (soil) and climatic factors will determine the presence or absence of various types of vegetation. Because carnivores eat herbivores, and herbivores eat plants, the distribution of plant species governs the distribution of animal species. The basic nutritional requirements of plants are carbon dioxide, water and minerals, so their distribution might be expected to be either uniform or random. Rainfall, light intensity, daylength and temperature are climatic factors which determine the pattern of the world's vegetation.

Vegetational zones are named after their climax vegetation – the natural vegetation which would grow if there were no human interference. The distribution of organisms therefore depends on:

1 Whether the species reaches the area
2 Suitable climatic and edaphic factors
3 An appropriate food source
4 Whether the species can survive competition or predation.

The ecological significance of soil The soil does not offer an inexhaustible supply of mineral ions to plants. Many essential elements may be in short supply within the community. During the millions of years of evolution, the soil and atmosphere would have become completely emptied of elements if it were not for recycling processes. The micro-organisms which use soil as a habitat are extremely important in recycling many elements, for example carbon and nitrogen.

The carbon cycle
Green plants absorb carbon dioxide from the atmosphere and utilize if for the manufacture of carbohydrates, proteins and fats. The plants are eaten by herbivores which digest and assimilate the foods originally synthesized by the plants. Carnivores continue the food chain, but eventually death ensures that all links of the chain are returned to the soil. No matter how long or complex the food chain/web, the carbon which was originally incorporated into the producers sooner or later forms part of dead organisms upon which saprophytes feed, or is returned to the atmosphere as carbon dioxide from respiration. The carbon in the dead organic matter becomes converted to carbon dioxide by putrefying bacteria and fungi. As a result of these processes, the concentration of carbon dioxide in the atmosphere remains fairly constant.

The nitrogen cycle
Soil-dwelling bacteria are some of the few organisms which can utilize atmospheric nitrogen. They are able to convert it to nitrates which can be absorbed by plants and used during protein synthesis. Herbivorous animals acquire their nitrogen from plants, whereas carnivores rely on animal protein for their nitrogen supply. The removal of nitrate from the soil is balanced by its return as a result of bacterial activity. The circulation of nitrogen within the community is

similar to that of carbon. Nitrogen is passed from the plants to herbivores and along a food chain. Saprophytes eventually decompose the tissues of all the organisms concerned and so liberate the nitrogen once more. They liberate it as ammonia which reacts with carbon dioxide and water present in the soil spaces. At this point in the cycle, two species of bacteria present in all soils play a vital part. One, called *Nitrosomonas,* is able to derive all the energy it needs from that released when ammonium carbonate is oxidized.

$$(NH_4)_2CO_3 + 3O_2 \longrightarrow 2HNO_2 + CO_2 + 3H_2O + ATP$$

The nitrous acid reacts with salts in the soil to produce nitrites. These are acted upon by *Nitrobacter* which derives its energy by oxidizing nitrites to nitrates.

Another method of nitrate production is via a species of bacterium called *Rhizobium leguminosarum,* which lives in swellings on the roots of plants belonging to the family Leguminosae. They can incorporate atmospheric nitrogen into their protoplasm. The host plants absorb some nitrogenous molecules from the bacteria. Furthermore, the whole community eventually benefits from their activity because, when the host plant dies and is decomposed, a greater amount of nitrogen is released in the soil than would otherwise be the case.

Link topics

Section 6.1 Autotrophic nutrition (photosynthesis)
Section 6.3 Heterotrophic nutrition (saprotrophs and parasites)

Suggested further reading

Barnes, R. S. K., *Estuarine Biology,* Studies in Biology No. 49, 2nd ed. (Arnold 1984)
Brafield, A. E., *Life in Sandy Shores,* Studies in Biology No. 89 (Arnold 1978)
Brehaut, R. N., *Ecology of Rocky Shores* Studies in Biology No. 139 (Arnold 1982)
Edwards, P. J. and Wratten, S. D., *Ecology of Insect-Plant Interactions,* Studies in Biology No. 121 (Arnold 1980)
Etherington, J. R., *Wetland Ecology,* Studies in Biology No. 154 (Arnold 1983)
Hassell, M. P., *The Dynamics of Competition and Predation,* Studies in Biology No. 72 (Arnold 1976)
Phillipson, J., *Ecological Energetics,* Studies in Biology No. 1 (Arnold 1966)
Solomon, M. E., *Population Dynamics,* Studies in Biology No. 18, 2nd ed. (Arnold 1976)
Townsend, C., *The Ecology of Streams and Rivers,* Studies in Biology No. 122 (Arnold 1980)

QUESTION ANALYSIS

1 The land snail, *Cepaea,* is common in a variety of habitats, for example woodland, hedgerow and rough grassland.

The shell of the snail shows variation in colour, for example yellow, pink or brown. The shell may also be banded, that is striped with a number of darker bands.

The banding on the shell and its colour are both genetically determined.

Most of the snails living in a predominantly green habitat are yellow, whereas most of those living in an environment such as decaying leaves are pink and brown. Most of the snails living in a habitat with a uniform background have unbanded shells, whereas most of those in a habitat with a more broken background have banded shells.

Living snails were collected from five sites in each of two habitats, A and B. The percentages of yellow shells and of banded shells obtained from each of the sites in the two habitats are given in the table below.

	% Yellow	% Banded
	72	100
	68	95
Habitat A	75	90
	78	75
	84	65
	10	40
	12	38
Habitat B	5	20
	8	15
	20	0

(a) Comment on these results and suggest the likely nature of the two habitats A and B. (4)
(b) Birds such as thrushes prey on snails by cracking open the shells on a stone. Broken shells were collected over a period of time from around such a stone situated in an area of rough grassland and most of them were found to be pink and unbanded. However, most of the living snails in the same area were found to have yellow and banded shells. Suggest an explanation for this finding. (4)
(c) In an area of grassland which is heavily grazed, snails with pink shells and snails with yellow shells

were found in approximately equal numbers. Suggest a reason for this.   (2)
Mark allocation 10/200   Time allowed 8 minutes   *London Jan 1987, Paper 1, No. 13*

This data analysis question requires the ability to comprehend and interpret written and numerical information. While some application of general biological principles is needed, the answers are mostly contained within the information provided, so read it carefully. For part (a) analysis of the table shows that yellow banded snails predominate in habitat A. From the written information it can be seen that this suggests habitat A is mainly green, and broken in background, e.g. rough grassland. Similarly the pink and brown unbanded snails found in habitat B suggest this is a woodland.

To answer part (b) the candidate must realize that pink, unbanded snails are not camouflaged in rough grassland. Being more conspicuous they are more easily seen by thrushes than the yellow banded snails. The pink unbanded snails are therefore more likely to be eaten by these birds. This leaves a preponderance of yellow banded snails who, due to their selective advantage, gradually become the common variety.

The reason required in part (c) is that grazing results in shorter vegetation and consequently reduced cover for the snails. The yellow banded ones thus become as conspicuous as the pink unbanded ones and both types are predated upon equally, resulting in approximately equal proportions of each.

2 (a) Describe what is meant by the following terms:
  (i) community (ii) ecosystem (iii) food chains.
  (b) Evaluate the use of studying food webs, rather than food chains in ecology.
  (c) Consider the trophic levels of a pyramid of numbers and illustrate how energy is lost in passing through the levels.
  (d) Briefly describe three ways in which nitrogen is incorporated into a food chain.
  Mark allocation 20/100   Time allowed 30 minutes   *In the style of the Oxford and Cambridge Board*

This is a structured question. The answer requires applying basic knowledge of ecological principles in a critical way.

The descriptions in (a) should be as concise as possible. Carefully worded definitions of the terms would be adequate to gain full marks for this part of the question.

 (i) A community is a localized group of several populations of different species interacting with one another and the physical/chemical factors of the environment.
 (ii) An ecosystem is a localized group of communities and their physical environment.
 (iii) A food chain is a series of stages through which energy passes, always beginning with chemical energy incorporated in plant tissues.

The answer to (b) will require an assessment of the use of food webs rather than food chains to illustrate feeding patterns in a community. One could say that it is more useful to identify and construct food chains within a food web to reduce the complexity of representing theoretical feeding relationships. However, there are occasions when food webs are more useful, i.e. they give a more accurate picture of the natural feeding relationships. It is extremely rare to have linear food chains because few animals confine themselves to a single type of food and few plants serve as the only food for one type of herbivore. The answer should be augmented by referring to specific examples, preferably those studied during your field work.

Considering trophic levels (c), it must be remembered that by no means all of the stored chemical energy present in plant tissues is appropriated by herbivores. Indeed the proportion seems surprisingly low. It has been estimated that in some cases, as little as 0.1% of the chemical energy stored in plants is ever converted into available energy in the body of a herbivore. A similar wastage occurs between subsequent stages of the food chain. The size of the animal population within a community is related to the productivity of the plant population. All the organisms within a food chain will lose energy as heat during respiration, just as they will lose heat in the form of faeces and excretory matter and through decay when they die.

The ways in which nitrogen enters food chains are as follows (d):

1 With the aid of *Nitrosomonas*, a species of bacterium which lives in soils, ammonium compounds can be oxidized to nitrites which in turn are oxidized to nitrates by another bacterium called *Nitrobacter*. The nitrate ions are then absorbed by plants and their nitrogen is incorporated into plant protein which enters the food chain when the plants are eaten. The starting point for the sequence has been ammonium compounds in the soil. It is important to remember that these have been formed as a result of putrefaction of nitrogenous excretory products such as urea, which has been produced by animals.

2 Electrical and photochemical fixation of atmospheric nitrogen results in the formation of nitrogen oxides which then form nitrates and can be absorbed by plants.

3 Nitrogen-fixing bacteria, *Rhizobium*, in the root nodules of leguminous plants incorporate atmospheric nitrogen into their protoplasm. The host plants then absorb some nitrogenous molecules from the bacteria and incorporate them into their protein.

3 The figures on page 241 represent pyramids of biomass and number of organisms in the same ecosystem, each of which is not drawn to scale.

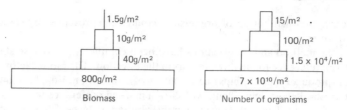

Fig. 11.6 Pyramids of biomass and numbers

(a) Explain why the relationships between the various trophic levels are different when comparing the two figures.

(b) What further kind of pyramid could be constructed to give additional information about the four trophic levels?

(c) Why are there seldom more than four trophic levels in each pyramid?

Mark allocation 5/100   Time allowed 6 minutes

Associated Examining Board November 1980, Paper I, No. 12

Besides testing the ability to recall basic ecological principles, this question assesses skills of application and deduction.

(a) The trophic levels of the biomass pyramid give no indication of numbers of individuals, i.e. a single plant could be the producer and a few thousand insects could be in the first consumer level. The pyramid of numbers gives no indication of biomass at each level.

(b) A further kind of pyramid could be an energy pyramid, giving the number of kilojoules at each trophic level. Such a pyramid would give an idea of energy flow.

(c) The answer to this depends on the Law of Conservation of Energy, i.e. at each trophic level, there is less available energy than in the previous one. There is so much energy loss, that the ecosystem rarely can support more than four trophic levels.

4 (a) What is meant by a soil?   (3)

(b) Explain why a loam soil is a fertile soil.   (8)

(c) What are the effects of adding to a soil   (i) animal manure   (ii) lime?   (9)

Mark allocation 20/100   Time allowed 35 minutes   London Board June 1980, Paper II, No. 3

The question is divided into three parts, all of which test recall.

Part (a) requires a simple statement of definition of a term but note that there are few marks for this part. The points to remember are that:

1 Soil is the upper layer of the earth's crust in which plants grow.

2 Soil is a mixture of organic and inorganic materials which forms an ecosystem for micro- and macro-organisms and provides minerals for plant growth.

Part (b), which carries a large proportion of marks, should be answered in continuous prose. It depends on knowledge of loam and fertile soil, but these terms must be related in the answer and not considered in isolation. The main points to remember are:

1 Loam is a mixture of sand, clay and humus.

2 The proportions of sand and clay are such that they combine the necessary water-holding properties, without reducing aeration, and they also prevent water-logging.

3 The water acts as a solvent for mineral salts and the air provides oxygen.

4 The humus consists of the remains of dead organisms. It acts as the base of a decay chain. Saprophytes break down detritus to minerals which can be used by plants.

5 A fertile soil is one that supplies all the factors that plants need for full growth.

A description of the role of putrefying bacteria in the decay of animal manure is required in (c) (i). Briefly, the manure provides nitrogenous material which is converted to ammonium compounds. These are acted on by nitrifying bacteria to form nitrite, which in turn, forms nitrate by the action of another group of nitrifying bacteria. The point of adding manure is to provide an indirect source of nitrates for plant growth.

Lime is added to soil to reduce its acidity and to aid flocculation (ii). The former condition results from poor drainage and the inability for complete breakdown of organic material. The latter condition means that the soil requires better drainage and aeration. Flocculation makes the clay particles clump together in larger crumbs which will have larger air spaces between them.

Probably, most of the nine marks allocated for part (c) will be given to part (i) because this will require the most detailed treatment of the nitrogen cycle.

5 With reference to a habitat you have studied give an account of the interrelationships between the biotic, edaphic and climatic factors and show how these affect the distribution of the organisms.

Mark allocation 20/100   Time allowed 35 minutes   In the style of the London Board

This is a question requiring recall of ecological principles in relation to practical field work techniques.

Your answer will depend on the actual habitat which you have studied and should include an account of a quantitative method used to record distribution, e.g. a belt transect, line transect, the use of a quadrat.

After giving a detailed account of your method of recording distribution, using illustrations where appropriate, the distribution of organisms should be represented by some graphical means, e.g. a histogram or map.

Once the method and observations have been completed, a concise statement as a conclusion is needed with reference to the three factors in the question. The biotic factors may show predator–prey or grazer–producer relationships on the pattern of distribution. Edaphic (soil) factors will be important in any terrestrial ecosystem. Plants will show a marked preference for certain types of soil and a non-tolerance of others. This will show in their distribution patterns. Climatic factors, which affect the soil, vegetation, and animals to be found in any habitat, are foremost in determining the climax of any distribution pattern in a habitat. It is much easier to illustrate the effect of climate on the distribution of organisms within a biome than in a smaller unit, but the amount of incident light falling on plants, and the accumulation of rain water in a poorly drained area could be considered as effects of climate. The climatic factor may be used to 'tie' the whole conclusion together because the other two factors are dependent on it.

11.2   Man and his environment

Assumed previous knowledge Ecological principles. Community structure. Succession. Genetic resistance to pollutants.

Underlying principles

Primitive man affected his environment by hunting, fishing and removing trees for fires and shelters. He was nomadic and therefore any effect he had on an area was temporary and the environment had adequate time to recover. About 11 000 years ago man began to cultivate his own crops and thus communities settled in one place in order to harvest and store the crops. The use of tools and later domestication of animals led to more efficient agriculture, the ability to produce more food and hence support a larger population. Demands were created on the environment and trees were felled to provide shelters and more land for cultivation. The cultivation of crops was aided by domesticated animals. The energy therefore came from the food they ate and this limited its efficiency. With the advent of fossil fuels all this changed and little of the energy in the crops grown went back into growing the crops. Instead machines driven by fuels helped cultivate the crops, the majority of which were available to support a larger population which in turn required more efficient agriculture to support it. Fertilizers, pesticides and rapid transportation all developed and brought with them pollution problems in addition to those caused by burning the fossil fuels. Such fuels are a finite source of energy and hence there was pressure to develop new energy sources. Nuclear energy was developed with its consequent environmental problems.

Points of perspective

Answers to questions on this topic are all too often vague and lacking in specific detail. High marks will be obtained by a candidate who can give:

1 The precise names of pollutants
2 Figures concerning lethal doses or the quantities of pollutants produced
3 The exact nature of the effect of a pollutant
4 A wide range of examples.

It would be of benefit to have some knowledge of the effects of pre-industrial man on the environment. A good candidate will understand the principles and methods of conservation.

Essential information

Man's effects on the environment are largely a result of two factors:

1 The need to produce human food supplies
2 The population explosion.

Human food supplies

Agriculture

Probably the first attempts at cultivation began with the deliberate use of fire to clear large areas of land for planting; by doing so, man began to alter the structure of communities. Climax vegetation was destroyed and replaced by monoculture, e.g. a field of cereal crops replaced a forest of oak. It was soon discovered that if the same crop was grown for several successive seasons, the gross yield fell because crops have specific demands on the soil's resources. The problem could be avoided by crop rotation or by replacing lost materials with fertilizers. The latter solution raises another problem because excess use of fertilizers often causes pollution of rivers due to leaching and draining.

Ecologically, agriculture is contrary to the natural development of communities. Mono-culture results in large areas planted with a single species of plant instead of the climax community of many species. As a result, weeds and pests compete with the single species for environmental resources. When man attacks the competitors with herbicides and pesticides, he creates additional problems of pollution of soil and water by toxic chemicals. Agriculture provides enormous food resources for those herbivores which share the same food as man. With many plants of one species grown close together, there is no problem of individuals finding new food plants at the dispersal stage. Thus, large numbers of herbivores can build up at the expense of crops. Whole ecosystems are therefore affected by the change in the original climax community.

The population explosion

There has been a rapid increase in human population since the seventeenth century and this has inevitably had its effect on the environment. Ancient agricultural races have been replaced in certain parts of the world by populations founded on the industrial and technological revolutions. The results have been an increase in the use of non-replaceable resources, e.g. fossil fuels, and an increase in pollution of land, sea and air by products of fuel combustion. Solutions to these problems rely on co-operation between ecologists, politicians and economists. A cost-effective method of reducing pollution is usually not easy to find and conservationists must argue their case against industry and other cost-conscious sections of the community. There is no dispute about the scientific aspects of the ecological situations involved.

An increase in population imposes greater strains on food supply and its distribution. There are basically two contrasting points of view relating to the food supply problem.

1 Many agriculturalists believe that, with the use of fertilizers, herbicides, and pesticides, the world can produce enough food for the predicted growth rate of population.
2 In contrast, many ecologists disagree and argue that, already at least half the world's population are under-nourished and they suggest that it will be impossible to feed adequately the probable population forecast for the year 2000.

There are three main methods of improving this situation:

1 To stabilize the human population.
2 To increase the efficiency of food production and utilization.
3 To develop new food sources.

At present we lose a great deal of food during its transport from the source of production to the consumer. This is due to competition between man and other organisms, e.g. rats, mice, insects and moulds. Improvement in techniques for combating these is feasible.

Some food is lost for economic reasons because of high transport costs from regions of over-production to regions of need. There are instances of countries burning surplus food or allowing it to rot while third world countries have populations suffering from malnutrition and starvation.

It is very easy to confuse biological and political problems in a discussion concerning human food supplies. The biological problems relate to the efficiency with which different types of food can be produced in different environments. Since energy is lost at each link of the food chain, the most efficient way for an omnivore like man to tap solar energy is to eat plants. The highest primary production comes from tropical crops, e.g. sugar-cane which can grow all the year round but most of the plant's body is not eaten by man. Usually only parts, such as fruit, seeds, leaves or roots are used so that much primary production is not used.

Plant proteins differ from animal proteins in the proportions of amino acids. Herbivores have to eat large quantities of food to obtain certain essential amino acids. Eating animals is economical in terms of the bulk of food consumed although it is wasteful of solar energy. Of the present supply of protein for human food 70% comes from vegetation and 30% from animals. Intensive livestock rearing is the most efficient way of producing those animal proteins which are traditional foods. The mass production of chicken and veal relies on processed diets which incorporate parts of organisms normally discarded as inedible. Occasionally antibiotics and hormones are added to the diets with subsequent dangerous side effects to the secondary consumers.

Ideally, protein should form about 10% of the total food intake in terms of energy; this constitutes the main difference between diets of well-fed and under-fed races of the world. An adequate diet should contain about 44 g per day protein. In the USA, the average consumption is about 64 g per day, whereas some third world countries have people surviving on 9 g per day.

In 1962, there were 0.5 hectares of arable land per person and it is estimated that the figure will be 0.3 hectares per person by 2000 AD. This implies that for agriculturalists merely to maintain the present levels of nutrition would mean doubling the productivity of arable land. It is possible that marine production could be doubled and terrestrial protein production increased by about 50%.

Other possible sources of food could depend on advances in biotechnology. Certain bacteria can be used to convert petroleum into protein and this so-called single-cell protein production could be an answer to the world's food shortage. However, these sources can only be exploited through energy-consuming industrial processes and the amount of energy needed to produce edible protein may be so great that the source is totally uneconomic. Nuclear energy or the use of solar energy will probably be essential if a large human population is to survive in the future, despite the potential harmful effects of the disposal of nuclear waste.

Pollution Pollution is a difficult term to define. It is derived from the Latin word *polluere* which means 'to contaminate any feature of the environment'. It may be broadly said to be 'adding to the environment a potentially hazardous substance or source of energy at a rate faster than the environment can accommodate it'. In this broad sense the definition includes not only man's activities but also certain natural processes.

Air pollution

This form of pollution is largely a result of burning fossil fuels such as coal, coke, or fuel oil. Smoke contains carbon particles, carbon dioxide, sulphur dioxide and fluoride. Domestic combustion is of importance in those areas which have not introduced smokeless fuels. Smoke is a major health hazard as it affects the respiratory tract, increasing the incidence of bronchitis, and may even be a carcinogen. Normally, the cells lining the bronchial tubes produce mucus to moisten the surface and trap bacteria. When cells are diseased, they cannot function properly and are susceptible to bacteria. Whenever fog builds up in air heavily polluted with smoke, a condition called smog is produced. The dampness, combined with the effects of suspended carbon in smoke, increases the incidence of pulmonary disease.

The problem of reducing industrial smoke is complex; it has been tackled in a number of ways. In 1956, the British government introduced the Clean Air Act which made it illegal to discharge smoke in many areas. This meant that households and factories had to use smokeless fuels, e.g. electricity, gas, oil or solid smokeless coals. The introduction of electric, diesel, and gas turbine locomotives also reduced air pollution from the level it was when steam engines were in use. Modern technology allows fuels to be burnt more efficiently, and washing processes can be installed to remove soluble and heavy particles before the gases are sucked up the chimney where sieves or electrostatic precipitators are also installed.

To allow efficient combustion of petrol, lead 'anti-knock' agents are added and these lead compounds are released into the atmosphere along with carbon monoxide and various nitrogen oxides. The lead is easily absorbed through the lungs and may cause mental retardation in children. Non-lead 'anti-knock' agents are available but are more expensive.

When fossil fuels are burned sulphur dioxide is discharged into the atmosphere. Coal and coke contain about 1–2% sulphur. Much of this can be removed but the cost is high. Control of this form of pollution depends on high chimneys to carry the fumes away. In some areas, problems have arisen when it was washed out from the waste gases by rain as sulphuric acid. This corrodes buildings. Plants are particularly vulnerable to sulphur dioxide pollution because the gas is absorbed through stomata and is absorbed into the leaf cells in lethal doses.

Fluoride is a waste-product of certain industrial processes involved in the manufacture of pottery, bricks, steel and aluminium. It is absorbed by plants and becomes concentrated in leaves. Cattle and sheep grazing on the plants become affected over a period of time. Bones become soft and joints stiffen. Methods used to get rid of fluoride depend on building higher chimneys. Other methods have proved very expensive.

Aquatic pollution

Freshwater pollution Man soon realized that rivers are the easiest and cheapest means of transport. Towns grew alongside them, and the simplest way to get rid of unwanted material, including sewage, was to put it into the river. Rivers became the first sewers and, before the population explosion of the industrialized communities, were efficient as such and still remained healthy. However, when the bulk of the waste became too much, bacteria thrived and began to deplete the oxygen from the rivers causing a **biological oxygen demand (BOD)**. In still waters the effects have been marked with many lakes in the world 'dying' as a result. Mass epidemics have spread via water systems because of pollution by pathogens. It was not until the 1800s that sewage works were constructed to treat the waste before discharge. Toxic wastes from industrial sources are numerous and varied. The main danger is from organic chemical and fertilizer manufacturers. Heavy metals act as non-competitive inhibitors of enzymes and build up via food chains once they enter the environment. Pesticides, herbicides and fungicides drain away from fields, enter waterways, and are concentrated through food chains so that top carnivores become poisoned. Excessive use of fertilizers can also give rise to a build-up of toxic by-products by leaching and draining. **Thermal pollution** becomes a problem where large quantities of heated water are discharged, e.g. from power stations. The consequent rise in temperature not only kills organisms directly, but reduces the solubility of oxygen in water, thus killing aerobic organisms. Anaerobes can thrive in these conditions and compete

Table 11.4 Summary of major pollutants

Medium affected	Pollutant			
	Natural	*Man-made*		
		Domestic fires	*Internal combustion engine*	*Industrial wastes*
Atmospheric	Sulphur dioxide in volcanic smoke Radioactivity due to cosmic rays	Grit Smoke (carbon particles) Tars Carbon dioxide Heat (thermal)	Carbon monoxide Nitrogen oxides Ozone Lead compounds Various organic compounds Noise	Sulphur dioxide Hydrogen sulphide Ammonia Fluorine Sulphuric acid Heavy metals, e.g. zinc, copper Radioactive fallout from bomb tests and nuclear power stations Noise
Aquatic	Eutrophication by leaching of minerals Putrefaction in small ponds due to leaves Silting in lakes and rivers Leaching of toxic ions, e.g. copper, iron	Inert, e.g. washings from mines and quarries Putrescible, e.g. domestic sewage Toxic, e.g. copper, zinc, lead, mercury from industrial processes Tainting, e.g. dyes Eutrophication, e.g. nitrates from fertilizers, phosphates from detergents Thermal, e.g. cooling water from power stations Radioactive, e.g. nuclear power station wastes Oil, e.g. washings from oil tankers, accidents at sea		
Terrestrial	Local accumulation of toxic minerals, e.g. ferrous salts Radioactive minerals	Dumped wastes, e.g. slag heaps, spoil heaps, domestic rubbish Derelict buildings and land, e.g. disused houses, factories and mines, old railway lines and canals Pesticides (overuse, drift and persistence), e.g. inorganic, di-nitro, organophosphorus, mercuric and other compounds, chlorinated hydrocarbons, hormone weed killers Radioactivity, e.g. wastes from nuclear power stations		

successfully with other forms of life for environmental resources. Detergents are rich in phosphates and these provide nutrients for the vegetation in rivers and lakes. Toxic by-products from algal blooms result in the death of many organisms

The prevention of these forms of pollution depends on treating waste-products at the source so as to render them harmless. Cost seems to be the major problem because the expense of treatment must be met by increasing the cost of the end-product manufactured or by reduction of profit margins.

Marine pollution
Oil pollution of the sea is one of the most emotive forms of pollution probably because its effects are direct and easily visible. Illegal washing of oil tanks at sea, together with accidental loss of oil by collision, wrecks, and oil rigs provide the biosphere with one of the most unsightly forms of pollution. The oil directly affects vertebrates which come in contact with it, e.g. sea birds, seals and fish, with the physical effect of preventing gaseous exchange. Planktonic invertebrates are also poisoned and so whole food chains become affected.

Methods of treating oil pollution have had limited success. They include absorbent methods involving covering oil with an absorbent material such as straw and collecting the oil-soaked material. Another approach is to spray the oil with a solution of plastic. The plastic hardens to form a coating over the oil. Plastic foam and sawdust have also been used but the problems with these methods include cost effectiveness and the difficulty of spreading the absorber evenly and then collecting it at sea. Alternatively, heavy materials may be added to the oil causing it to sink. Bacteria break down the oil quite rapidly when it is on the sea bed. Sand, treated chemically to make oil cling to the grains, has been used to sink oil and because very small amounts of chemicals are needed to treat the sand, it is cheap and safe to marine life. Calcium sulphate and pulverized fuel ash, when treated with silicone, are other effective materials for sinking oil. Detergents have been used to reduce the surface tension of oil but unfortunately may damage wildlife when ingested.

Radioactive pollution

One of the unavoidable consequences of the use of nuclear energy is the production of radioactive waste. Highly reactive, long-lived wastes are usually concentrated, sealed into lead containers, and dumped in the deep sea. Wastes with a lower level of activity are dispersed by the sea, which dilutes them to acceptably safe levels. The amounts which can be discharged are strictly controlled. One of the most dangerous effects of radiation is the damage it causes to human chromosomes. Mutations often result. These are often passed from one generation to the next before they exert their often damaging effects.

In considering the dangers posed by a radioactive substance a number of factors need to be taken into account. These include the half-life of the substance, its possible concentration in food chains and its role in an organism. ^{90}Strontium for instance is particularly hazardous to humans:

1 It is produced by nuclear explosions and is therefore constantly produced.
2 It has a half-life of 28 years and therefore persists.
3 It is readily absorbed by grasses and concentrated in cows' milk which is consumed in large quantities by humans, especially babies who are particularly vulnerable because their cells are rapidly dividing.
4 It becomes concentrated in human bones, which contain the bone marrow which is rapidly dividing to produce blood cells. This division may therefore become disturbed and leukaemia can result.

A radioactive substance released at a relatively harmless level may become accumulated to dangerously high levels along a food chain. The levels of the phosphorous isotope ^{32}P in a North American river were found to be:

1 In the water – 1 (arbitrary unit)
2 In phytoplankton – 1000
3 In aquatic insects – 500
4 In fish – 10
5 In ducks – 7500
6 In duck eggs – 200 000 (shells contain calcium phosphate)

Here again the main danger is that the highest levels are found next to the developing embryo which comprises rapidly dividing cells.

Terrestrial pollution

There are two distinct types of terrestrial pollution: waste materials that are dumped and pesticides. Waste heaps, derelict mines and buildings present the following problems:

1 They are unsightly.
2 They waste possibly useful land.
3 They may create a danger to life, e.g. children trapped in mines, the Aberfan slag heap disaster of 1966.
4 They may be a health hazard by attracting vermin.
5 They cause urban decline, causing people to move away which creates social decay.

Pesticides are numerous. It is essential that they are properly used. Often however they 'drift' from their targets or are leached into streams, rivers and lakes. A good pesticide should:
1 be specific only to the pest it is directed at
2 be rapidly broken down and not persist in the natural environment
3 not be accumulated through food chains.

Link topic

Section 11.1 Ecology

Suggested further reading

Arvill, R., *Man and Environment,* 3rd ed. (Penguin 1973)
Bradshaw, A. D. and McNeilly, T., *Evolution and Pollution,* Studies in Biology No. 130 (Arnold 1981)
Cornwell, A., *Man and the Environment* (Cambridge University Press 1984)
Edwards, R. W., *Pollution,* Oxford/Carolina Biology Reader No. 31 (Packard 1972)
Mellanby, K., *The Biology of Pollution,* Studies in Biology No. 38, 2nd ed. (Arnold 1980)
Ottaway, J. H., *The Biochemistry of Pollution,* Studies in Biology No. 123 (Arnold 1980)
Royal Commission on Environmental Pollution, 1st report (HMSO 1971)
Spellerberg, I. F., *Ecological Evaluation for Conservation,* Studies in Biology No. 133 (Arnold 1980)

6 Discuss the impact of man on the environment, with reference to the use of pesticides, artificial fertilizers, overfishing and the disposal of radioactive materials.

Mark allocation 20/100   Time allowance 40 minutes   *London Board 1983, Paper II, No. 8*

Following a short general introduction, each of the four aspects should be dealt with separately (and in isolation). Take care not to get carried away or become emotional about the subject. Try to be objective and scientific in your answers by relating the facts in a clear and logical manner. Distribute your time equally between the four aspects, spending ten minutes on each. If possible always give specific examples of substances and organisms. The answer plan below gives some ideas for an answer but is rather general because the actual examples used in an answer will depend upon which ones have been studied by the candidate. Where specific groups or individual organisms are mentioned these represent well-known examples.

Answer plan

Use of pesticides – used to remove unwanted organisms – examples of pests – to increase crop yields or remove vectors of human diseases – often not specific and may kill beneficial organisms, thus disrupting food webs – may be concentrated along food chain, killing animals at the top of the chain – may affect products of animals e.g. produce thin shells in birds' eggs – may be slow to break down, and may have long-term effects – can be washed off, or blown, and so affect other areas – over-use leads to resistance in pest.

Artificial fertilizers

Used to increase yield of crops – may be substituted for organic fertilizers, and so affect soil texture – can run off into rivers, causing eutrophication and algal blooms (algae die and decay, leading to oxygen deficiency) – may be harmful to plants other than those they are aimed at – may destroy beneficial soil micro-organisms.

Overfishing

Reduces fish populations – affects fish distribution – disrupts food webs due to increase in plankton following reduction of fish predators – other animals feeding on plankton may thus increase in numbers and the populations of their predators in turn increase – fish predators e.g. seals decrease in number – may ultimately cause extinction of fish species.

Disposal of radioactive materials

Depends on half life, but most have very long-term effects – no absolutely safe means of disposal – burial in old mines may lead to seepage – storage in holes in rocks is only safe until geological movements occur – dumping at sea in containers still allows slow escape – direct contamination may cause ionisation of chemicals in cells and hence death – radioactivity may be accumulated along food chains (e.g. ^{32}P) – may be concentrated in specific parts of an organism (e.g. ^{90}St).

7 Write a brief essay on the influence which man has on natural habitats.

Mark allocation 25/60   Time allowed 45 minutes

Southern Universities Joint Board June 1979, Paper II, No. 13

There are several ways of approaching this essay but both positive and negative aspects of the effects of man on the environment should be emphasized. The essay should begin with an introduction in which ecology and conservation should be put into historical perspective relating the increase in population which has occurred over the centuries to improvements in agriculture and its environmental consequences. The long-range effects of deforestation, pesticides and pollution could provide material for detailed consideration. The essay could end with an account of the ways in which man is taking positive steps to redress the balance of his past actions and how he is attempting to conserve natural habitats. The following points should provide an outline for the essay:

1 Populations fluctuate in a normal stable environment, but if man interferes with nature, they may rise rapidly and then crash. Specific examples should be given to illustrate this principle.

2 The enormous increase of human population during the last 500 years can be related to agricultural advances, with the consequent use of land normally considered as natural habitat. However, the growth of population of some countries has fallen in recent years and this could be seen as an opportunity to conserve remaining natural habitats in these countries.

3 The increase in population has affected natural habitats because of deforestation and over-grazing of livestock leading to soil erosion. Inefficiencies of previous agricultural practice in various parts of the world have ruined the landscape, and the introduction of alien species which have become feral has had profound influences on the native fauna and flora.

4 Mechanization of agriculture, including heavy machinery and the cultivation of large areas of a single crop, has introduced the problems of weeds and pests. The herbicides and pesticides used to combat these have often had side effects on natural habitats. Perhaps the best way of controlling pests is by biological control, but the agents of control may sometimes become pests themselves. Examples of successful biological control have been the introduction of the cactus moth and the myxomatosis virus into Australia. However, greatest success is often achieved when a biological agent is combined with a chemical.

5 The misuse of pesticides is an example of pollution. Pollution of air, water and land has affected natural habitats and should be critically examined.

6 The essay should conclude with suggestions for the future development of ideas related to conservation of natural habitats. We must control the growth of our population, try to produce food more efficiently without destroying natural ecosystems, and ensure that our wildlife resources are used as wisely as possible.

8 Indicate those activities of man that have resulted in air and water pollution and discuss some of the measures which may be used to prevent further pollution.
Mark allocation 20/100   Time allowed 35 minutes   London Board June 1981, Paper IIB, No. 10

This question requires description of the negative and positive effects of man on the environment. In all questions of this type the overriding need is to give precise facts supported by examples.

The 'activities' referred to are the increase in population and the industrialization of the world. It is advisable to take the two types of pollution first and leave discussion of preventative measures until later. Examples of air pollution should include not only combustion of fossil fuels and industrial air pollutants, but also the spraying of herbicides. Try if possible to refer to specific incidents, e.g. the release of dioxin from a factory in Seveso in Northern Italy. Water pollution should be discussed with reference to domestic wastes such as sewage, agricultural runoff, industrial wastes including radioactive and thermal pollution and the problem of oil.

It is impossible to include all pollutants of air and water, but the essential thing is to give a few examples of each that cover a full range (see answer plan for suggestions). In each case state clearly the source of the pollutant and its effect on the environment.

The discussion of the preventative measures should be more than just a list. The word 'discuss' suggests that examiners are looking for some comment on the measures, for instance its effectiveness or economic feasibility. Again it is most important to give precise detail. It is inadequate to talk about the desirability of legislation; instead the candidate should discuss the legislation that actually exists. No other topic seems to arouse such emotion in candidates with the result that many get carried away. Essays are frequently biased and unscientific. The examiner is often so bombarded with the personal views of candidates that bear no relation to what is on their mark schemes that candidates obtain little or no credit. Keep to the facts, support your answers with sound scientific principles and examples and point out clearly how each measure works to 'prevent further pollution'.

A list of the actual measures is given in the answer plan. Note that the question states 'some' of the measures. Do not attempt to cover all but discuss about six in some detail.

This type of Paper II question for the London Board has specific marks for the way in which the question is answered as well as the content. This fact is clearly stated in the rubric and is true of questions by some other boards. Candidates should therefore write accurately, fluently and keep to sound, well-reasoned scientific argument rather than personal bigotry.

Answer plan

Activities

1 Industrialization
2 Population explosion

For air and water pollution choose a total of about 12 examples. Give the source and effect in each case.

Air pollution

1 **Domestic** (fossil fuels): carbon, carbon dioxide, carbon monoxide, ozone, lead, nitrous oxides
2 **Industrial:** sulphur dioxide, various toxic gases, cement dust, asbestos, fluorine
3 **Agricultural:** pesticide sprays
4 **Radioactive:** fallout

Water pollution

1 **Domestic:** sewage, detergents
2 **Agricultural:** fertilizer and pesticide runoff
3 **Industrial:** toxic wastes, e.g. mercury, thermal pollution, radioactivity, oil

Preventative measures: choose six to eight examples from:

1 Population control: contraception
2 Smokeless zones: 1956 Clean Air Act
3 Alternative energy sources: solar/wind/water power
4 Improved fuels: fewer toxic products
5 Improved technology: fewer polluting by-products, e.g. electric cars
6 Positioning of towns, factories, airports to give minimum disturbance to the environment
7 Smoke filters
8 Non-lead petrol additives
9 Control of effluent discharge
10 Legislation, national and international agreement on pollution, e.g. discharge of oil from tankers and nuclear bomb trials.

9 Read the passage and answer the questions which are based on it.

'In 1949 synthetic organic pesticides were widely introduced into the Canete Valley of Peru in an attempt to control cotton pests in this important agricultural area. Many ecologists and evolutionary biologists argued against this approach, but their voices went unheeded, and massive applications of these pesticides began. Initially the program was very successful. Application of chlorinated hydro-carbons increased yields from 440 lb per acre in 1950 to 648 lb per acre in 1954. The entire valley was blanketed with these insecticides. They were applied from airplanes, and in many places trees were chopped down to facilitate the process. There were several fairly early repercussions from this campaign. When the trees were removed, many birds that nested in them disappeared, as did a number of other animals. Over the years the frequency of insecticide treatments had to be increased, and each year they had to be started earlier.

Soon the real troubles began. In 1952 the chlorinated hydrocarbon BHC failed to stop aphid infesta-tion. In 1954 the chlorinated hydrocarbon toxaphene failed against the tobacco leafwork. In 1955 there was a major infestation of boll weevils. New previously unencountered pests began to appear in the valley, and old pests began to show very high levels of DDT resistance.

In 1955 synthetic organophosphates were used in place of the chlorinated hydrocarbons, and the frequency of treatments was increased from two or three times monthly to once every three days. In that one year, despite this intensive effort, yields dropped over 300 lb per acre. In 5 years of intensive warfare against insect pests, all of man's initial successes had been wiped out, and he was worse off than when he began.'

(Reproduced from 'The new Higher Paper II Interpretation Question' Arnold, B., Mills, P. R., Aberdeen College of Education Newsletter 34 November 1979 by permission of the editor.)

(i) What is a pesticide? (1)
(ii) What evidence is there that the use of pesticides was initially successful? (1)
(iii) How might the removal of trees lead indirectly to an increase in the number of pests? (2)
(iv) Why did the whole valley have to be blanketed with insecticide and not just the cotton fields? (2)
(v) Why weren't all the aphids killed in 1949? (1)
(vi) In what way did the changes occurring as a result of the addition of the pesticides to these populations of insects illustrate the principle of natural selection? (3)
(vii) Why was it that new, previously unencountered pest species were able to invade the Canete Valley after 1955? (2)
(viii) Describe the levels of pesticide you would expect to find in a top carnivore in this ecosystem. Give a reason for your answer. (2)
(ix) In what ways were the farmers worse off after five years of using pesticides? (3)
Mark allocation 17/125   Time allowed 20 minutes   *In the style of the Scottish Higher II*

Questions (i), (ii), (iii), (v), (viii) and (ix) test for understanding of the passage but questions (iv), (vi) and (vii) make demands which go beyond the passage itself.

The candidate needs initially to read and understand the passage and answer direct questions which test his comprehension of it. Other questions expand this understanding by requiring the candidate to apply biological principles and specific knowledge to points arising from the passage.

Increasingly in the scientific world articles and papers are written which must be understood by others if they are to be of value. The ability to comprehend such material is therefore an essential prerequisite of any good scientist. For this reason this style of question is proving increasingly popular with examination boards. Candidates should read the passage initially to obtain the basic ideas and principles involved. A second reading should be made during which more specific detail and under-standing should be absorbed. Only then should the questions be attempted. After reading each question the candidate should return to the passage, if not for the specific answer, at least to give some idea of the information required from his own knowledge. An outline of acceptable answers to the questions is given below:

  (i) A chemical used to kill pests.
  (ii) Increased yields.
  (iii) Fewer predators. Less predation, or numbers of pests not controlled, or equivalent.
  (iv) The pests may occur outside the cotton fields. The pests may not have their full life cycle in the cotton fields.
  (v) Some were resistant.
  (vi) An idea of how this particular example demonstrates: variation, change in the environment, survival of those best suited to the new conditions, spread of this character through the whole population (any three points).
  (vii) Ecological niches available. Lack of competition. Immigration (any two points).
  (viii) Levels would be high. Idea of accumulation as pesticide passes through the food chain.
  (ix) Overall decrease in yields. Increase in number of pests. Increase in different kinds of pests. Pesticides are now ineffective (any three points).

Index